The Captain's Guide to Hurricane Holes

The Bahamas and Caribbean

by
Captain Dave Underill
and
Stephen J. Pavlidis

Seaworthy Publications, Inc. • Cocoa Beach, Florida

The Captain's Guide to Hurricane Holes
The Bahamas and Caribbean
Copyright © 2018 by Captain Dave Underill and Stephen J. Pavlidis

Published in the USA by:
Seaworthy Publications, Inc.
6300 N. Wickham Rd
Unit 130-416
Melbourne, FL 32940
Phone 310-610-3634
email orders@seaworthy.com
www.seaworthy.com - Your Bahamas and Caribbean Cruising Advisory

All rights reserved. No part of this book may be reproduced, stored in a retrieval system, or transmitted in any form or by any means, electronic, mechanical, photocopying, recording, or by any storage and retrieval system without permission in writing from the Publisher.

CAUTION: The authors have taken extreme care to provide the most accurate and reliable charts possible for use in this edition, nevertheless, the charts in this guide are designed to be used in conjunction with DMA, NOAA, and other government charts and publications. The Authors and Publisher take no responsibility for the misuse of the charts in this edition. All charts are Copyright © 2017 Stephen J. Pavlidis unless otherwise noted.

All aerial photos of The Bahamas courtesy of Capt. Paul Harding, Aerial Imaging. All aerial photos of the Virgin Islands courtesy of Todd Duff (a special thank you for the updated Virgin Island information too). All aerials of Puerto Rico courtesy of Bob Greiser. Guatemala photo by Bongo Bob Meredith. All photos of Mexico, Belize, and Cuba are courtesy of Capt. Dave Underill. All other photos courtesy of Stephen J. Pavlidis.

Book design: Stephen J. Pavlidis, Nightflyer Enterprises, Melbourne, FL

Library of Congress Cataloging-in-Publication Data

Names: Underill, Dave, Captain, 1958- author. | Pavlidis, Stephen J., author.
Title: The captain's guide to hurricane holes : the Bahamas and Caribbean / by Captain Dave Underill and Stephen J. Pavlidis.
Other titles: Captain's guide to hurricane holes, the Bahamas and Caribbean
Description: Cocoa Beach, Florida : Seaworthy Publications, Inc., [2018]
Identifiers: LCCN 2017045408| ISBN 9781892399960 (pbk. : alk. paper) | ISBN 1892399962 (pbk. : alk. paper)
Subjects: LCSH: Pilot guides--Caribbean Sea. | Pilot guides--Bahamas. | Hurricanes. | Marine meteorology. | Heavy weather seamanship.
Classification: LCC VK971 .U53 2018 | DDC 623.89/22365--dc23 LC record available at https://lccn.loc.gov/2017045408

Dedication

This book is dedicated to Scott Bell, who has been my best friend for 55 years, and his father, the late Larry Bell, who got me hooked on sailing a long time ago.

Capt. Dave Underill

This book is dedicated to my parents, Elizabeth and Basil, for introducing me to the beach, the sea, and boats.

Stephen J. Pavlidis

Table of Contents

Irma and Maria 12

Introduction 14

Foreword 15

Hurricane Formation 17
    Storm Surge 18

Hurricane History 20

Hurricane Preparation 22

The Northern Bahamas 25
    Grand Bahama: *West End* 25
    Grand Bahama: *Freeport, Bradford Yacht* 26
    Grand Bahama: *Xanadu Channel* 26
    Grand Bahama: *Sunrise Channel* 26
    Grand Bahama: *Silver Cove, Ocean Reef Yacht Club* 27
    Grand Bahama: *Bell Channel Bay* 28
    Grand Bahama: *Grand Lucayan Waterway* 29
    Grand Bahama: *South Riding Point Harbour* 31
    The Bight of Abaco: *Randall's Creek* 31
    The Bight of Abaco: *Basin Harbor Cay* 32
    The Bight of Abaco: *Mores Island* 32
    Abaco: *Grand Cays* 33
    Abaco: *Double Breasted Cays* 35
    Abaco: *Carter's Cays, Hogsty Harbour* 35
    Abaco: *Allan's-Pensacola Cay* 36
    Abaco: *Green Turtle Cay, Black Sound* 37
    Abaco: *Green Turtle Cay, Bluff Harbour* 37
    Abaco: *Green Turtle Cay, White Sound* 37
    Abaco: *Mariposa* 38
    Abaco: *Treasure Cay Marina* 38
    Abaco: *Baker's Bay Marina* 39

Abaco: *Leisure Lee and Hill's Creek* 39
Abaco: *Marsh Harbour* 41
Abaco: *Man-O-War Cay* 42
Abaco: *Hope Town Harbour* 43
Abaco: *White Sound* 44
Abaco: *Tilloo Pond* 44
Abaco: *Snake Cay* 45
Abaco: *Little Harbour* 45

The Central Bahamas 48

The Biminis: *North Bimini* 48
The Biminis: *South Bimini* 48
The Berry Islands 49
Berry Islands: *Great Harbour Cay Marina* 49
Berry Islands: *Little Harbour* 50
Berry Islands: *Chub Cay Marina* 51
Andros: *Kamalame Cay Marina* 52
Andros: *Fresh Creek* 52
New Providence: *Nassau Harbour* 54
New Providence: *Lyford Cay Marina* 54
New Providence: *Albany Marina* 55
New Providence: *Coral Harbour* 55
New Providence: *Palm Cay Marina* 56
New Providence: *Rose Island, Salt Pond* 56
Eleuthera: *Davis Harbour Marina* 57
Eleuthera: *Powell Point* 57
Eleuthera: *Hatchet Bay* 59
Eleuthera: *Spanish Wells* 60
Eleuthera: *Royal Island* 60
Cat Island: *Springfield Bay* 61
Cat Island: *Hawk's Nest Creek* 62
Cat Island: *Bennett's Harbour* 63
Cat Island: *Orange Creek* 63

The Exumas 65

Highborne Cay Marina 65
The Pond at Norman's Cay 66
Warderick Wells: *South Warderick Wells* 67
Compass Cay Marina 69

    Sampson Cay 70
    Staniel Cay 70
    Cave Cay: *Safety Harbor Marina* 72
    Rudder Cut Cay: *The Pond* 73
    Great Exuma: *Emerald Bay Marina* 74
    Great Exuma: *Stocking Island* 74
    Great Exuma: *Crab Cay and Red Shanks* 75

The Southern Bahamas 77
    Long Island: *Joe Sound* 77
    Long Island: *Stella Maris* 78
    Long Island: *Dollar Harbour* 79
    Long Island: *Deadman's Cays* 81
    Long Island: *Little Harbour* 81
    The Jumentos: *Ragged Island* 82
    The Jumentos: *Boat Harbour* 82
    Crooked/Acklins District 83

The Turks and Caicos 84
    Provo: *Turtle Cove Marina* 84
    Provo: *Leeward Going Through-Pine Cay* 86
    Provo: *Discovery Bay* 88
    Provo: *Caicos Marina* 88
    West Caicos: *West Caicos Marina* 89
    Grand Turk: *North Creek* 91

The Dominican Republic 93
    Luperón 93
    Cofresi: *Ocean World Marina* 94
    Samaná: *Puerto Bahía Marina* 94
    Bahía de San Lorenzo 95
    The Southern Coast: *Barahona* 95
    The Southern Coast: *Puerto de Haina* 95
    The Southern Coast: *Santo Domingo* 96
    The Southern Coast: *Casa de Campo Marina* 96
    The Southern Coast: *Cap Cana Marina* 97

Puerto Rico 98
    Western Shore: *Puerto Real* 98
    Western Shore: *Boqueron* 98

Southern Shore: *La Parguera* 100
Southern Shore: *Bahía de Guanica* 101
Southern Shore: *Bahía de Guayanilla* 101
Southern Shore: *Salinas* 102
Southern Shore: *Jobos* 102
Eastern Shore: *Palmas del Mar* 103
Eastern Shore: *Puerto del Rey Marina* 105
Eastern Shore: *Fajardo* 105
Northern Shore: *San Juan* 106

The Spanish Virgin Islands 107
Culebra: *Ensenada Honda* 107
Vieques: *Ensenada Honda* 107
Vieques: *Puerto Ferro* 109
Vieques: *Puerto Mosquito* 110

The United States Virgin Islands 111
St. Thomas: *Mandahl Bay* 111
St. Thomas: *Flamingo Bay* 111
St. Thomas: *Krum Bay* 112
St. Thomas: *Sapphire Bay Marina* 112
St. Thomas: *Benner Bay* 113
St. Thomas: *Charlotte Amalie* 114
St. Thomas: *Brewer's Bay* 114
St. John: *Enighed Pond* 114
St. John's: *Hurricane Hole* 114
St. John: *Coral Harbor* 116
St. Croix: *Christiansted* 116
St. Croix: *Green Cay Marina* 118
St. Croix: *Salt River Bay* 118

The British Virgin Islands 120
Tortola: *Road Town* 120
Tortola: *Paraquita Bay* 121
Tortola: *Maya Cove* 121
Tortola: *Trellis Bay* 122
Tortola: *Hannah Bay* 122
Tortola: *Sea Cow Bay* 123
Virgin Gorda: *Virgin Gorda Yacht Harbour* 124

  Virgin Gorda: *North Sound* 124

The Leeward Islands 125
  St. Martin: *Anse Marcel, Radisson Marina* 125
  St. Martin: *Oyster Pond* 126
  St. Martin: *Simpson Bay Lagoon* 127
  Antigua: *St. John's* 128
  Antigua: *Jolly Harbour* 128
  Antigua: *Falmouth Harbour* 128
  Antigua: *English Harbour* 129
  Antigua: *Indian Creek* 130
  Antigua: *Mamora Bay* 130
  Antigua: *Ayres Creek* 131
  Antigua: *Parham Harbour* 132
  St. Kitts: *St. Kitts Marine Works* 132
  St. Kitts: *Christophe Harbour Marina* 133
  Guadeloupe: *Marina de Rivière Sens* 133
  Guadeloupe: *Marina Bas du Fort* 134
  Guadeloupe: *Rivière Salée* 134
  Guadeloupe: *Marina de St.-François* 136
  Marie Galante and Desirade 136
  St. Barths: *Gustavia* 137

The Windward Islands 138
  Martinique: *Cohe du Lamemtin* 138
  Martinique: *Cul-de-Sac du Marin* 140
  Martinique: *La Marina Hâvre du Robert* 141
  Barbados: *Port St. Charles* 142
  Barbados: *Bridgetown* 143
  St. Lucia: *Rodney Bay* 143
  St. Lucia: *Marigot Bay* 144
  St. Vincent: *Ottley Hall* 145
  Canouan: *Glossy Bay Marina* 145
  Carriacou: *Tyrrel Bay* 146
  Grenada: *St. George's* 147
  Grenada: *True Blue Bay* 148
  Grenada: *Prickly Bay* 148
  Grenada: *Mt. Hartman Bay* 149
  Grenada: *Clarke's Court Bay* 150

Grenada: *Port Egmont, Calvigny Harbour* 150
Grenada: *St. David's Harbour* 151

Trinidad and Tobago 152
Trinidad: *Scotland Bay* 152
Trinidad: *Winns Bay* 153
Trinidad: *The Carenage* 153
Trinidad: *Port of Spain* 153
Tobago: *Bon Accord Lagoon* 154

The Northern Coast of Jamaica 156
Port Antonio: *West Harbour* 156
Bogue Lagoon 156

Honduras and the Bay Islands 159
Bay Islands: *Guanaja. El Bight* 159
Bay Islands: *Guanaja. West of Bonacca* 160
Bay Islands: *Roatán, Old Port Royal* 160
Bay Islands: *Roatán, Calabash Bight* 161
Bay Islands: *Roatán, Hog Pen Bight* 161
Bay Islands: *Roatán, Bodden Bight* 162
Bay Islands: *Roatán, Second Bight* 162
Bay Islands: *Roatán, CoCo View Marina* 163
Bay Islands: *Roatán, Brooksy Point Marina* 164
Bay Islands: *Roatán, Old French Harbour* 164
Bay Islands: *Roatán, French Harbour* 164
Bay Islands: *Roatán, Brick Bay* 165
The Bay Islands: *Utila, Puerto Este* 165
Mainland Honduras: *Puerto de Cabotaje* 166
Mainland Honduras: *Laguna el Diamente* 167
Mainland Honduras: *Puerto Cortes* 168

The Cayman Islands 169
Grand Cayman: *North Sound* 169
Grand Cayman: *Governor's Harbour* 170
Grand Cayman: *Harbour House Marina* 171

Guatemala 173
Bahía de Graciosa 173
Puerto Barrios: *Ensenada San Carlos* 174

Puerto Barrios: *Amatique Bay Marina* 175
Rio Dulce: *Río Tatin, Río Lampara* 175
El Golfete: *Laguna Salvador, Laguna Calix* 175
El Golfete: *Bahía de Tejano (Texan Bay)* 176
El Golfete: *Unnamed Bay* 176
El Golfete: *Bahía Buenavista (Gringo Bay)* 177
El Golfete: *Río Chacon Machaca* 177
El Golfete: *Laguna Quatro Cayos* 177
Rio Dulce: *Laguna Escondida* 177
Rio Dulce: *Río Ciénaga* 178
Rio Dulce: *La Bacadilla* 178
Rio Dulce: *Monkey Bay Marina* 178
Rio Dulce: *Fronteras* 179
Rio Dulce: *La Joya del Rio Marina* 179
Lago Izabal: *Puerto Refugio* 179
Lago Izabal: *Río Oscuro* 179
Lago Izabal: *Bocas de Bujajal* 179

Belize 182
Cucumber Beach Marina 182
Sapodilla Lagoon 183
Hakim's Boatyard 184
Mango Creek 184
Ycacos Lagoon 185
Big Creek 185
Other Marinas and Boatyards 186

Mexico's Caribbean Coast 187
Bahía del Espiritu Santo 187
Bahía Ascension 188
Puerto Aventuras Marina 189
Cancún: *V&V Marina* 189
Isla Mujeres 190
Isla Blanca: *Laguna Chakmochuk* 191

Cuba 193
Northern Coast: *Canal de Barco* 193
Northern Coast: *Esperanza* 194
Northern Coast: *Cayo Morillo* 194

    Northern Coast: *Bahía Honda* 194
    Northern Coast: *Marina Hemingway* 195
    Northern Coast: *Varadero, Marina Gaviota* 196
    Northern Coast: *Cayo Cruz del Padre* 197
    Northern Coast: *Bahía de Vita* 197
    Southern Coast: *Bahía de Baitiquiri* 198
    Southern Coast: *Santiago* 198
    Southern Coast: *Puerto de Casilda* 199
    Southern Coast: *Cienfuegos* 199

Weather Broadcasts 200
    TV and Internet 200
    HF Weather 200
    Local Weather Broadcasts 201
    The Bahamas 201
    The Turks and Caicos Islands 202
    Puerto Rico and the Spanish Virgin Islands 202
    The United States and British Virgin Islands 202
    The Leeward Islands 202
    The Windward Islands 202
    Trinidad and Tobago 203
    Mexico 203
    Belize 203
    Cuba 203
    Guatemala 203
    Honduras 203
    Jamaica 203

List of Charts 204

Chart Legend 211

List of Haul Out Yards 212

List of Marinas 216

Index 226

About the Authors 232

Bahamas/Caribbean Hurricane Tracking Chart 233

Irma and Maria

by
Stephen J. Pavlidis

THIS GUIDE WAS WRITTEN IN THE SUMMER OF 2017, AND WAS BEING EDITED while we were witnessing the destruction wrought by Hurricane Irma and Hurricane Maria in The Bahamas and Caribbean. I was shocked at the amount of utter devastation these storms left behind and how some of the holes, so favored by both cruisers and charter fleets for hurricane protection, lost nearly every vessel present, while other holes escaped with little or no damage. I have gone through this guide again since Hurricane Irma and Hurricane Maria passed through the islands and annotated the text to reflect how some of these places survived. The one good thing to take from this is that the people affected will rebuild, it is their way, they have done this for centuries, but the damages from Irma and Maria will set them back for months, some for years.

Irma left a huge path of destruction from Barbuda and Antigua through the Virgins and then right up the middle of Florida. The eye of Irma went over Barbuda which is now little more than a ghost town; everybody has been evacuated off the island and who knows when they will return. Irma then leveled St. Martin, St. Barth's, Anguilla, and the U.S. and British Virgin Islands. However, *North Sound Boatyard* located on Crabbs Peninsula on Antigua, about 25 or so miles south of Barbuda, suffered minor damage and all the boats there were fine. *Jolly Harbour Boatyard* on the western shore of Antigua also suffered little damage.

Irma taught much about the holes that people have been using for years in the Virgins. Paraquita Bay, the safe hole for most of the BVI charter fleet, the safe hole that charges for moorings and lines up the charter vessels in nice, long rows, was decimated and has set the Virgin Islands' charter industry back who knows how long. Nearly every hole in Irma's path suffered with few exceptions. Nanny Cay was wiped out, both the docks and the other infrastructure but they are rebuilding already (they have ordered new docks which should be installed in early 2018).

In North Sound, Virgin Gorda, the *Bitter End Yacht Club* is in total ruins and closed for rebuilding.

In the USVI, St. John and St. Thomas were hit hard but Benner Bay (especially the area at the head of the bay known as "The Lagoon"), Flamingo Bay, and Mendahl Bay survived with just a few losses. The small cove north of the airport runway and south of Brewer's Beach on the west side of St. Thomas also proved a valuable hole with its mangroves and 7' depth where boaters survived both Irma and Maria. *Sapphire Bay Marina* suffered some boat losses as well as dock destruction. All in all, St. Thomas had a better survival rate for boats than did Tortola and Virgin Gorda where gusts to 200 mph and tornadoes laid those islands to waste, there was no truly safe place there.

IGY stated that *Blue Haven Marina* on Provo in the Turks and Caicos Islands, *Yacht Haven Grande* on St. Thomas, USVI, *American Yacht Harbor* at Red Hook, *USVI, Yacht Club* at Isle del Sol, St. Martin, and *Simpson Bay Marina* are all closed for repairs, when they will open is anyones guess. Most marinas in Simpson Bay and Marigot were heavily damaged and will be closed for a while. Gustavia suffered a lot of damage but should be up and running by the time this guide is published. Christophe Harbour on St. Kitts seems to have made it through with little damage. As did *St. Kitt's Marine Works.*

Irma skirted the northern coast of Cuba, heavily damaging the marina and boatyard at Gaviota but leaving *Marina Hemingway* virtually untouched. A few marinas in The Bahamas were damaged but all were up and running within a week of Irma's passing.

Maria appeared to be following in Irma's wake beginning her path of destruction by leveling Dominica and then hitting St. Croix hard before crossing Puerto Rico and knocking out ALL power on the island (even snapping concrete power poles) and leaving few vessels unscathed. *Puerto del Rey Marina* suffered minimal damage and most boats survived with little harm. The damage to local boats in Puerto Rico is sad as many Puerto Rican boaters usually keep an eye out for strong storms and many will simply head south for three days to the ABCs and return after the storm has passed Puerto Rico.

The eye of Maria then passed approximately 35 miles to the east of Grand Turk and North Creek faired as well as can be expected with little damage.

So what have we learned? We have discovered that some of the best holes are not as safe as many claim them to be while other, perhaps not so well-known holes did their jobs in two major hurricanes. Bear these lessons in mind when you seek refuge.

All in all, the best protection is not to find yourself in the hurricane zone during hurricane season, call it avoidance. You might wish to consider Panama or Venezuela's offshore islands.

Introduction

Even though the IDEAL hurricane hole might not exist, if I were cruising and faced with the threat of an imminent hurricane, I would want to know where the most secure location in my area was located. This is what this guide intends to do. What choices does one have? In most areas within the hurricane belt, dry land options are taken up sometimes a year in advance.

For most of us, our boat is our home. It's where we live out our dreams. It's where our loved ones snuggle beside us. It's where we escape from the mundane life back on land. The goal of this guide is to give your boat the best chance possible to withstand a storm of utterly unimaginable magnitude. As for yourself, you should be off the boat and can celebrate with a local beer AFTER the storm passes. Unfortunately, the reality is that most of us will stay with the boat, attempting to 'save' it in 130 mile per hour winds. This guide, while definitely not advocating staying aboard when there's a monster outside literally trying to kill you, will help you as well, hopefully while wearing the proper survival gear!

The majority of hurricanes travel at approximately 10-15 mph. With a sailboat traveling at 5 knots and forecasters being relatively accurate as to the expected path 3 days before a storm hits, that gives one roughly 24-36 hours to find the very best shelter that you can find and be anchored properly, not trying to outrun it. We've tried to cover at least one 'best' for each of the described areas and if there just weren't any acceptable anchorages or marinas, the boat and YOU should not be in that area during hurricane season. With this guide, Stephen J. Pavlidis and I are going to show you the best of what's available in the area that you cruise as well as tips for protecting your property. I will repeat a number of details over and over. This is to stress the importance of each. Remember, the first goal is to stay alive, the second, to save your boat.

Capt. Dave Underill

Foreword

by
Stephen J. Pavlidis

To begin with, let me just reassure you by stating that there is no such thing as a hurricane hole. There is no anchorage, marina, or boatyard so secure that it cannot be decimated by a direct hit from a strong hurricane with a high storm surge (as we learned in 2017 with Hurricane Irma and Hurricane Maria; for more information see that chapter). There are no guarantees, there is no Fort Knox to hide in when a named windstorm threatens. Now, with that out of the way we can discuss how to protect yourself in those special places that offer the best hurricane protection.

This book will not go into great detail about the intricate state of conditions that create a hurricane, instead the purpose of this book is to pass along information that Dave and I have gathered first hand in our years of cruising. I'm sure that many readers will not agree with some of the places listed here, but perhaps they are missing the point. The material presented here is offered as an aid for the skipper seeking shelter from an oncoming storm. They might find themselves in a poor position for shelter and may not be able to get to a prime location and maybe they will find an alternative in these pages. I hope so. Being able to assist just one boater to save their vessel is worth the hard work involved in producing this publication. A good example is *Winns Bay* on the island of Trinidad. By no means is it an ideal spot for protection from a hurricane, however, if your time is short and it is the only place you can get to, it will have to suffice, but only if the winds are not forecast to be from any southerly direction.

I cannot stress enough to make and implement your plans EARLY! You will see this mentioned in these pages several times, EARLY! Some of the holes we offer here are small with only room for a couple of boats and you don't want to show up late and find there is no room for you. And never forget, the locals know the best places and they WILL be there early!

We will show where to haul out, hide, or get a slip, the choice is yours. Not every listing in this book is an ideal hurricane hole, many are not, however we have included them as they do offer some sort of shelter and may do if you cannot get to a more favored position. It is up to the skipper to decide his or her own hurricane plan and implement it with as much knowledge at hand as they can muster. It is our endeavor to bring that knowledge to you so that you can make a smarter, safer decision.

Our charts are meant to supplement government issued charts and should be used in conjunction with government produced charts. We have also included a *Bahamas/Caribbean Hurricane Tracking Chart* in the back of this book.

Now what can you do to help yourself? The primary thing you can do, besides making a move EARLY, is to always know where the eye is forecast to track and the direction that the wind is expected to blow. As I mentioned in the chapter on "Hurricane Preparation," a neat trick to determine wind direction and the location of the eye of the hurricane is to stand facing into the wind. Extend your right arm out from your body and point. As a general rule, that is where the eye lies, approximately 90 degrees off the wind. The forecast track and expected wind direction are your primary tools for choosing protection.

I suggest that when you get to an area where you plan to be for a while, take a day and familiarize yourself with the nearest hurricane protection with an eye for how to get in and where you would secure your boat dependent on the wind forecast. If you seek a slip or haul out, check with the marina or yard managers concerning their hurricane policy, time is of the essence.

Now let's discuss what you might be able to do with the information presented in this publication. As mentioned, the choice to stay aboard or not is a personal one but most will leave your boat after securing it. In this book we will give you information about where to get a slip, where to haul out, and where to anchor if that is your choice. In some of the places we present you might find that you will be the only boat there and have no way off your vessel, make sure you want this before making a move.

We will show places that offer protection for shallow draft as well as deep draft vessels, controlling depths will be shown either in the text or shown on the accompanying chart. Where we have them, aerial photos are included to give you a better view of the protection offered by a particular hole.

Many of the boatyards that do hurricane haul outs do so for the season with an intricate cradle and tie down system. This is not cheap and can run into the thousands of dollars but then again it IS for the season, not just till the end of one storm. A haul out in the Caribbean for a season with a pit (a keel hole) and tie downs may cost you between 2K and 5K (U.S. Dollars). This is not to say that you cannot find a boatyard to haul you just for the duration of one storm, but expect to be at the end of the line of those that paid more for their protection.

Some marinas require vessels to leave with the approach of a storm, so call ahead to verify their hurricane policy. In a marina with floating docks, the docks can float off pilings with a high storm surge and you could get caught by a whole dock blowing down on you. (Ft. Pierce City Marina, 2004, in Florida comes to mind.) Slamming down on a piling or another boat is no way to spend a hurricane.

Most of the places we offer you in this publication are good for Cat 1 - Cat 3 storms, if the storm is a Cat 4 or 5 there is not much else to do except ride it out if you plan to stay aboard. At those wind strengths you probably won't be able to work on deck adjusting lines or dealing with a deck problem.

Chapter 1

Hurricane Formation

by
Capt. Dave Underill and Stephen J. Pavlidis

THE BAHAMAS AND CARIBBEAN TRAVEL BROCHURES SHOW THE IDYLLIC, postcard pictures of a quiet lagoon with barely a ripple on the surface of the water. On the white sand beach stands a thatched roofed hut nestled amongst the palm trees. There we are, the lone sailboat anchored in this paradise with mounds of lobster and conch just waiting on the sea-bottom for us to dive in and catch in the 85-degree water. Fast forward to the reality of a 135 mile per hour storm, ripping the hut to shreds, shooting coconuts at us like cannonballs, waves pummeling against the boat so hard that it makes the bow launch skyward every two seconds. The thunder of a locomotive blasting in our ears almost drowns out the sounds of the straining and stretching anchor lines that desperately try to hold us from being crushed on the surrounding rocks. The horizontal rain is pelting us so hard that it feels like nails being hammered into our skin and seems to go on forever....

Hurricanes need the tropics. It's what CAUSES theses monsters to form. The combination of warm water of at least 80° Fahrenheit (26.6° Celsius), moist air, spin, and low wind shear all contribute to the formation of these great storms.

Just like on a warm, sunny day, when we touch our skin it feels warmer than the air around us. This is because we absorb the heat. In a similar manner, the earth and its seas absorb the heat from the sun and thus the temperature of the air closest to the land and water is higher than in the upper atmosphere.

When air heats up, it's molecules move further apart and become less concentrated. Less concentrated means less weight so the air rises where the pressure on it from the atmosphere is less. In the tropics, as this warm moist air rises, water vapor condenses from the evaporation of salt water, forming storm clouds and rain. (Think of the dew on your boat every morning when you wake up). The pressure difference, or gradient, between the low-pressure air rising and the higher pressure air that is filling in beneath it creates wind. (Think of the low and high-pressure on a sail that LIFTS you up into the direction of the wind, instead of just sailing downwind). The eye of a hurricane is not fully understood by meteorologists. Most of the up drafts from the rising air flows up and outward but some ends up flowing up and inward towards the center. This sinking air contains the lowest pressure and actually creates a storm free zone but with the highest winds in the 'wall' that mark its borders. This eye is typically 20-40 miles (30-65km) in diameter.

Spin comes from the Coriolis effect. That is the phenomenon that molecules track to the right of their target in the Northern Hemisphere and the left in the Southern Hemisphere because of the Earth's rotation that is from west to east. At the equator the Coriolis effect is zero.

Wind shear simply relates to the upper atmospheric winds that can disrupt the existing wind pattern from rising low-pressure and falling high-pressure. High wind shear, combined with dry air, is the reason not all low-pressure systems form into hurricanes. Also contributing is *La Niña* (a phenomenon associated with the cold currents of the *Pacific Ocean*) which lessens the likelihood of hurricane formation in the *Atlantic* by increasing wind shear from the *Pacific* crossing over into the Atlantic. *El Niño* has the opposite effect, decreasing high level wind shear and thus increasing the likelihood of hurricane formation

So, there you have the perfect mixture of ingredients in our idyllic little harbor in the tropics for the making of a hurricane. Warm water from the equatorial region being closest to the sun (as opposed to the north and south poles which are farther from the sun), moist air from the humidity and evaporation of the warm salt water of the ocean, spin from the Coriolis effect, and low wind shear.

And don't forget, the quadrants of a hurricane. The one NE of the eye is the deadliest while the one SW of the eye offers the least damage, primarily because the eye has already passed you. The best location in any hurricane is to the west of the forecast track of the eye.

Storm Surge

Statistics prove that, by far, the greatest number of deaths in a hurricane occur due to storm surge (up to 90% of deaths). The highest recorded storm surge was in Australia in 1899, a surge of 42'!

Storm surge can be defined simply as the increase in the height of water that occurs as a result of an approaching storm or even during the storm. Think of standing in a pool. Now take your arm and swing it across your body, slightly underwater. This wave that you have created is the surge.

There are numerous factors involved with the creation of storm surge: the wind strength of the storm, the barometric pressure of the storm, the actual size of the storm, the forward speed of the storm, *El Niño* and *La Niña*, the coastal geography, the present high or low tide caused from the lunar cycle, the ocean floor topography, and rising seawater levels associated with climate change.

Hurricane Ivan's 165 mph winds and 8'-10' storm surge flooded much of the island of Grand Cayman. Ivan produced 23' heights in Jamaica only to be surpassed by Hurricane Allen's 190 mph winds and 39' surge on the north coast. Allen generated a 6'-12' surge in Cuba even though the storm passed just to the west of the island. The off-lying island of Cayo Largo experienced a 16' surge. Hurricane Allen also had the second strongest winds ever recorded in the Atlantic basin (190 mph). In contrast, Hurricane Earl had top winds of 80 mph and produced less than 4' of surge when it struck northern Belize.

The lower the barometric pressure, the greater the storm surge although this effect is less than the other factors involved. Since hurricanes are produced by areas of low-pressure, this results in less force being produced in a downward motion. Your barometer measures how much of this force is present at a specific location. Since hurricanes can have a pressure drop of 40 millibars (the measurement commonly used by barometers), this would result in a sea level rise of 40 cm (1mb=1 cm rise). Hurricane Wilma had a low-pressure of 882 millibars, at the time a record for the Atlantic basin, but only produced a storm surge of 3' in Key West when it passed just north of it.

The overall size of the storm influences the total amount of water moving towards a certain destination. Hurricane Gilbert had tropical force winds extending 575 miles in diameter. The storm surge penetrated over 3 miles inland in Cancún, Mexico.

The forward speed also affects the storm surge. Think of your arm moving just slightly under the surface of the pool. One time move it slowly, the next time move it rapidly. Hurricanes can move quickly, as demonstrated by the 60-70 miles per hour in the Long

Island express of 1938, or slowly, as in Hurricane Francis, which had no movement for 12+ hours over Grand Bahama.

Island geography helps lessen the impact of storm surge. Water is simply able to flow around islands. Contrasted to land masses such as Mexico's Caribbean shore where all the water being pushed forward has no escape route and must flow over land. Cancún reported winds from Hurricane Wilma of 140 mph but it produced a storm surge of 10' with waves reportedly battering the third floor (30' above the ground) of many hotels.

Lunar tides only add to the storm surge. It increases the total volume of water available to increase. Two feet of additional tides, as are common in the Caribbean, combined with a 6' storm induced surge, can mean the difference of water reaching over a mile further inland. The ocean floor topography plays a major factor in storm surge. Over deep water, the force of the storm is generated downward somewhat, lessening the height of the surge. In shallower water this energy can only be directed forward, thus increasing the surge.

Conversely, the steeper rise in the ocean floor can produce larger and more destructive waves; one cubic yard of seawater weighing close to one ton. Hurricane Ivan produced waves of 15' at Cienfuegos in central Cuba. The same Category 4 hurricane, as it passed over the *Naval Research Laboratory's* wave-tide gauges in the relatively shallow *Gulf of Mexico*, generated waves of 90'. The undeniable fact of a rising sea level can only add to surge heights in the same manner as the lunar tides do, only on a more permanent basis.

One last note. In 2017, we experienced a rarity in Florida. Hurricane Irma, a Cat 5 that hit land as a Cat 4 near Naples, Florida, and then skirted up the center of the state, brought a storm surge to both the east and west coasts of Florida, as well as portions of the southeast Georgia coast.

Chapter 2

Hurricane History

by
Capt. Dave Underill and Stephen J. Pavlidis

HURRICANE SEASON, OFFICIALLY DESIGNATED BY THE NATIONAL OCEANIC and Atmospheric Administration, begins June 1 and ends November 31. Mother Nature has no calendar, however, as hurricanes have been recorded in December, January, March, April, and May. Add to this the fact that the difference in wind speed between a Tropical Storm and a Hurricane is 1 mph (73 to 74 mph) which enlarges the parameters to include a significant storm every single month of the year. As examples, Tropical Storm Arthur in May 2008 caused in excess of $78 million in damages and five deaths. As recently as 2016, in January, Hurricane Alex formed near the Azores. Also in 2016, in the month of May, Tropical Storm Bonnie caused over $600,000 in damages and two deaths. According to Wikipedia, "...off-season cyclones are most likely to occur in the central to western Atlantic Ocean and most do not make landfall. Of the storms that did strike land, most affected areas surrounding the Caribbean Sea. Cumulatively, at least 398 deaths occurred due to the storms."

Hurricanes don't obey any 'rules.' Hurricane Lenny first formed in the western Caribbean on November 13, 1999. Instead of the 'normal' east to west movement, Lenny traveled from west to east for its entire existence. It first produced large waves that killed two people in northern Columbia and wave heights in the ABC islands off Venezuela reached 10'-20'. Heavy rainfall and flooding affected Jamaica, Haiti, the Dominican Republic, and Puerto Rico where a 544' freighter was washed ashore. Wind speeds of 112 mph were recorded in St. Croix where they experienced 15'-20' storm surge at Fredricksted. The storm continued its devastation to the British Virgin Islands and on to Anguilla and St. Martin/St. Marteen where it dropped 27" of rain. Lenny produced 16' waves in St. Bart's and in Saba winds of 167 mph were recorded before the wind gauge blew off. Major flooding occurred in Antigua and Barbuda. Twenty-foot waves struck the western shore of St. Kitts and Nevis and washed 600 feet inland. Montserrat, Guadeloupe, Dominica, Martinique, St. Lucia, St. Vincent and the Grenadines, Grenada, and as far south as Trinidad and Tobago were all affected.

Hurricane Ivan formed as a Cape Verde storm off the west coast of Africa on August 31, 2004. Its notoriety was gained as the southern-most hurricane to form on record as it became a Category 3 with winds of 120 mph when it hit the southern tip of Grenada, previously thought of by insurance companies as 'below the hurricane belt', where 39 people died. Every building in the capital of St. Georges was damaged or destroyed. It left 18,000 people homeless, and the entire island was without power. There were an estimated 650 boats in Grenada prior to the storm hitting. Roughly

20% of those that were hauled ended up toppled on the ground. Of those left in the water, approximately 1/3 ended up sunk or up on the rocks.

The storm gained Category 5 status as it passed south of the Dominican Republic, only to weaken slightly to Category 4 as it passed within 20 miles of the Cayman Islands. 95% of the buildings on Grand Cayman were either damaged or destroyed as a result. Ivan turned North to slam into Gulf Shores, Alabama bringing a 10'-15' surge of water and spawning over 100 tornadoes. After traveling through seven states, Ivan re-entered the Atlantic to again become a tropical storm and circled back to strike Florida's east coast, re-emerge on the west coast, and hit southern Louisiana as a tropical depression.

Hurricane Iris formed in 2001 in the central Caribbean. It passed just south of Jamaica and slammed into southern Belize. The entire 20-mile-long peninsula of Placencia was under water. 20 of the 28 deaths occurred as a result of the passengers and crew remaining on board the vessel m/v *Wave Dancer* that was flipped by the Category 4 winds. The thrust of the investigation following the loss of lives on *Wave Dancer* was centered around if the Captain failed to follow the hurricane plan on weather advice, whether the dock lines used were adequate (some polypropylene was used), whether the engines should have been running prior to storm impact, and whether the size of the boat in relation to the (shorter) dock contributed to the loss of life.

In 2015, the 790' container ship 'El Faro' was caught in the 30'-50'seas of Hurricane Joaquin off Crooked Island in the Bahamas. The ship lost its propulsion and was presumed to have sunk, taking all of its 33 crew with it. Three months later the disappearance mystery was solved as the hull was discovered in ocean depths of 3 miles.

Hurricane Martha was short lived but was significant because of how far south she formed. In 1969 it hit Panama which lies at only 9° above the equator.

The aftermath of the storm can sometimes be as devastating as the storm itself. No power for refrigerated food, flooding, cholera outbreaks, mosquitoes, looting, and downed live power lines can all occur and sometimes with deadly results. The third world countries can be slow to recover but even the United States can suffer the longer restoration process. Following Hurricane Jean in 2004, Sebastian, Florida was without power for 11 days. And that was 50 miles north of where the eye came onshore in the most modern country in the world!

Hurricane Katrina struck New Orleans on August 29, 2009 as a Category 4 storm packing winds of 120 mph. It had already devastated the Bahamas, Cuba, and South Florida. It is estimated that 1833 people died in Louisiana, with millions being left homeless from the flooding. Maximum winds reached out a distance of 30 miles and the storm surge was measured at 10'-28'! The floods did not recede for weeks.

As recently as 2016, Hurricane Matthew affected virtually every Island of the Caribbean and resulted in almost 1,000 deaths. In paths that were far below normal hurricane routes in the Caribbean, Hurricane Matthew skirted the waters north of the ABC's, while just after the end of hurricane season, Hurricane Otto, damaged numerous boats on the Caribbean coast of Panama and brought tropical storm force winds to the San Blas Islands, making landfall along the eastern coasts of Costa Rica and Nicaragua, the first hurricane to ever track so far south and hit Costa Rica.

One hurricane experience is one too many. Every sailor of worth has a tale of stormy horror. In 1984, Tropical Storm Klaus strengthened without much warning, approaching St. Thomas and putting some hundred-plus boats on the hard. 1989's Hurricane Hugo was a catastrophe in the Virgins, St. Croix was crushed. Sailors who had fled to what they hoped was safety in Culebra were unpleasantly surprised when the eye of the storm followed them. Then, 1995 brought the double barrels of Hurricanes Luis and Marilyn: Antigua, St. Maarten and St. Thomas were truly blown apart. In the following years Hurricanes Bertha and Georges made living on the water a worrying Hell.

And then in 2017 we met Irma, Harvey, and Maria. Irma was a horrendous Cat 5 storm that decimated Barbados, the Virgins and well as Florida. Irma, followed Harvey (50 inches of rain in Houston, Texas), and was quickly followed by Mara which leveled Dominica and knocked out ALL power on Puerto Rico. Avoid all Cat 4s and Cat5s if you are able.

Chapter 3

Hurricane Preparation

by
Capt. Dave Underill and Stephen J. Pavlidis

THE FIRST THING YOU MUST DO IS RESEARCH THE FORECAST TRACK FOR THE hurricane and decide which side of the track you will be on and how far from the eye your vessel will lie. This knowledge will be the primary tool in deciding your approach to storm survival by giving you a good idea about the conditions that will be affecting your vessel (bearing in mind that hurricanes are apt to change direction and strength at any time fooling even the best of forecasters). You might wish to find a nice, safe, mangrove lined hole in which you can anchor to hide from the coming storm, or you might wish to get a slip in a protected marina, or haul out in a boatyard that specializes in hurricane protection. I feel comfortable about most of the protection in this guide in hurricanes up to a Cat 3. However, if a hurricane is a Cat 4 or Cat 5, many of the places that we think are the safest may not be (see the chapter on Irma and Maria). In this guide will discuss how to protect yourself in those special places that offer hurricane protection. Let's begin by passing along a few hints as to how to secure your vessel while getting along with your neighbors, and then learn where to find the best protection.

To begin with, it is necessary to understand the quadrant structure of a hurricane and it all begins with the storm's eye. Plot the forecast track of the eye and determine where you are in relation to it. If you divide the hurricane up into four quadrants, two will lie on each side of the forecast track, think of the eye and the quadrants as a gun-sight. The most destructive quadrant (for referral purposes we shall pretend our hurricane is tracking north) will be to the northeast of the eye while the quadrant to the southeast of the eye will be less destructive as the winds there will be weaker as the eye moves along its path. The quadrant to the northwest of the eye will be not quite as vicious as the northeast quadrant but it will still be a deadly place, the southwest quadrant being less so. Ideally, you should strive to be to the west of the forecast path of the hurricane for best protection, and as far from the eye as possible.

Once you have decided your course of action, make sure your fuel is topped off and that you have enough food and water for an extended period. Also, make sure that you have enough cash to see you through as phone lines may be down for a while after the storm passes which might prohibit credit card usage. Once your tanks, lockers, and wallet are topped off, you can head for protection. Some skippers prefer to head to sea when a hurricane threatens. Some will take off at a ninety-degree angle from the hurricane's forecast path, those in the lower Caribbean usually head toward Venezuela or the *Rio Dulce* while boaters in Puerto Rico and the Virgins may head to Grenada or take a 3-day sail south to the ABCs.

We cannot advise you as to what course of action to take, that is up to each individual cruising boat and their own particular circumstances, but unless absolutely necessary, it is not advisable to gamble with racing a storm that is unpredictable (no matter what the forecasters claim). Whatever course you choose to take, the prudent skipper will make his or her move EARLY! For protection, most of us would prefer a narrow canal that winds deep into the mangroves where we will be as snug as the proverbial bug-in-a-rug. These canals are rare, and to be assured of space you must get there early. When a storm threatens, you can bet that everybody will soon be aware of it and the early birds will settle in the best places. Sure, those early birds might have to spend a night or two in the hot, buggy mangroves, but isn't that better than coming in too late and finding the best spots taken and your choices for protection down to anchoring in the middle of a pond with a bit of fetch and no mangroves to surround you like a security blanket? Hint number one...get to safety early, and secure your vessel.

So how do you secure your vessel? Easy! First, find a likely looking spot where you'll be safest from the oncoming winds, a spot with a short fetch and good holding. Try to deduce by the forecast path of the storm where the wind will be coming from as the storm passes and plan accordingly (remember that the winds blow counterclockwise around the center in the northern hemisphere). If your chosen spot is in a creek that is fine. Set out bow and stern anchors and tie off your vessel to the mangroves on each side with as many lines as you can, including lines off the bow and stern to assist the anchors. A neat trick to determine wind direction and the location of the eye of the hurricane is to stand facing into the wind. Extend your right arm out from your body and point. That is where the eye lies, 90 degrees off the wind.

Use plenty of chafe protection (I like old fire hose, leather, and if nothing better is available, towels secured with duct tape) as the lines lead off your boat and rig your lines so that they don't work back and forth on the mangroves as well. If chain can be used to surround the mangroves, that will help (not the mangroves of course). Mangroves are strong so try to tie to the largest and use several lines as backup.

If other boats wish to proceed further up the creek past your position, remove the lines from one side of your boat to allow them to pass. Courtesy amongst endangered vessels will add to the safety factor of all involved, especially if somebody needs to come to somebody else's aid. If your only choice is to head into the mangroves bow or stern first, always go in bow first; it stands to reason that if you place your stern into the mangroves serious rudder damage could result. I prefer to go bow-in as far as I can, until my boat settles her keel in the mud (trying to keep the bow just out of contact with the mangroves), tie off well, and set out at least two stern anchors (the largest ones you have) with as much scope as possible. It is probable that other boats will be tying off into the mangroves in the same manner on each side of you, courtesy dictates that each skipper assist the other in the setting of anchors (so that they don't trip each other) and the securing of lines in the mangroves (and don't forget to put out fenders). Work with other skippers to assure that everybody will have swinging room in the event of a wind shift.

If you must anchor in the open, away from the mangroves, place your anchors to give you 360° protection. The greatest danger to your vessel will likely be the other boats around you, and in the Caribbean, there's going to be a better than average chance that you'll be sharing your hole with several unattended boats, often times charter boats that are not secured as well as you would like them to be as well as being left unattended. A good lookout is necessary for these added dangers. I've seen some folks that put out three anchors 120° apart, whose rodes lead to a swivel. From the swivel, a chain leads over the bow roller to fasten strongly to the deck. This eliminates chafe at the bow roller.

If you plan to be in hurricane waters during hurricane season, you might want to carry several sand screws. These are large threaded rods that you will have to screw deep into the sand, they are similar to the anchors that secure the guys on power poles and can be incredibly strong. You might want to have SCUBA gear or a hookah rig aboard to assist you with air during installation. Once installed you can add a shackle and a line to the sand screw to help secure your vessel.

Some skippers prefer the safety of a narrow seawall lined canal as can be found in many finished and unfinished housing projects. Leisure Lee in Abaco, Bahamas, comes to mind. If you try to avail yourself of the protection afforded by such a place be sure to run anchors onto the land well past the seawall, don't count on the cleats on the seawall, or the seawall itself to hold your vessel. Seawalls have been known to crumble and fall into the canal separating from the land they are fronting.

There are differing opinions on whether to haul out for a hurricane, or to tie off in a marina. A lot of cruisers will tell you they've had success at both, but there's an equal number that will advise against it. On the hard, a domino effect can topple one boat after another, and slips in marinas, if the owners will let you stay for a blow, are often narrow and care must be taken to avoid contact with your neighbor. Note that floating docks are famous for going adrift in a strong storm surge and destroying all boats attached to those docks and others where the runaway dock lands.

Once secure, your next step is to strip everything off your boat and stow it below. Sails, bimini, awnings, rail-mounted grill, wind-generators, solar panels, jerry cans, and anything small and loose that can become a dangerous object should it fly away at a hundred miles an hour. Make sure that your neighbors do the same; their loose objects could be hazardous to your health. In addition, don't forget to secure your dinghy!

The decision to stay aboard is a highly personal one. Some of us that have insurance will head for a hotel or some other shelter ashore, while others, whose only insurance is their seaman's skills, will ride the storm out aboard. If you decide to stay aboard, pack all your important papers in a handy waterproof container, and in the most severe of circumstances, use duct tape to secure your passport, wallet, or purse to your body.

Plan ahead as you secure your vessel so that you will not have to go on deck if you don't absolutely need to, as it is most difficult to move about in 100-knot winds. Keep a mask and snorkel handy in the cockpit; you might need it to stand watch. Also, keep a flashlight and a sharp knife close at hand; you never know when you might need them.

There are some obvious things that we should do as boaters when a tropical storm or hurricane is approaching. Removing all sails, biminis, placing chafe gear on lines, and removing anything on deck are standard practices. But you should also consider:

Placing as many old tires as you can find around dock pilings

Seriously evaluating all lines and replacing as needed at the BEGINNING of hurricane season

A secondary bridle on catamarans

Maximum of two lines on any deck cleat

Topping off water tanks, propane tanks, and batteries

Strapping down microwaves and tv or setting them down on the sole

Taking dinghy, paddle boards, and kayak ashore and securing

Have life vests, waterproof flashlights, and a mask and snorkel within easy access

Having personal information (passports, licenses, etc.) in waterproof containers

Using winches for strength on anchor lines or mangrove lines (it's easier to adjust them from the cockpit with the winch)

Duct tape or quick tie running rigging together (sheets, halyards, furling lines)

Securing the roller furling system

Dropping the boom to the deck and securing

Empty the bilges and make sure your bilge pumps work and have backups for them.

Chain around dock cleats and pilings

Cell phones fully charged

Remove all canvas, sails, and anything that WILL fly off. Place chafe gear on each line as they pass through your chocks. Try to remember the point that sometimes, "It's not just how well you prepare your boat, but how poorly others around you prepare theirs." So, offer help to others.

Chapter 4

The Northern Bahamas

THE NORTHERN BAHAMAS, AND IN PARTICULAR THE ABACOS, IS ONE OF THE most popular destinations for cruisers to The Bahamas. We will show you where to find protection in a slip, hauled out, or at anchor on Grand Bahama, and even a couple of places you might find shelter in the *Bight of Abaco*. And of course, we will cover all the spots in the Abacos, some of which you may have not heard of before. One note, the reader may notice that we do not tout *Spanish Cay Marina* as hurricane protection and for good reason. The tiny breakwater offers little if any protection in heavy westerly seas.

Grand Bahama: *West End*

Although the *Old Bahama Bay Marina* seems to get trashed with a direct hit from a hurricane, many cruisers still like the protection and convenience of West End, Grand Bahama so it is included here (if a direct hit is forecast, and you still have time, you might want to find better protection along the southern shore of Grand Bahama or even across the *Gulf Stream* in Florida.

A waypoint at 26° 42.23' N, 79° 00.15'W, will place you approximately ½ mile west of the entrance channel leading into the marina as shown on Chart GB-1A (next column). From this position head straight in between the well-marked jetties and follow the channel around toward the fuel dock and slips. An alternative to the marina is the canals just west of the marina, but they will get trashed too.

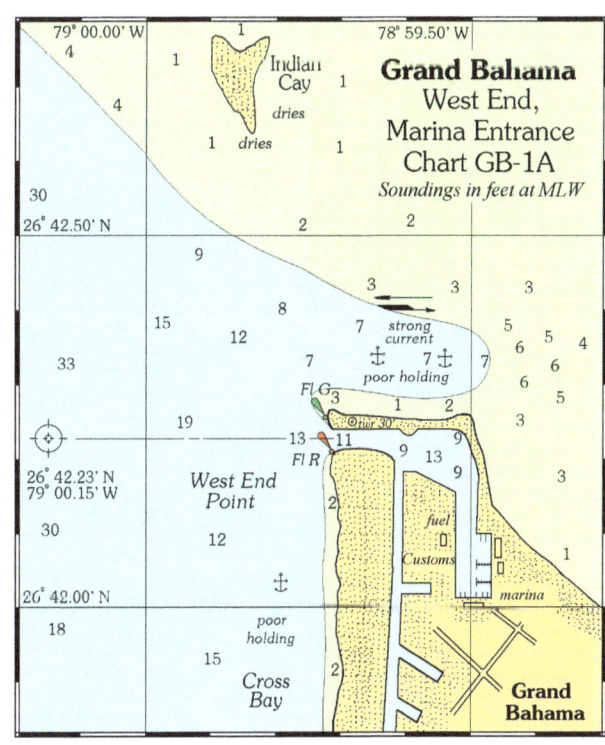

Grand Bahama: *Freeport, Bradford Yacht*

The only reason a recreational yacht would have for entering busy, heavily commercial *Freeport Harbour* in order to seek shelter from a hurricane would be to head for *Bradford Yacht and Ship* a short distance up *Hawksbill Creek* for a haul out or a slip.

As shown on Chart GB-2 (below), a waypoint at 26° 30.10' N, 78° 47.05' W, will place you approximately one nautical mile SW of the entrance channel to *Freeport Harbour*. From this position take up a course of approximately 021° magnetic and you will soon see the lighted (G) range inside the harbor. Keep an eye out for the huge offshore bunkers.

As shown on the chart, if you work your way up dredged *Hawksbill Creek* you will quickly come to *Bradford Yacht and Ship* (http://bradford-marine.com/), the only full-service yard on Grand Bahama (and primarily geared for larger vessels).

Bradford has a 150-ton *Travelift* (they even provide tie-downs), a 1200-ton floating dry dock that can haul boats to 235', and 20 slips with depths to 25'. You can reach *Bradford Yacht and Ship* by VHF on ch. 16, or by phone at 242-352-7711 or by fax at 242-352-7695.

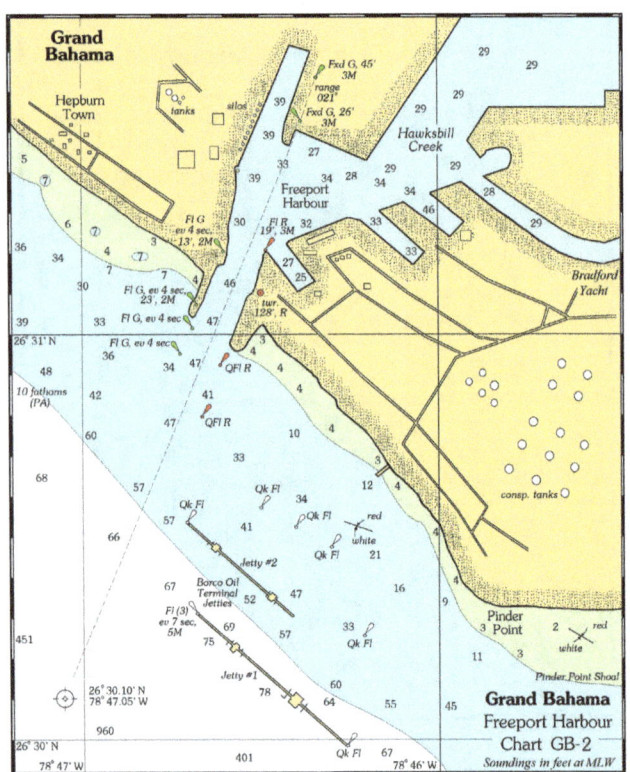

Grand Bahama: *Xanadu Channel*

Along the southern shore of Grand Bahama is a long and shallow barrier reef. Though there are quite a few shallow spots between *Xanadu Channel* and *Bell Channel* (mostly close to shore), the really shallow areas begin just west of *Bell Channel* so use extreme caution and keep a good lookout in these waters. It would be prudent to stay one mile offshore along this stretch of shoreline. All of the channels on the southern shore of Grand Bahama are dangerous to enter in strong southerly winds so be sure to arrive EARLY! The land here is low and flat and offers little resistance to the wind.

Beginning just a few miles east of *Freeport Harbour* are several channels leading to marinas and private developments. The first of these is *Xanadu Channel* named after the famed *Xanadu Hotel*. The marinas here are well protected and in some places one can even tie off in the more protected canals.

A waypoint at 26° 29.00' N, 78° 42.40' W (not shown on Chart GB-4 next page) will place you approximately ½ mile southwest of the entrance to *Xanadu Channel* between two marked jetties. The flashing red and green lights may not work, so bear that in mind if attempting to enter at night. From this position head towards the channel mouth between the jetties trying to remain parallel to the lay of the channel. If you stray a bit don't fret, it's fairly deep on either side of the channel with no hazards. Once inside the channel continue between the jetties and the marina and hotel will open up on your port side very soon.

Grand Bahama: *Sunrise Channel*

Less than a mile to the east of the *Xanadu Channel* is the channel leading into *Sunrise Marina* (formerly Runnin Mon Marina) and *Knowles Marina and Boatyard*. The channel has experienced shoaling in recent years and the controlling depth is now 5' at MLW.

A waypoint at 26° 29.00' N, 78° 41.31' W will place you approximately ½ mile southeast of the jetties at the entrance to the channel. From this position take up a course of approximately 340° to enter the

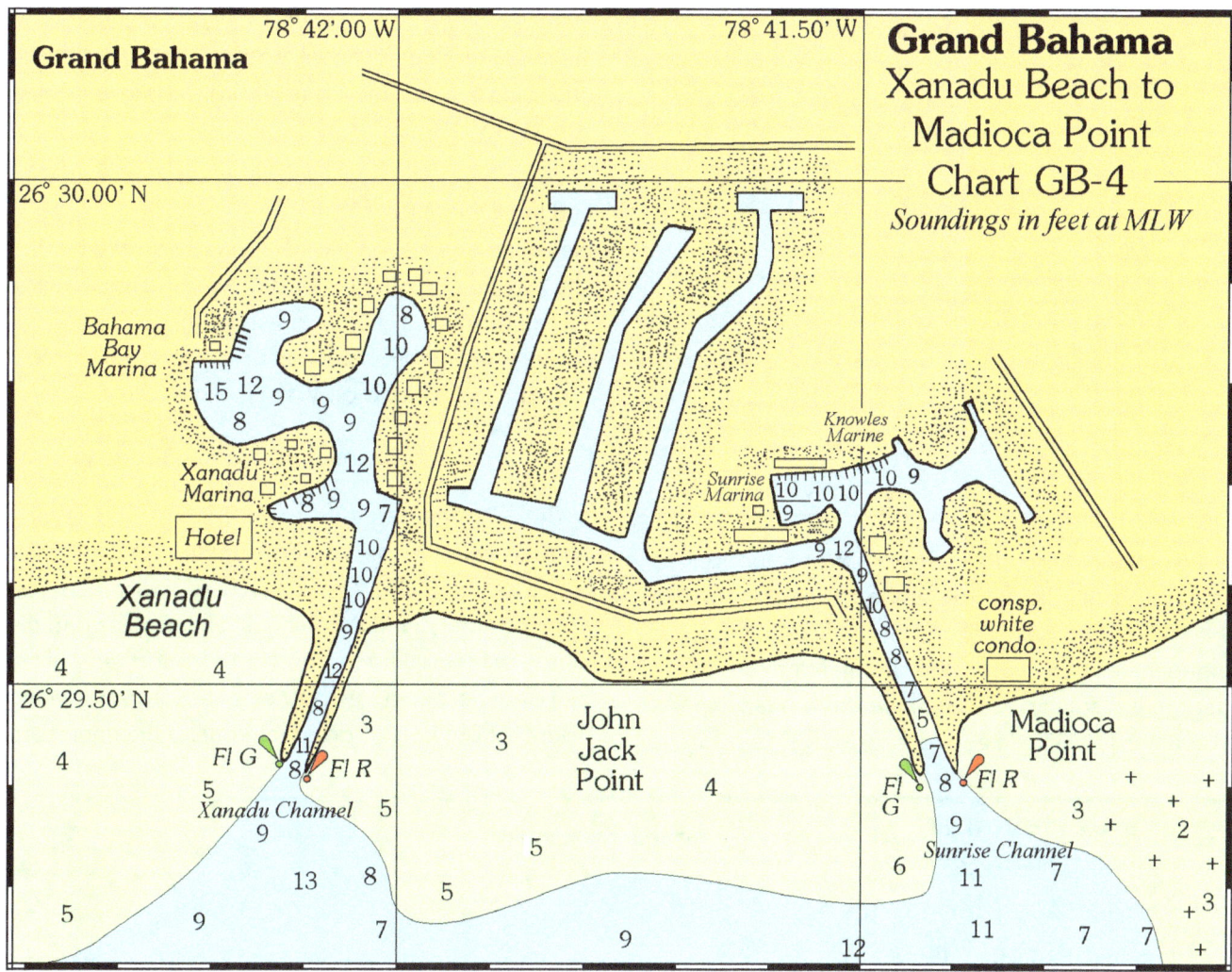

channel mouth between the two jetties shown on Chart GB-4 (above).

After you enter the channel you'll pass over a shallow spot with a depth of 5' at MLW so if you draw more, plan on using the tide to enter. The deeper water is to the west of center channel upon entry until past the shoal.

Once inside and past the shallow spot, forge ahead past the canal that works off to the west and *Sunrise Marina* will open up to port ahead while to starboard is *Knowles Marine*. Here you can get a slip or haul out. *Sunrise* has 70 floating slips that can accommodate yachts up to 110 feet. *Knowles Marina* has a 40-ton lift as well as stern-to and side-tie slips. At the end of the small canal to port upon entry is *Sea Breezes Marina* with 400' of frontage accommodating vessels to 50' LOA and drafts to 6'.

Grand Bahama: *Silver Cove, Ocean Reef Yacht Club*

Silver Cove is the name of the man-made canal network that lies approximately 2 miles east of *Xanadu Channel* and approximately 1½ miles west of *Bell Channel*. *Silver Cove* is home to the *Ocean Reef Yacht Club and Marina* and the *Hawksbill Yacht Club*.

You won't be able to haul out here but *Hawksbill Yacht Club* (242-373-1513) has 12 slips for vessels to 45' LOA, while the *Ocean Reef Yacht Club* (http://www.oryc.com/, 242-373-4661) has 50 slips and can accommodate vessels to 120' with a 6' draft though some slips are a bit deeper.

A waypoint at 26° 29.00' N, 78° 39.35' W will place you approximately ¾ nautical miles southeast of *Silver Point* and the jetties as shown on Chart GB-5 (see next page). Steer approximately 340° to enter

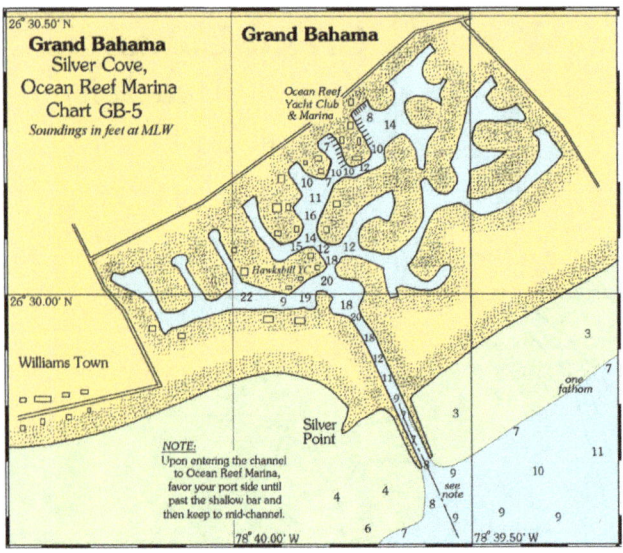

the channel. The course here is not as important as staying parallel to the lay of the channel inside the jetties as shown on the chart. The shallower water lies along the eastern side of the channel and to a minor degree just south and east of the eastern jetty tip. As you enter the jetties keep to your port side a bit, as there is a small bar along the eastern jetty just inside the channel. You will have 7' at MLW around this bar and once past the end of the jetties the canal becomes very deep, 18'-20' in most places.

The *Hawksbill Yacht Club* lies to port just inside. To find the *Ocean Reef Yacht Club and Marina* keep in the center channel as it winds around past two curves. The marina and office will open up on your port side; you can't miss it. If you are having any problems finding your way inside or to the marina, contact *Ocean Reef Marina* on VHF ch. 16.

Grand Bahama: *Bell Channel Bay*

Bell Channel Bay offers two marinas and several canals where a person can anchor. A waypoint at 26° 29.95' N, 78° 37.30' W will place you approximately ¾ mile southeast of the jetties and the entrance to *Bell Channel* and just a bit northeast of the *Bell Channel Sea Buoy* (Fl W ev 2 sec). If approaching from offshore keep a good lookout for the unmarked

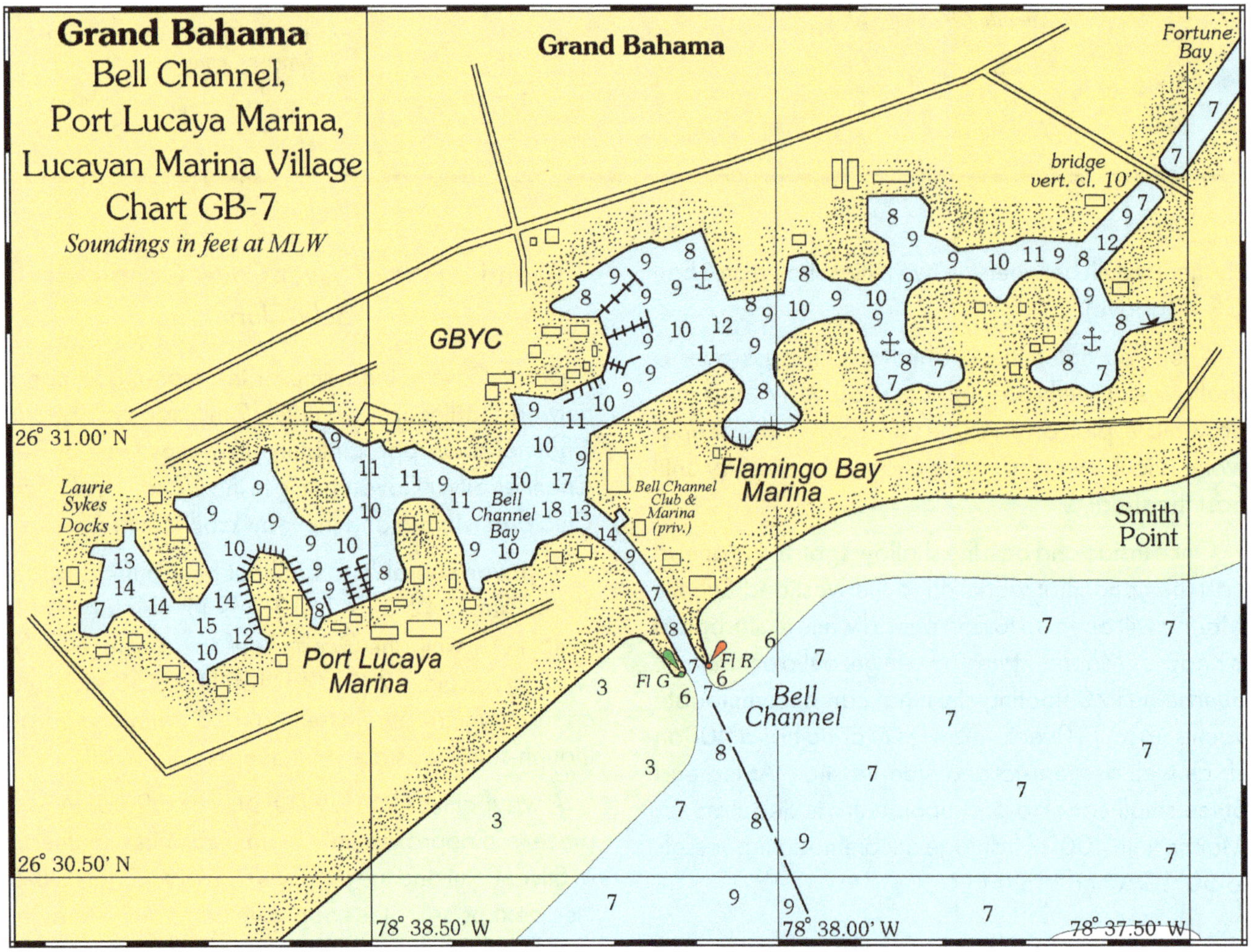

28 • THE CAPTAIN'S GUIDE TO HURRICANE HOLES

buoys to the west of this waypoint; they are for cruise ship use and constitute a hazard to the small boater. At night a flashing amber light marks their field but the individual buoys (7) are not lit. From the waypoint take up a course of approximately 340° towards the entrance between two flashing red and green markers and the lit jetties.

On the starboard side of *Bell Channel* upon entry is the small and private *Bell Channel Marina* (http://www.bellchannelclub.com/index.html), it is only for those who own condos on the adjacent property. However, once inside *Bell Channel Bay*, as shown on Chart GB-7 (previous page), if you turn to port you will find the spacious and upscale *Port Lucaya Marina* (http://www.portlucayamarina.com/, 242-373-9090), with 160 slips that can accommodate a vessel to 190' LOA, and vessels with a beam up to 40'.

Proceeding west past *Port Lucaya Marina* you will find *Laurie Sykes Docks* where owner *Laurie Sykes* has 15 slips. The docks can accommodate vessels to 100' LOA with beams to 20'. You can contact Laurie at 242-373-1344. This is a family run marina and very low key and friendly.

Proceeding inward and to starboard after the *Port Lucaya Marina*, you will come to the *Grand Bahama Yacht Club* which has just reopened (formerly *Lucayan Marina Village*). In a cove to starboard you will find the *Flamingo Bay Hotel and Marina* (http://www.flamingobaymarina.com/). The marina can accommodate 25 vessels to 100' LOA.

If you wish to anchor here, work your way into some of the coves to the east of the marinas and pick a spot (see photo below).

Grand Bahama: *Grand Lucayan Waterway*

If you need shelter the *Grand Lucayan Waterway* offers excellent protection from seas for vessels with drafts of over 7', but if you draw that much you can only enter the waterway from the south.

Approximately 3.5 miles northeast of *Bell Channel* lies the southern entrance to the *Grand Lucayan Waterway*, a man-made canal that effectively splits Grand Bahama in two. The entire waterway from the *Casuarina Bridge* southward is concrete walled with many small coves leading off the main canal.

Port Lucaya, *Bell Channel*

As shown on Chart GB-8, (next column) a waypoint at 26° 31.30′ N, 78° 33.33′ W will place you approximately one mile south of the entrance to the *GLWW* between two lit jetties. From the waypoint head north, keeping the sea buoy to starboard, and enter between the jetties to continue up the canal in good water

For shelter there is a nice cove that lies on the north side of the first canal that branches off to port (to the west as shown on Chart GB-8 and Chart GB-9 (below) as you head northward. This anchorage is known as *Top Gallant Basin.* Although the bottom was dredged when this area was built decades ago, the holding is very good in a sand/mud bottom and almost every little cove along the waterway offers a good anchorage with seawalls all around you, many are lined with vacant lots. If you plan to anchor off the *Grand Lucayan Waterway*, remember; since the lots all are privately owned, never tie up to someone's dock without permission.

Grand Bahama
Fortune Point to
The Grand Lucayan
Waterway
Chart GB-8
Soundings in feet at MLW

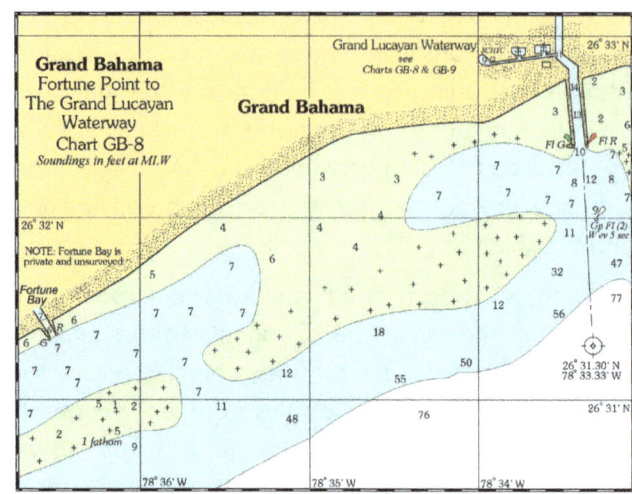

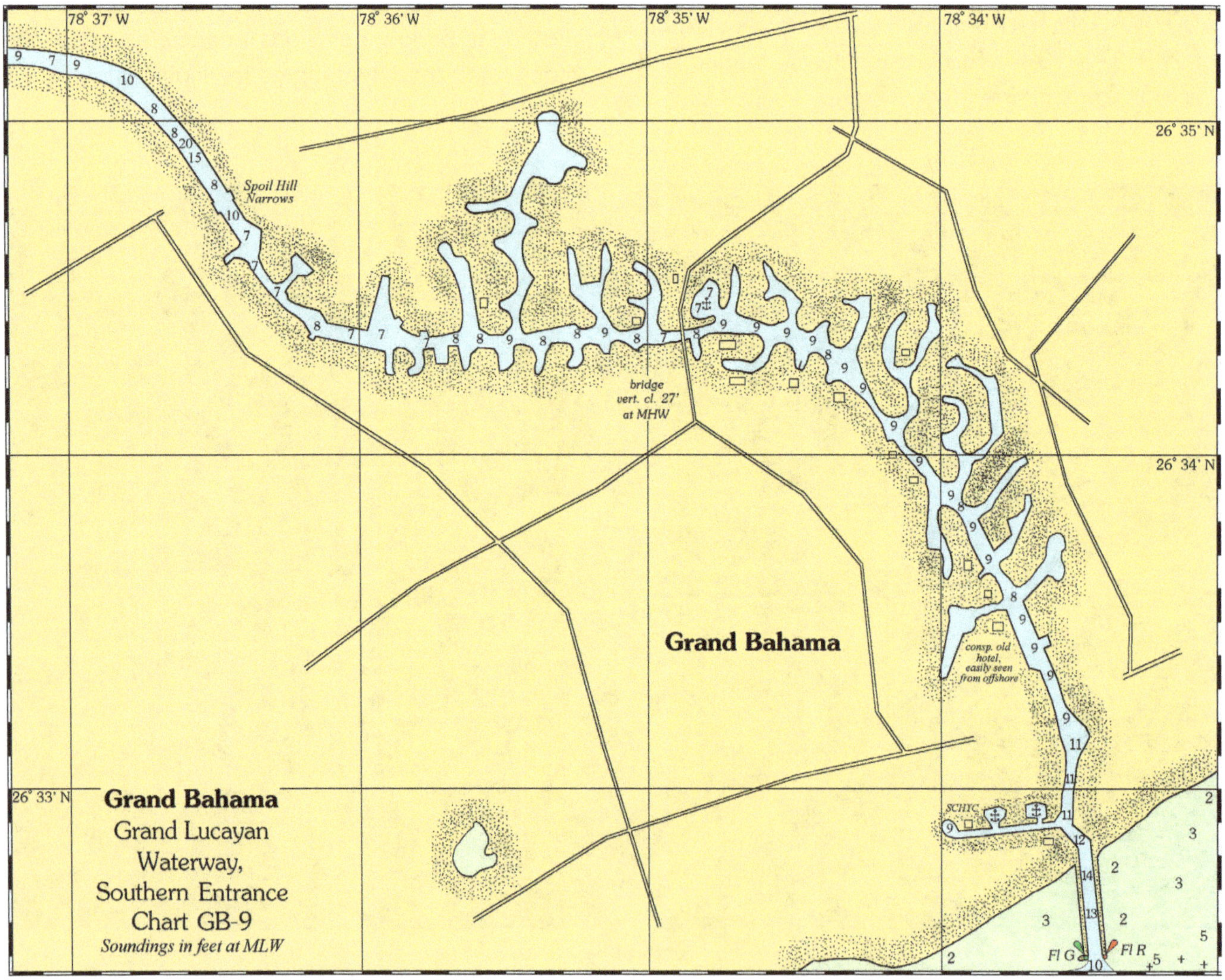

A few miles up the waterway is the *Casuarina Bridge* (see Chart GB-9 below), a fixed bridge with a 27′ vertical clearance at high water. This is the controlling height for the waterway and bars sailboats

Grand Bahama
Grand Lucayan
Waterway,
Southern Entrance
Chart GB-9
Soundings in feet at MLW

and tall powerboats from utilizing the shortcut through to the *Little Bahama Bank*.

If you're approaching the *Grand Lucayan Waterway* from the north, from Mangrove Cay, you must first clear the point of a sandbank that lies S/SW of Mangrove Cay and is marked by a lit, though not working, piling and light (FL W).

A heading of 200° from Mangrove Cay for ten miles will take you to this position. From here, head to a waypoint at 26° 38.40' N, 78° 39.70' W, which sits in 7' of water in *Dover Sound* as shown on Chart GB-10 (below).

From this position take up an approximate course of 152° to the first set of lit markers that will lead you in to the *Grand Lucayan Waterway* proper via a dredged channel. The water shallows to 3' here at MLW and the tide must be used if you draw more (note that the tide here is 2 hours behind the tide at Freeport or Lucaya).

Once between the two markers, split the rest of the pilings and you will come to a lit pair at the end of the channel at the entrance to the *Waterway*. Turn to starboard here keeping the two single pilings to starboard. The depths in the waterway itself run from 6'-12' and there is a bit of current that you will really notice at the *Casuarina Bridge* and at the conspicuous Spoil Bank Narrows where the *Waterway* narrows in width.

Grand Bahama: *South Riding Point Harbour*

Located approximately 35 miles east of Freeport and 15 miles east of Peterson Cay, there is a safe harbor to be found at the South Riding Point port facility. This is a commercial facility and I only mention it here in case someone needs a safe harbor in an emergency, especially a deep draft vessel. The harbour authority may tell you that you cannot enter but by Bahamian law nobody can deny shelter to a vessel in the event of a hurricane.

When approaching *South Riding Point Harbour* be sure to give the offshore bunkering terminal a wide berth. As shown on Chart GB-12 (below), a waypoint at 26° 36.60' N, 78° 12.75' W will place you approximately ½ nm SSE of the entrance channel. From the waypoint head NNW into the marked entrance channel, and follow it into the turning basin. Before entering the channel you must hail the terminal on VHF ch. 16 (*South Riding Point Shore Control*) for permission and instructions.

The Bight of Abaco: *Randall's Creek*

A waypoint at 26° 50.00' N, 77° 31.50' W, will place you approximately ¼ mile west of the mouth of *Randall's Creek* as shown on Chart AB-BI-5 (next page). A shallow draft vessel of 4' or less can work its way up *Randall's Creek* at the south end of Randall's Cay for protection from a major windstorm, but be warned that this anchorage has a lot of current and a rocky/marl bottom in many places.

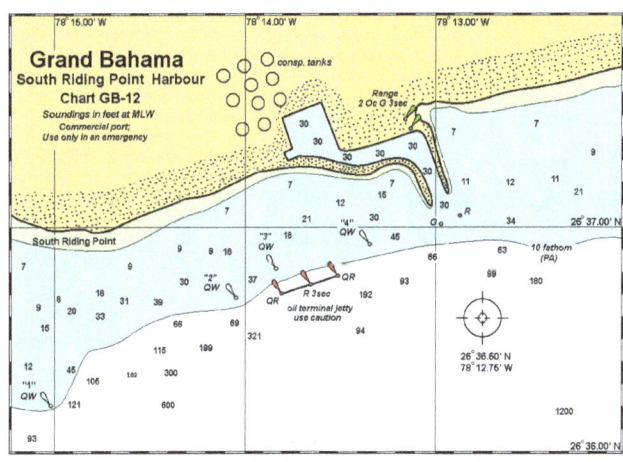

If seeking shelter, you might be better off two miles further south at Basin Harbour Cay.

The Bight of Abaco: *Basin Harbor Cay*

Basin Harbour Cay, besides being one of the prettiest cays in the *Bight of Abaco*, offers some of the best protection in nearly all weather conditions. The high rocky bluffs covered in cactus lead to an excellent anchorage that is only open to the southwest.

For all around protection it is best to head for the center of Basin Harbour Cay and its anchorage. As shown on Chart AB-BI-5 (see below), a waypoint at 26° 47.75' N, 77° 30.10' W will place you approximately ¼ mile southwest of the entrance to this anchorage. From this position enter the harbor and anchor where your draft allows. You will find that the harbour rapidly shallows to a fairly uniform 5' or so and gradually shallows as you approach the shore.

The anchorage itself is open to the southwest and west so if southwest winds are forecast it is best to anchor in the southwestern corner of the anchorage for the best protection. After the winds move into the west, and if you can, it would be better to move to the northern side of the anchorage where the water is a little deeper. In strong southeasterly through southwesterly winds, this anchorage develops a surge as swells work their way into the harbour around the southern tip of the harbour mouth. In normal conditions his surge is not dangerous, only uncomfortable, but in a hurricane it could be disastrous. Know the wind forecast before you choose to hide here. The bottom in this harbour is marl in places and the holding here is tricky at times, make sure your anchor is set well.

The Bight of Abaco: *Mores Island*

At the northern tip of Mores Island you will find two small creeks, the first, the westward one, is far

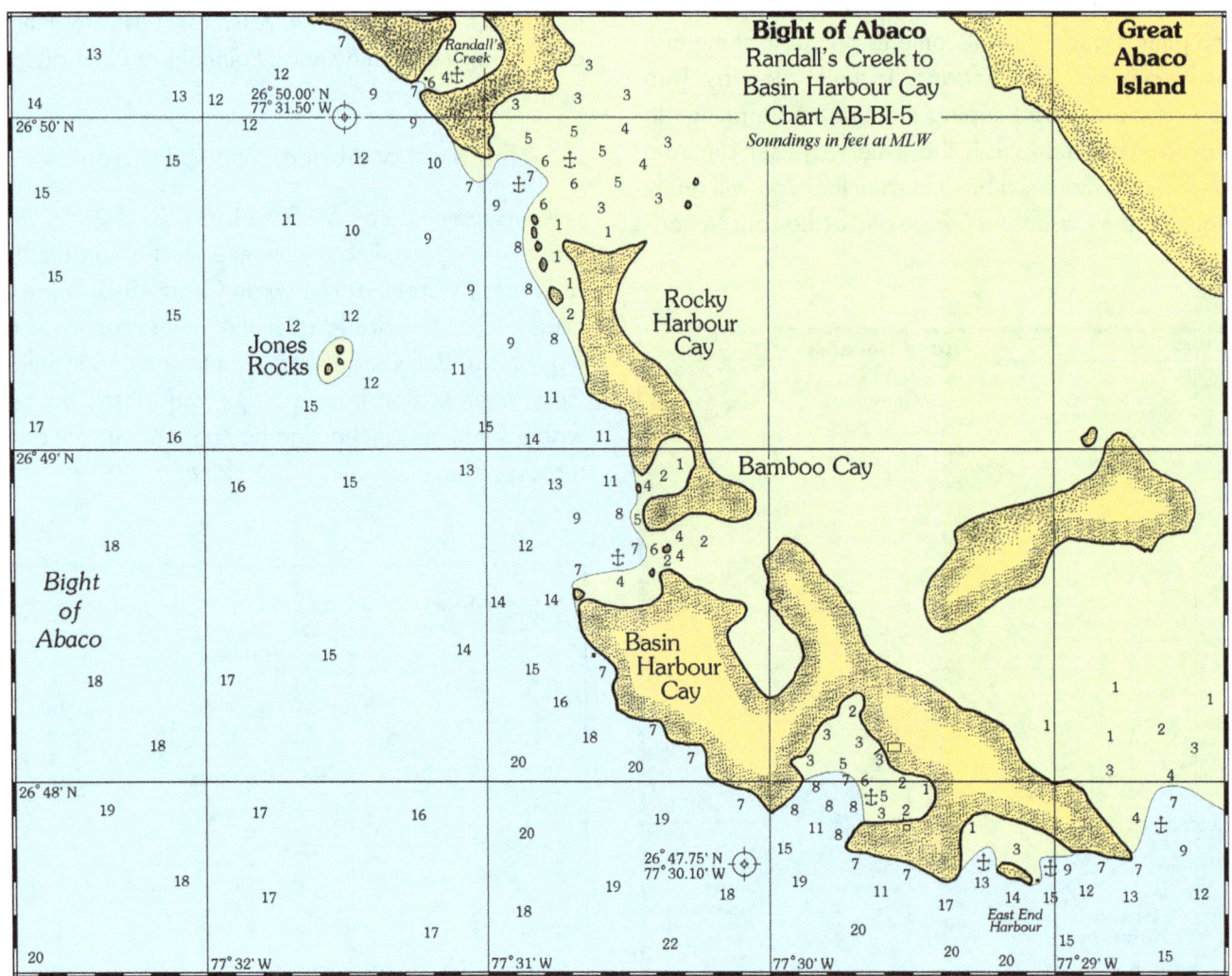

32 • THE CAPTAIN'S GUIDE TO HURRICANE HOLES

too shallow for anything of any real draft, perhaps a Gemini catamaran but that's about all. The second creek that you come to as you head eastward around the northern tip is quite a bit deeper, 5'-7' at MLW in spots, of course there are a few places where it's 2'-3' in spots also but this the only place in the vicinity to give you any sort of shelter from a major storm if your draft is not too deep.

Approaching Mores Island head for a waypoint at 26° 20.55' N, 77° 35.20' W (as shown on Chart AB-BI-13 below) which will place you about ¾ mile northwest of the northern tip of Mores Island. From here simply head for the creeks at the northern tip of Mores Island

The bar at the mouth of the creek will take drafts of less than 5' at high tide, and 5' on a very high tide. This is one of those places you should check out with the dinghy first.

The entrance channel is straightforward, steer for the middle of the opening parallel with the lie of the creek itself and squiggle around any shallow spots that you come upon.

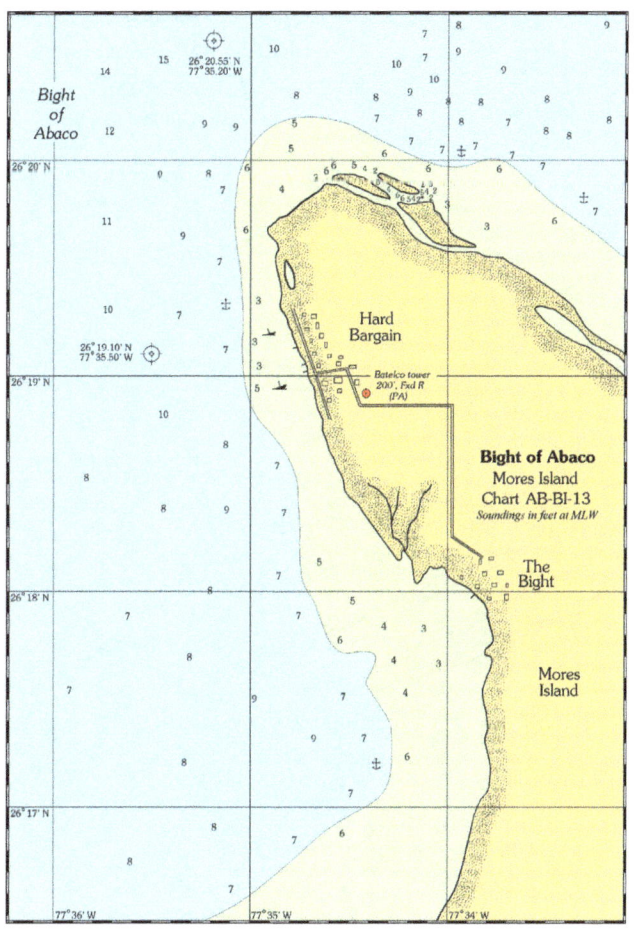

Once over the bar, about 3'-4' at MLW, you will be in 5'-6' at MLW. You can venture farther up the creek but the channel shifts around some shallow, grassy bars of about 2' at MLW. Five feet can be carried all the way to the end of the creek and you can even pass between the small mangrove islets and the large cay to the northeast. But that channel will end in a 2' grassy bar as shown on the chart. It should go without saying that you'll need two anchors set in a Bahamian moor here as there is a lot of tidal current. The creek is open to the northwest but even a strong northwest wind sends very little seas into the creek, the bar breaks most of it.

Abaco: *Grand Cays*

The Abacos are probably the most popular islands as far as the number of visiting boats, and they are no stranger to hurricanes. The flat land offers little protection from the wind so one must concentrate on finding a hole that gives good shelter from the seas while maximizing protection from the wind. We will discuss the Abacos from north to south.

Starting in the north, Grand Cays offers minimal protection and is only mentioned here as an option, not as a prime hurricane hole. The local marina, *Rosie's Place*, has wooden docks that are not in the best of shape and we cannot recommend getting a slip here to ride out a hurricane, instead, we suggest that you anchor in the harbour as it offers fair protection although the holding is iffy in spots.

To enter the harbour at Grand Cay, head for a waypoint at 27° 12.60' N, 78° 18.80' W. At this point you will be approximately ½ mile south of the entrance channel to the protected anchorage between Little Grand Cay and Sea Horse Cay as shown on Chart AB-4 (next page).

From the waypoint head north and then northwestward as you pass south of Felix Cay and the light on its western tip. The channel here carries about 6' at MLW and winds its way west/northwest to the anchorage area off Little Grand Cay. The deepest water is just as you enter, and it shallows the further in you go to the west and north.

Make sure your anchor is set well in the harbor at Grand Cay, the holding is tricky (it is marl in spots).

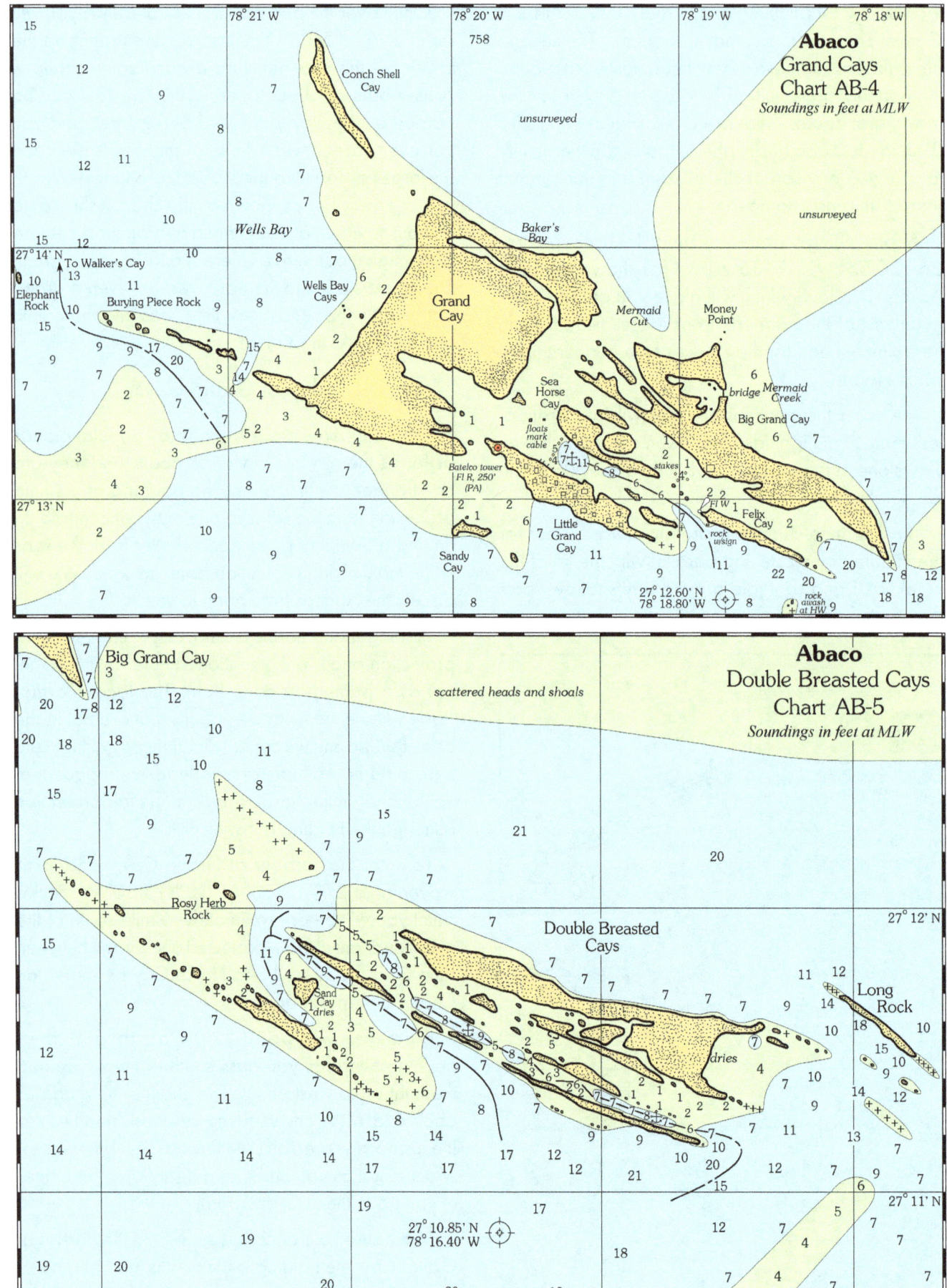

Abaco: *Double Breasted Cays*

Like the Grand Cays, Double Breasted Cays offers minimal protection and is only mentioned here as an option, not as a prime hurricane hole. The best protection lies in the creek between the cays at Double Breasted Cays.

A note about anchoring in the creek between the cays at Double Breasted Cay: the creek has great holding, but you must use two anchors in a Bahamian moor, there is a lot of current here. Don't be careless with your scope, you can't afford to have a lot of slack in it. If you're anchored NW/SE and a SW/NE wind should appear, it will push you to the side of the channel even though your anchors are holding. If you're expecting to ride out a blow here, you must place anchors in those directions also (don't worry too much about seas though, nothing builds up in this narrow channel).

A waypoint at 27° 10.85' N, 78° 16.40' W will place you approximately ½ mile south of the cays as shown on Chart AB-5 (previous page). You must be careful when entering this creek and once in it, the farther you go up the creek, the more hazards you must avoid. From the waypoint, steer towards the southeastern tip of Double Breasted Cay as shown on the chart. Pass around this tip and immediately turn to port, and head up the creek paralleling the lie of the channel. The entrance has a shallow bar of 6' at MLW with an even shallower spot just north of the channel, so be careful when entering and leaving. You might wish to check it out by dinghy first.

Once over the first shallow spot you'll find yourself in water from 7'-9' deep. You can anchor here if you like but better protection lies further up the channel as shown on the chart. If you desire to proceed up the channel you must first make sure the sun is not in your eyes or you'll never see the bars that you must steer around. Conversely, you cannot leave too early in the morning as the sun is again in your eyes and you'll have a devil of a time trying to trace your entrance route.

As you proceed up the creek from the entrance you'll notice that I've tried to mark on the charts the locations of the shoals you will find. For the most part these are easily seen as you approach them if the sun is overhead. They are mostly sand and grass, but they are very shallow, 1'-2' in places. You can work your way through the entire creek with the tide, but it's far safer to enter and leave via the southeastern tip of Double Breasted Cay.

Abaco: *Carter's Cays, Hogsty Harbour*

Between Top Cay and Old Yankee Cay is the entrance to *Hogsty Harbour*, sometimes called *Safety Harbour*, the local hurricane hole. The entrance is straightforward from the north over a 3' bar at MLW. Inside you'll find room for two good sized boats in water 7'-10' deep.

As shown on Chart AB-7 (see below), a waypoint at 27° 03.80' N, 78° 01.15' W, will place you about a mile southwest of Gully Cay in water from 7'-9' in depth. From here, the closer you get to the Carters Cays the shallower the water becomes. If you're approaching from Great Sale Cay, once you clear Little Sale Cay in Barracouta Rocks Channel you can head directly for the waypoint at Carters Cays.

If you are approaching from the SE, from Angelfish Point, Fox Town, or perhaps Allan's-Pensacola, you must avoid the large sandbank south of Moraine Cay and the Carters Bank as you approach the waypoint.

From the waypoint southwest of Carters Cays you have two options to access *Hogsty Harbour*. The traditional route is to put the diamond shaped range on Gully Cay on your bow and approach it on a course of approximately 40°. When you're only about 50' off Gully Cay turn to port, roughly to the northwest,

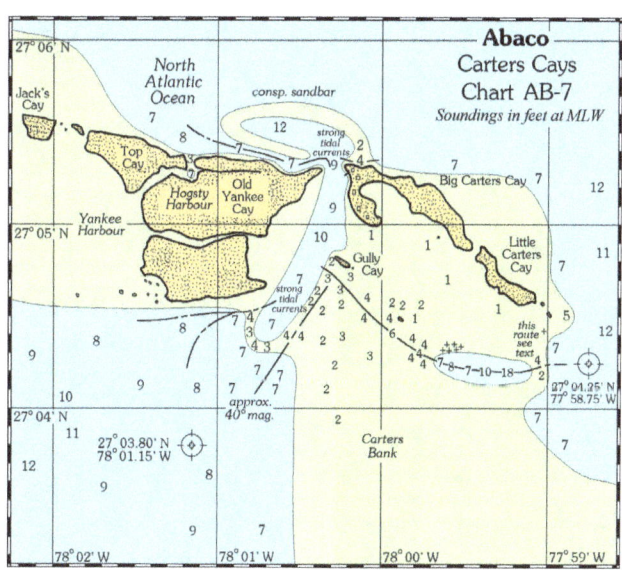

and cross over a shallow bar into the conspicuous deeper water.

This route used to work fine, but over the last few years this area has filled in and the controlling depth now is about 2'-2½' at MLW here. As you approach Gully Cay you'll notice lots of long marks in the bottom where people have drug keels and damaged props. This is no longer a boater friendly route for deeper draft vessels (over 4'). But you can use a different approach.

From the waypoint you'll notice the deeper blue water that lies southwest of Gully Cay and you should be able to discern the shallow bank that lies between you and that deeper water as shown on Chart AB-7. When used with the tide, this route saves a lot of wear and tear on your boat as well as your nerves. You can also cross the same bar at the extreme southwestern tip of the deeper water.

Between Top Cay and Old Yankee Cay is the entrance to *Hogsty Harbour*, sometimes called *Safety Harbour*, the local hurricane hole. The entrance is straightforward from the north over a 3' bar at MLW. Inside you'll find room for two good sized boats in water 7'-10' deep.

Abaco: *Allan's-Pensacola Cay*

Allan's-Pensacola Cay boasts one of the best hurricane holes in this section of the Abacos. Once two cays, Allan's Cay and Pensacola Cay, they were united during a hurricane in the not too distant past and remain as one today. At the eastern end of Allan's-Pensacola Cay is the aptly named Hurricane Hole. As shown on Chart AB-12 (below), the entrance is between two spits of land with your course favoring the southern spit. Don't stray north here as the water shallows quickly and the bottom is rocky. It's best to sound this route by dinghy first. This route will take 5' at high water and the shallowest spot is just inside

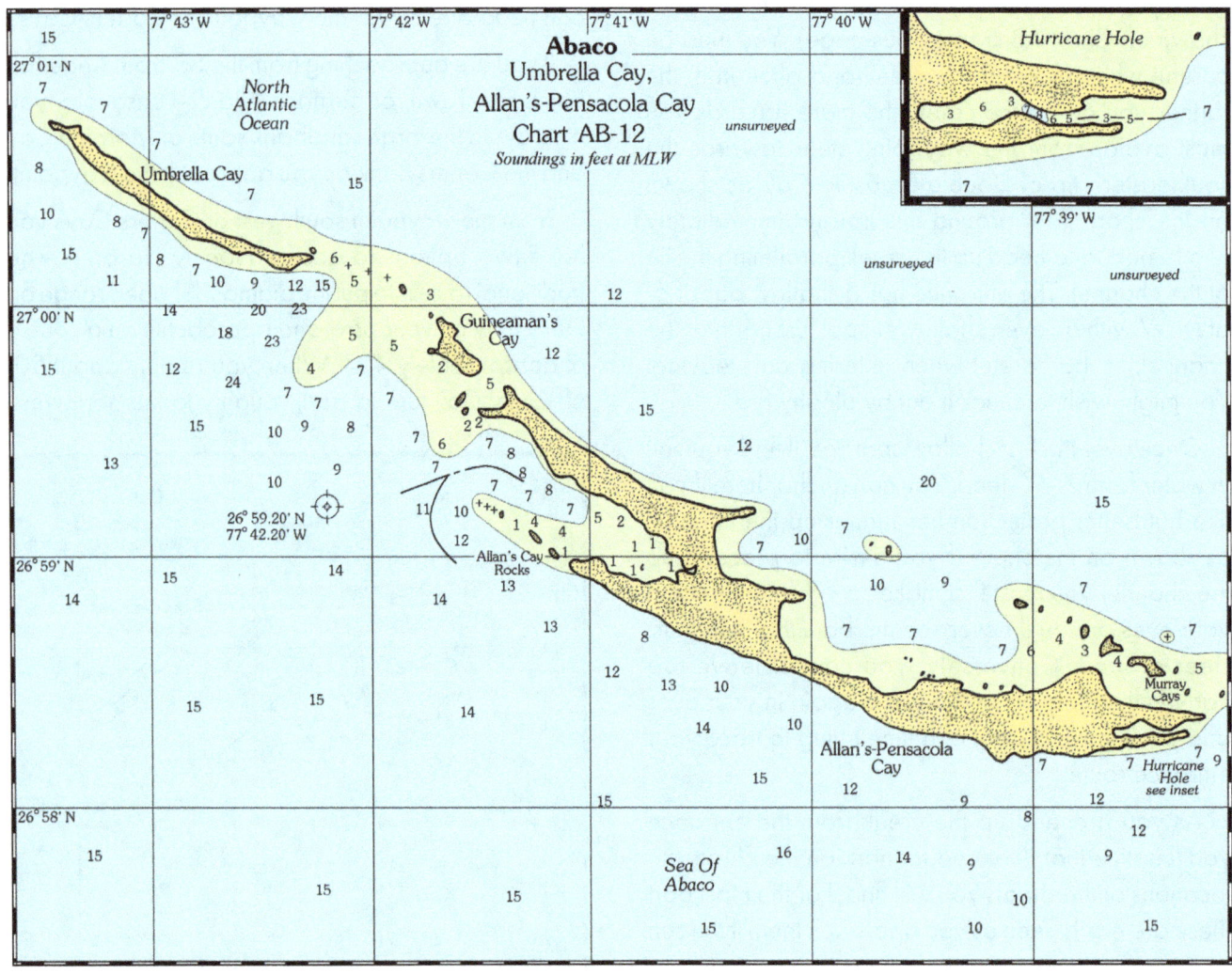

the end of the southern spit with about 3' at MLW. From here the water gets progressively deeper, 7'-9' in spots in the narrow creek that leads back to the small, shallow pond. The creek is lined with mangroves on both sides and if a hurricane threatened it would behoove you to tie up in the creek rather than in the small pond.

Abaco: *Green Turtle Cay, Black Sound*

For hurricane protection, Green Turtle Cay offers *White Sound* and Black Sound (both have all-around protection), and to a lesser extent, *Bluff Harbour* as shown on Chart AB-19 below. *White Sound* is highly recommended although there is a bit of fetch to allow seas to build up. *Black Sound* is a bit smaller, has a grassy bottom and a few concrete mooring blocks scattered about.

The entrance to *Black Sound* lies just north of the town of New Plymouth as shown on Chart 19. To enter *Black Sound* from the lee anchorage west of New Plymouth you must pass between the outer red (FL R) and green markers that lie between the mainland to the north and the point of land to the south (as shown on Chart AB-19A next column). Split the rest of the markers as you pass over the shallowest spot (4' at MLW) and continue past the large Abaco Yacht Services (AYS) yard on your port side and the *Other Shore Club* dock on your starboard side. When you're

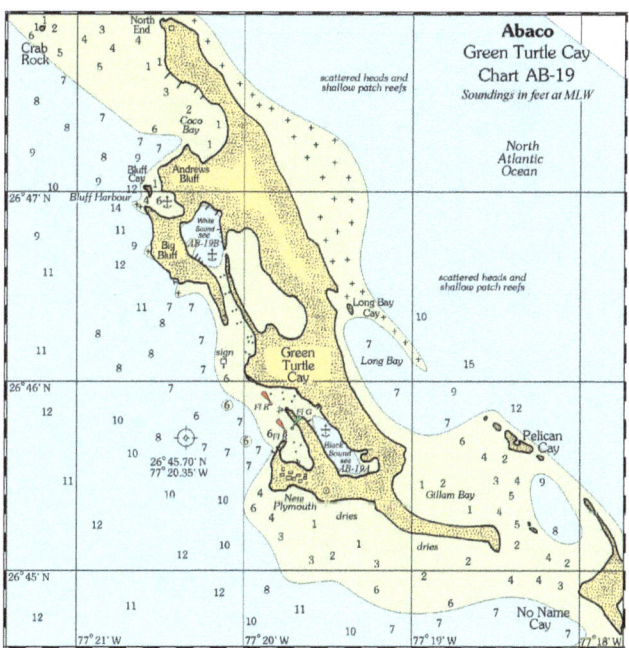

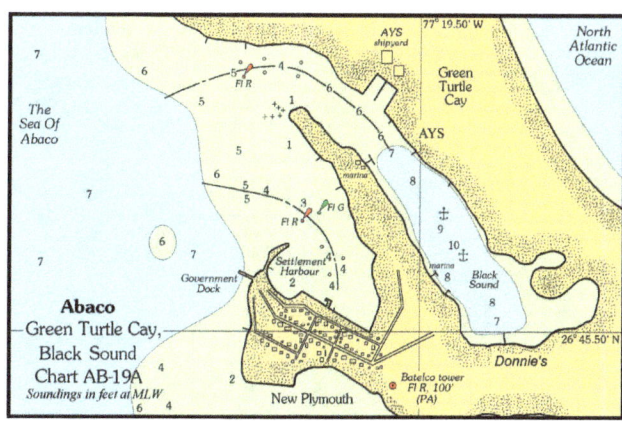

back in deeper water anchor wherever your draft allows. The bottom is quite grassy here so make sure your anchor is set well. If you seek to haul out, talk to the folks at *Abaco Yacht Services*.

Abaco: *Green Turtle Cay, Bluff Harbour*

A small, but very protected anchorage lies in tiny *Bluff Harbour* as shown on Chart AB-19 and Chart AB-19B (next page). The entrance lies south of Bluff Cay between Bluff Cay and the mainland of Green Turtle Cay and north of the entrance to *White Sound*. The entrance bar has almost 4' over it at MLW and you'll find 6'-7' and more in a small pocket inside. Watch out for the submerged rocks off the point to the south of the entrance channel. As protected as this anchorage appears, it is no place to be in strong west to northwest winds due to the seas they bring. Be advised that if you don't move EARLY local boats will fill up this tiny harbour.

Abaco: *Green Turtle Cay, White Sound*

The entrance to *White Sound*, as shown on Chart AB-19B (next page), lies a bit north of the entrance to *Black Sound* (as shown on Chart AB-19), between the two conspicuous signs, one of which sits well out into the water. The entrance is via a long, narrow (dredged to 30' wide), shallow (4½'-5'), but well-marked channel where you'll need the help of the tide.

To enter you will need to pass south of the outer sign, north of the inner sign (obviously!), and split the outer green and red buoys. You'll split a couple of pairs of markers here, and then as the channel curves to the north you'll want to favor the eastern side of the channel, hugging the stakes as you proceed northward into *White Sound*.

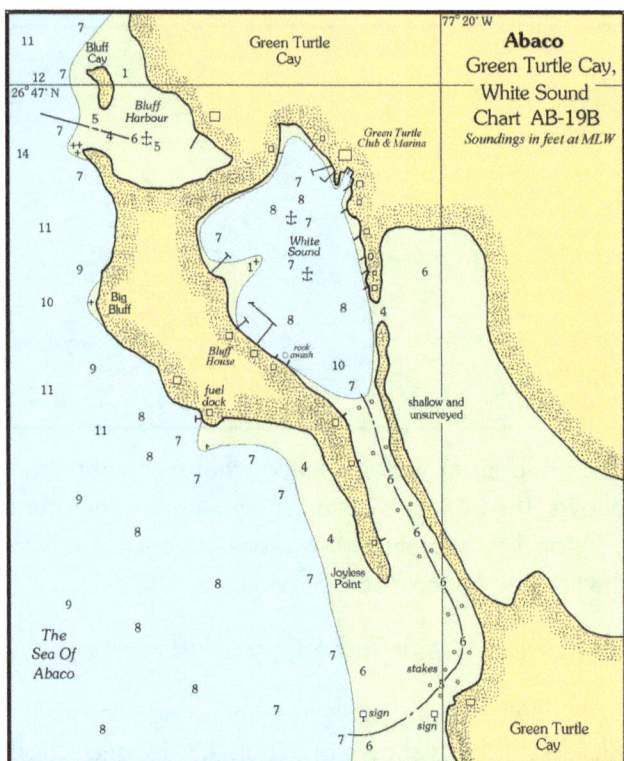

Mariposa, Abaco

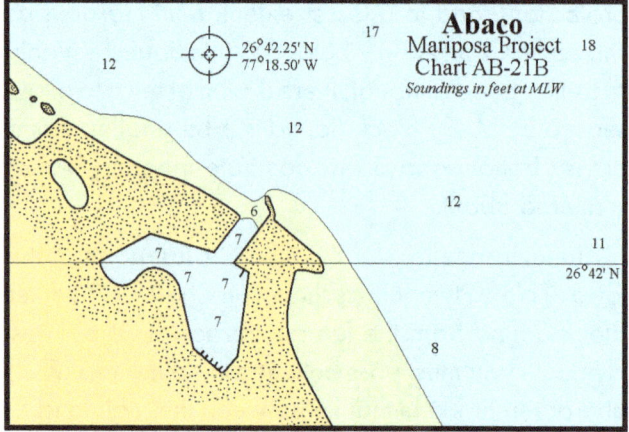

White Sound is a good anchorage as testified to by the 15 boats that road out Hurricane Bertha there in 1996. They had several hours of 115 mph winds and only two boats were damaged.

Abaco: *Mariposa*

Just north of Treasure Cay on the mainland is a jetty protected entrance into what was once a small lake called *Jackson Hole* (see Chart AB-21B at the top of the next column with aerial image). Today it appears to be a marina ready to open but is actually part of a private development by *Mariposa*. Some have suggested it as a possible hurricane hole, but it can be a bit surgy in northerly winds and seas. If you plan to enter here to find shelter, a waypoint at 26° 42.25' N, 77° 18.50' W, will place you approximately ¼ mile NNW of the entrance channel as shown on the chart. The lake was dredged and bullheaded a few years ago, then all development stopped, and its future is uncertain, use this harbour at your own risk.

Abaco: *Treasure Cay Marina*

Some of the best protection in the Abacos lies in and around *Treasure Cay Marina* where you can get a slip or anchor in the narrow canals surrounding the marina complex (see photo above). The marina now has a cabled boom system to restrict entry to the channel north of the fuel docks between 2300-0500. Security can be reached on VHF ch. 16 or by phone at 242-365-8899 to provide emergency access.

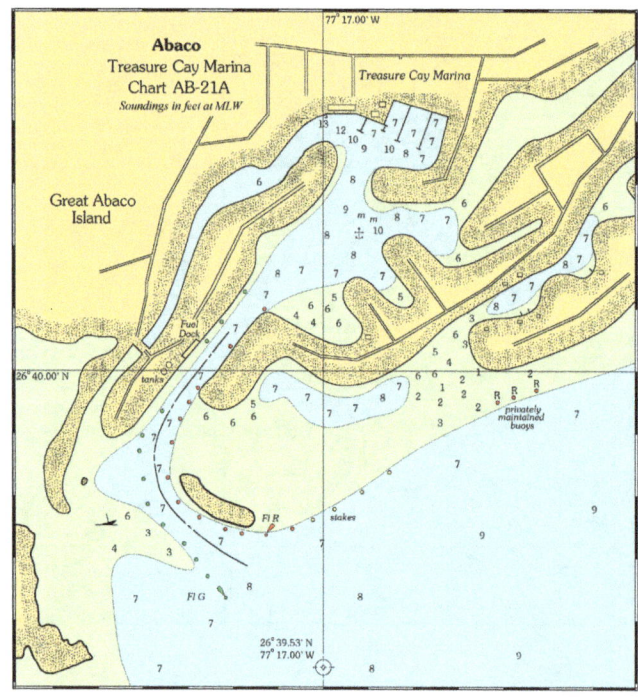

38 • THE CAPTAIN'S GUIDE TO HURRICANE HOLES

Treasure Cay Marina complex from the north, Abaco

Baker's Bay Marina, Great Guana Cay, Abaco

The entrance to *Treasure Cay Marina* lies along the southern shore of Treasure Cay just south of *Whale Cay Passage* and west of Great Guana Cay. As shown on Chart AB-21A (bottom of previous page), a waypoint at 26° 39.53' N, 77° 17.00' W, will place you approximately ¼ mile southeast of the entrance channel to the *Treasure Cay Marina* complex. From the waypoint you will see a string of markers leading past a small spoil island with a few trees on it. The entrance channel is to the west of this small spoil island as shown on the chart. Enter between the two rows of markers and follow them as they curve around to the northeast where you'll parallel the shoreline past the fuel dock and into the marina and canal area. If you don't want to anchor in the canals and refer a slip, Treasure Cay Marina (http://treasurecay.com/) offers 150 slips with a minimum depth of 7' at MLW.

Abaco: *Baker's Bay Marina*

Located at *Baker's Bay* on Great Guana Cay, *Baker's Bay Marina* (see photo above; http://bakersbayclub.com/) is geared towards providing service for members and condo owners although 150 slips are available for vessels to 250' LOA with drafts to 12.5'. You won't be able to anchor here but a slip gives pretty good protection in all but the worst storms.

If you are approaching *Bakers Bay* from the south, from *Marsh Harbour* for instance, you can head for a waypoint at 26° 40.50' N, 77° 09.80' W. This position lies approximately ¾ mile south of Bakers Bay as shown on Chart AB-22 (bottom of previous column). If you're approaching *Bakers Bay* from *Guana Cay Harbour*, keep an eye out for a pair of shoals, one only 6' deep at MLW, the other not much better at 7' at MLW (as shown on Chart AB-22).

From the waypoint, head generally north until the marked channel opens up to starboard and you can enter the marina basin. The northern and eastern parts of marina offer the most protection.

Abaco: *Leisure Lee and Hill's Creek*

There is a man-made canal complex called *Leisure Lee* lying just south of Treasure Cay on Great Abaco. Here you will find excellent protection from seas in 7'-8' but you will have to tie off to the trees along

THE NORTHERN BAHAMAS • 39

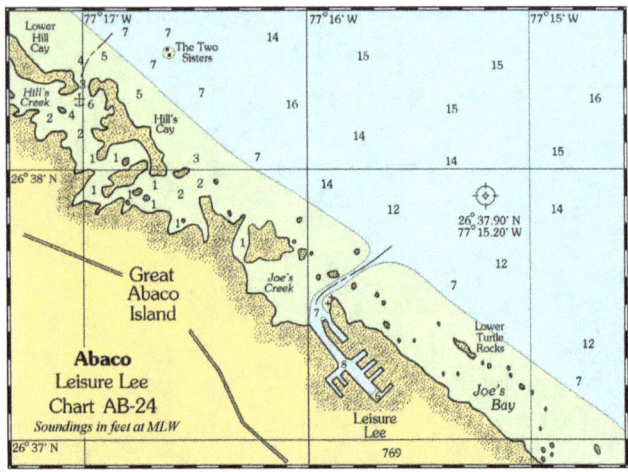

the shore as the entire complex is dredged and the holding is not good.

As shown on Chart AB-24, (above) a waypoint at 26° 37.90′ N, 77° 15.20′ W, will place you approximately ¼ mile northeast of the entrance channel into *Leisure Lee*. A range that sits northwest of the conspicuous condos marks the entrance channel to *Leisure Lee*. The range consists of a white "X" in front and a white rectangle behind it.

As you enter keep the rock with the small tree on it to port and head in on the range. As you approach the point of land northwest of the condos give it a wide berth if you can; there are submerged rocks just off the point. Round this tip of land to port and you can enter the canal complex. Quite a few of the lots have not been sold and there are plenty of places to secure your vessel in the event of a hurricane.

You will probably have to set your anchors ashore as the bottom has been dredged and the holding is not very good. Please don't tie up your vessel so as to block homeowners here and don't block or tie up to their docks without permission.

There is small and seldom used anchorage north of Leisure Lee at Hill's Creek. The entrance is between Hills' Cay and Lower Hill Cay and will barely take a 5′ draft at high water. Inside is a small area of deeper water in which to anchor. I strongly urge you to check

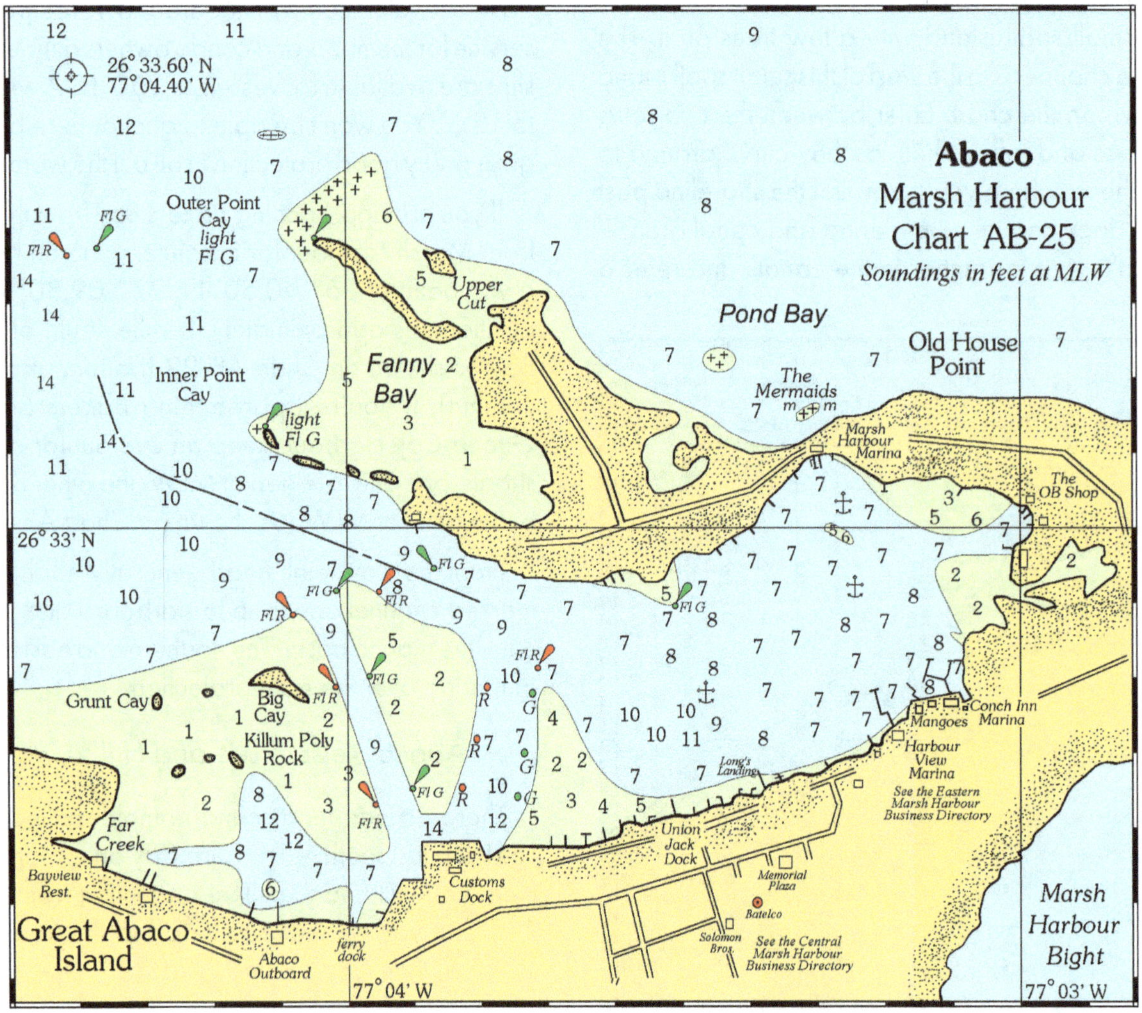

40 • THE CAPTAIN'S GUIDE TO HURRICANE HOLES

out this route by dinghy first before entering. This would make a fair hole for a shallow draft vessel.

Abaco: *Marsh Harbour*

Marsh Harbour (see photo below), although open to the west, has a wonderful sand/mud bottom that anchors so love. When Hurricane Fran blew through in 1996, the wind came from the west and 6' seas in the harbour were the norm (very rough in only 50 mph winds!) and most boats survived though the marinas took a lot of damage. The harbour is open to any westerly wind so keep that in mind when selecting a spot to anchor or a slip in which you can secure your vessel.

As shown on Chart AB-25 (previous page), the entrance to *Marsh Harbour* is fairly easy, although with the construction of the marked channel leading to the *Customs* docks, newcomers are often confused, following the marked channel to the turning basin off the dock instead of entering the harbor proper.

As shown on the chart, a waypoint at 26° 33.60' N, 77° 04.40' W, will place you approximately ¼ mile northwest of the shallow reef off Outer Point Cay. From the waypoint you can head south until you clear Inner Point Cay and can parallel the shoreline eastward into *Marsh Harbour*, this is where newcomers often get confused. The channel to the *Customs* dock is well-marked with buoys beginning just south of the waypoint given. Don't follow the markers to the *Customs* dock; follow the course shown on the chart to your chosen anchorage or marina.

Most of the harbor is only about 7'-8' deep at MLW but there is a deeper trough where the water is 10'-14' deep in places. This area lies to the east of the large shoal lying north of the *Customs* dock and just north of the *Union Jack Dock*, the large dock with all the dinghies tied to it.

Marsh Harbour as seen from the west; *Boat Harbour Marina* upper right

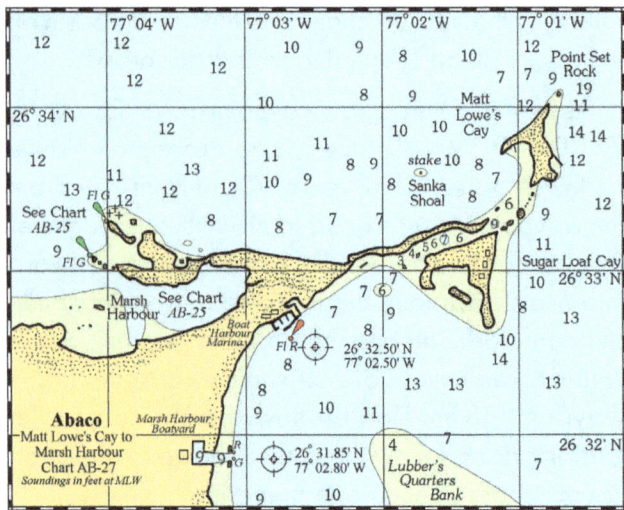

At the eastern end of *Marsh Harbour*, just south of the old *Sunsail* docks, is the shallow entrance to a small creek. This creek is good for drafts of 4' at high water only and is an excellent hurricane hole for shallow-draft vessels up to 40'. If you need this protection better plan on getting there very, very EARLY or the charter boats may beat you to it.

You can anchor in the harbour, get a slip at any of the marinas, and if you head to the south side of the peninsula that *Marsh Harbour* sits on, you will find *Boat Harbour Marina* (where you can find a protected slip) and *Marsh Harbour Boatyard* (where you can get hauled) as shown on Chart AB-27 above.

Abaco: *Man-O-War Cay*

Man-O-War Cay offers excellent all-around protection, a quality marina, and a place to haul out (although if a hurricane threatens you'd better be on the list a long time before its arrival.

As shown on Chart AB-28A (top of next column), a waypoint at 26° 35.30' N, 77° 00.22' W, will place you approximately 250 yards south of the entrance channel.

From the waypoint enter the channel midway between the two points of land as shown. The point of land to port is actually the southern tip of Dickie's Cay and is marked by a white flashing light. If you're heading into the *Eastern Harbour*, once you split the first pair of markers, you can turn to starboard and keep the next marker to port as you avoid the shoal area off the northern shore of the harbor and head into *Eastern Harbour* proper. *Eastern Harbour* (see

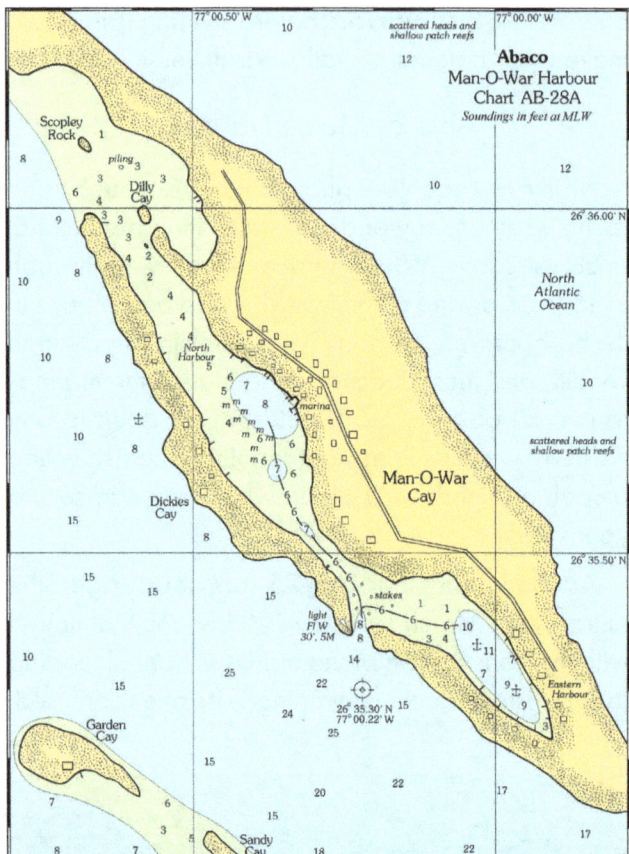

Eastern Harbour, Man-O-War Cay, Abaco

photo above), sometimes called *American Harbour*, is primarily used for vessel storage and there are a lot of moorings here. Do not pick up a mooring here with a boat name on it, or one that has a "reserved" sign on the float. Anchor wherever your draft permits and beware of unattended vessels during the storm if you stay aboard.

If you intend to enter *North Harbour*, enter the main channel as described above and instead of turning to starboard to head into *Eastern Harbour* bear more to port splitting the next pair of markers and following the channel into *North Harbour*. You'll find that *North Harbour* is usually crowded and the best idea (if you can't haul out) is to get a slip at *Man-O-War Marina*. The northern entrance to the harbor is shallow and should be avoided by all but the shallowest draft vessels and small outboards.

Abaco: *Hope Town Harbour*

Just to the south on Elbow Cay, *Hope Town Harbour* (see photo below) boasts very good protection. This harbour is usually very crowded with crewed and unattended vessels so bear that in mind before making the decision to come here.

If you arrive early enough and your draft is shallow enough you may be able to work your way up the creeks in the southern part of the harbour for better protection. There is an old hurricane chain stretched across the harbour to which you may be able to secure your vessel. Ask any local where to find the chain, perhaps they'll know and will tell you. There are several marinas where you might find a slip, and although the harbour prohibits anchoring (cruisers must take a mooring), nobody should say anything about anchoring during a threatening hurricane. You can also inquire about a haul out at *Lighthouse Marina*.

The entrance to Hope Town is fairly shallow and should be done on a high tide if you draw over 4'. From Point Set Rock you can head north of the Parrot Cays (the northernmost cay is marked by a flashing

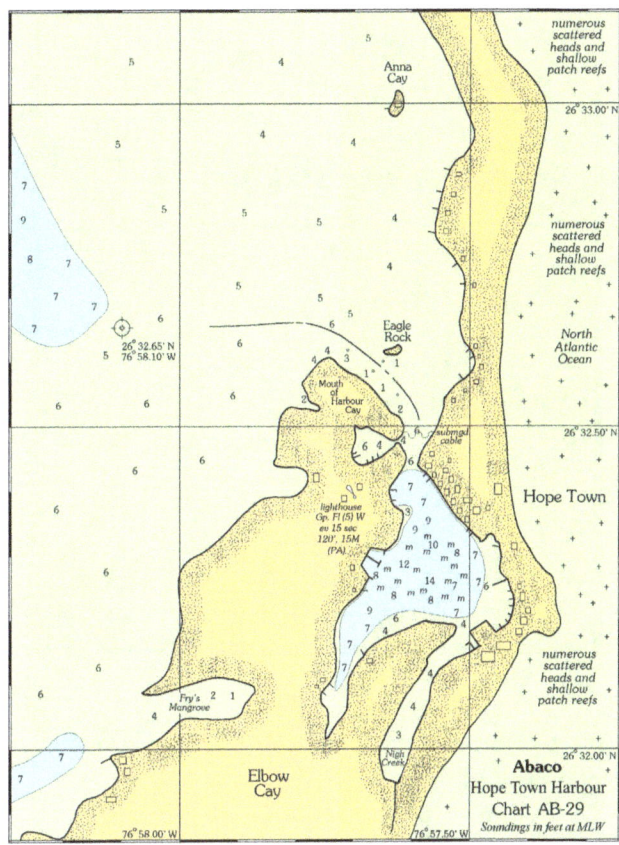

red light) for a waypoint at 26° 32.65' N, 76° 58.10' W, which will place you approximately ½ mile west of the entrance channel to the harbor as shown on Chart AB-29 (above).

From the waypoint head eastward as if to pass north of Eagle Rock. Use caution in this area as it's pretty shallow...deeper draft vessels can proceed towards the conspicuous quarry on the western shore of Mouth of Harbour Cay before turning northeast towards the entrance channel to avoid some 5' at MLW spots.

As you approach the channel between Eagle Rock and Mouth of Harbour Cay, take a look down the channel towards the shoreline of Elbow Cay where you'll see a concrete road leading away from the water's edge. Steer your vessel as if to head straight down this road (148° mag.), and you'll spot a range on shore consisting of two white poles with red reflectors. Favor the Eagle Rock side of this channel; there may or may not be some privately maintained markers here to show you the channel. When *Hope Town Harbour* opens up to starboard, turn and enter the harbour mouth staying approximately mid-channel. Before you actually enter *Hope Town Harbour* you'll see a small cove to starboard, this is

Hope Town Harbour, Elbow Cay, Abaco

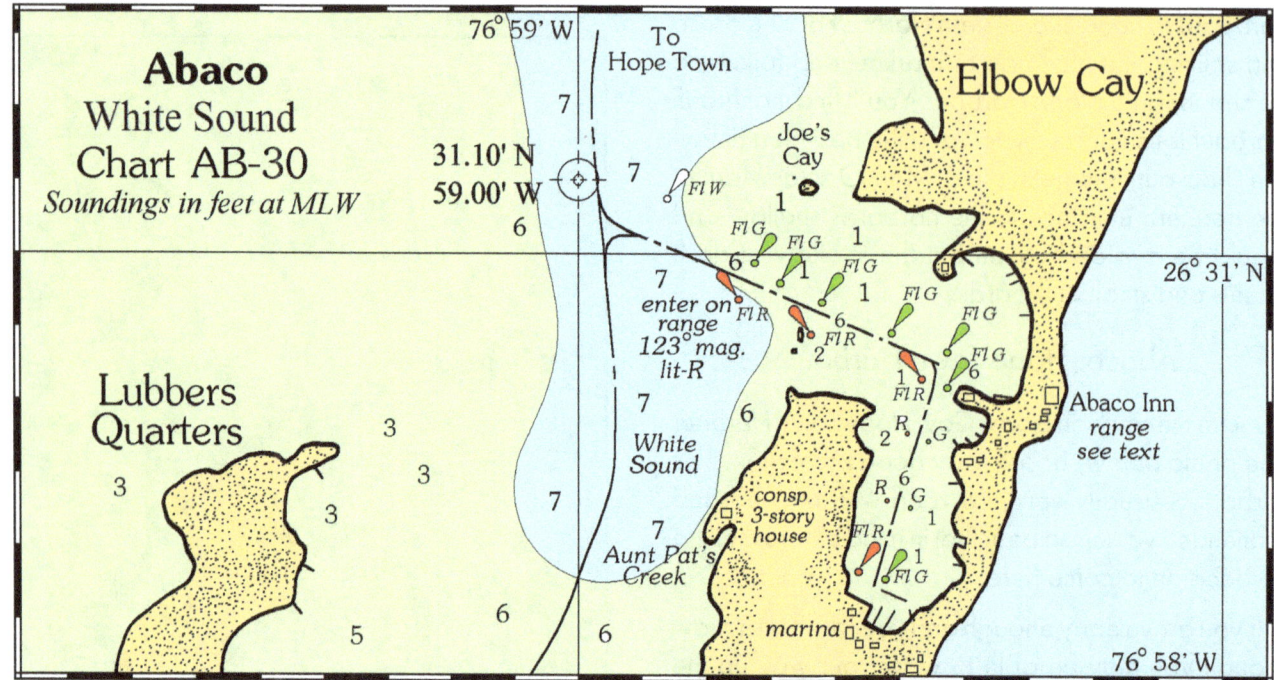

used primarily for storage of vessels and haul outs at *Lighthouse Marina*.

Abaco: *White Sound*

White Sound (see photo next column) lies about a mile north of the southern end of Elbow Cay and *Tilloo Cut*. Here you'll find a fairly straightforward entrance channel as shown on Chart AB-30 (above) that leads in to a protected marina. Be forewarned, a direct hit by Hurricane Floyd (Cat 4) in 1999 cut Elbow Cay in half, while Hurricane Jeanne (Cat 3) in 2004 breached the land to the east of White Sound and caused a temporary inlet to the Atlantic Ocean.

A waypoint at 26° 31.10′ N, 76° 59.00′ W, will place you approximately ¼ mile west/northwest of the entrance channel to *White Sound*. East of the waypoint you'll see a turning mark indicating the beginning of the dredged channel (6′ at MLW) into *White Sound* as shown on Chart AB-30. The entrance channel lies south of the light shown on the chart, and a range at the *Abaco Inn* (two red disks-lit red at night) will lead you in on a bearing of 123°M.

There really isn't any room to anchor here, you can't anchor in the channel and the rest of *White Sound* is too shallow, but the *Abaco Inn* offers complimentary dockage for guests of their restaurant and bar. Dockage here is Med-style, drop a hook and back

White Sound, Elbow Cay, Abaco

in to tie up stern to. Cruisers wishing to anchor can drop the hook south of the entrance as shown on Chart AB-30 in prevailing east/southeast winds.

If a slip is more to your liking, you can take the dredged channel south in *White Sound* to the *Sea Spray Resort and Marina* located at the extreme southern end of *White Sound*. The well-marked channel turns sharply southward (as shown on the chart) and as you approach the marina you will pass to the west of a jetty that protects the marina.

Abaco: *Tilloo Pond*

Tilloo Pond is a wonderful, well-protected little anchorage about halfway down the western shore of Tilloo Cay. *Tilloo Pond* will accept boats with drafts of less than 6′, and vessels drawing more than 3½′-4′ will have to wait for the tide to enter and leave.

This anchorage is a good spot to ride out a minimal hurricane, especially if you can duck in far enough behind Shearpin Cay as shown on Chart AB-32 (top of next column). However, If the hurricane is forecast to be greater than a Cat 1 you might not wish to hide here.

A strong west wind will create a bit of a surge in here as it funnels in through the entrance and bounces around inside so keep that in mind if the wind is forecast to come out of the west (which usually means the eye is passing north of you). The docks in the pond are all private so don't tie up there.

If approaching *Tilloo Pond* from the north, from the northern end of Tilloo Cay or Lubber's Quarters, you can parallel the shoreline of Tilloo Cay southward to a waypoint at 26° 27.00' N, 77° 00.50' W, as shown on Chart AB-32. This will place you approximately ¼ mile west of the entrance into the pond. If approaching *Tilloo Pond* from the south, from Middle Channel at Tilloo Bank, once clear of Tilloo Bank you can head for the waypoint at the entrance to the anchorage.

From the waypoint, the entrance to *Tilloo Pond* is fairly straightforward. Heading east, the entrance lies between Shearpin Cay and the small rock north of the cay. At the time of this writing there was a green marker just south of the small rock and if that is there when you arrive, keep it to port. The entrance channel lies roughly halfway between the small rock and Shearpin Cay. Once over the bar turn to starboard to anchor behind Shearpin Cay, do not turn to port as the water shallows almost immediately there.

Holding can be tricky in *Tilloo Pond* as the bottom is very grassy, make sure your anchors are set well.

Abaco: *Snake Cay*

Southwest of *Tilloo Pond* lies Snake Cay which has excellent protection in its mangrove lined creeks. In fact, Snake Cay's best attribute is that it is one of the best, and least known, hurricane holes in the Abacos.

If approaching from the north, from Witch Point or *Lubbers Quarters Channel*, you can head straight for a waypoint at 26° 26.90' N, 77° 02.70' W, which will place you approximately ¼ mile southeast of the southeastern tip of Snake Cay as shown on Chart AB-33 (next page) If approaching from the south, from

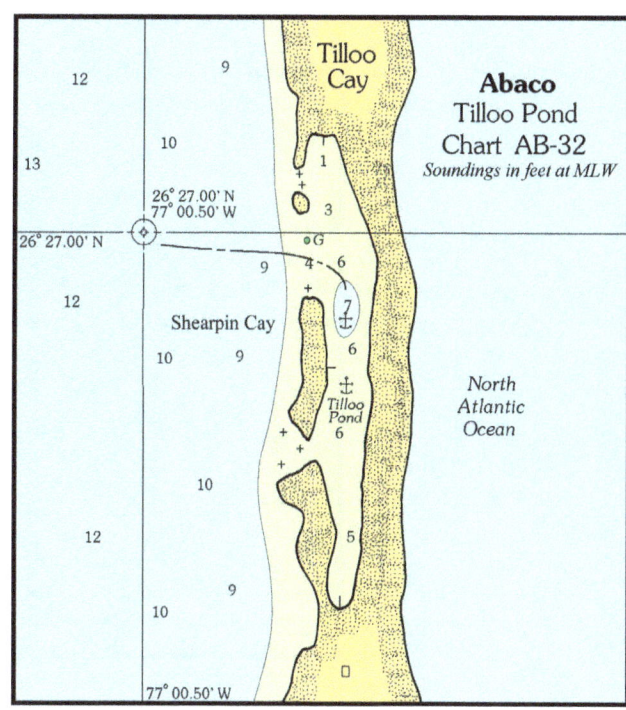

Middle Channel at *Tilloo Bank*, once clear of *Tilloo Bank* you can steer straight for this waypoint, Snake Cay will be clearly visible and bearing 310°.

As mentioned, Snake Cay is one of the best hurricane holes in the Abacos. There is a small, deep channel close in to the southern shore of Snake Cay that will take you back between Snake Cay and Deep Sea Cay and, if your draft allows, to a small deep pocket to the west of Deep Sea Cay. To access this area, follow close in along the southern shore of Snake Cay as shown on the chart. You'll pass between the shoal that lies northeast of Deep Sea Cay and the southern shore of Snake Cay in good water from 8'-20' in depth. There is a bit of current here so be prepared. You can anchor in the deep cut between Snake Cay and Deep Sea Cay or, if your draft allows, cross over the 3' bar (MLW) to head south/southwest to the conspicuous blue patch of water (7' at MLW) that lies west of Deep Sea Cay. Use caution here as the shallows surrounding the patch of deeper water are rocky, don't get blown up onto those areas.

Abaco: *Little Harbour*

Farther south you might consider *Little Harbour* (see photo at the bottom of the next column) though it is open to the north with a 4' bar across the mouth.

Approaching from the north or from *Little Harbour Bar*, head to a waypoint at 26° 20.10' N, 76° 59.95'

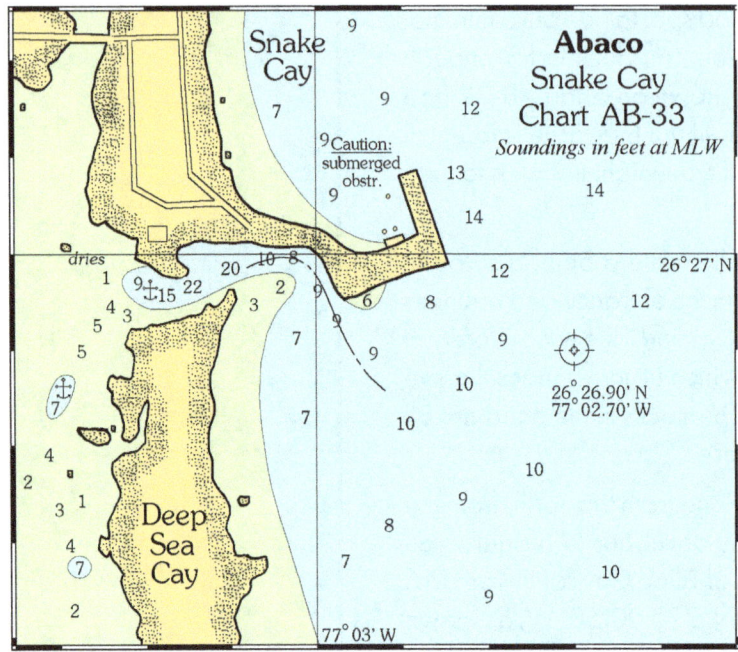

W, which will place you approximately ¼ mile north of the marked entrance channel into *Little Harbour* as shown on Chart AB-35 below.

From the waypoint, head generally south/southwest past Tom Curry's Point as shown on the chart. Here you'll be in deep water that gradually shoals to about 3½' in the marked channel just before you enter the deeper water of *Little Harbour*. This passage will require a rising tide for most boats.

The entrance channel into *Little Harbour* is marked (at the time of this writing) by four pairs of red and green markers and is easy to follow. It now has a new breakwater and the channel deepened. Once inside *Little Harbour* anchor wherever you desire. It is not advisable to take a mooring in the event of a hurricane, so when you anchor please make sure that your swing radius does not cause you to collide with any moored boats.

There are plans in the works to open a huge 44 slip marina in *Little Harbour* that can accommodate Mega-yachts so keep your eyes open for this, it may offer an alternative to anchoring in the bay for those who prefer a slip.

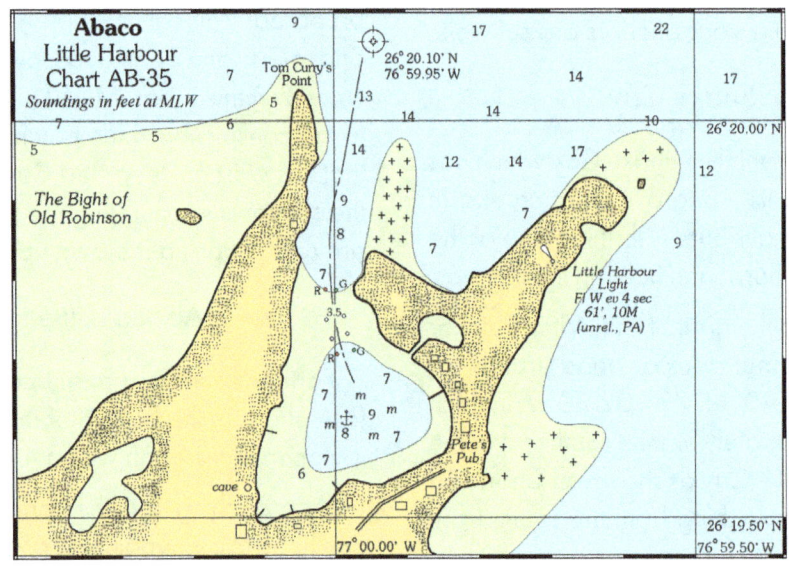

Little Harbour, Abaco

Aerial Imaging ©
For all your Aerial Photography!
Private Yachts or Island Projects

Fabulous Bahama Islands aerial photos for sale - great gifts!

Contact Us: 1-(242)-557-1813
Email: seaplanepilots@gmail.com
Also Yacht Provisions flown to your nearest airport!

Chapter 5

The Central Bahamas

THE CENTRAL BAHAMAS COVERS A LARGE AREA AND SEVERAL ISLAND CHAINS such as the Biminis, the Berry Islands, New Providence, Andros, Eleuthera, and Cat Island. These are wonderful cruising grounds and some places even offer fair hurricane protection whether you wish to anchor, haul out, or get a slip in a marina. In the case of Eleuthera and Cat Island, we will cover these from the south to the north. People may be surprised that we don't mention *Rock Sound* in Eleuthera but it is far to large and there is simply too much fetch to consider it as hurricane protection. You also might not want to read what we have to say about *Hatchet Bay* as well. In Andros, if you plan to avail yourself of Kamalame Cay you must be inside the barrier reef, and that is not something that you can do at night.

The Biminis: *North Bimini*

Bimini has several marinas if you would rather stay at a slip, the best protection being at one of the two marinas at the *Resorts World Bimini* at the northern end of the dredged entrance channel whose southern entrance is shown on Chart BI-3 at the top of the next page.

On the north side of South Bimini, also shown on Chart BI-3, is an entrance to some small canals with a 4' bar at the entrance from the harbour at North Bimini. Take into consideration that these canals have plenty of wrecks lining the shores along with old rotten pilings jutting up here and there. The surrounding land is very low, and the canals may become untenable in a high storm surge.

These canals could be considered as a possible hurricane hole although adequate protection is questionable due to the very low height of the surrounding land.

The Biminis: *South Bimini*

The finest hurricane hole in the Biminis is up the creek (sometimes shown as *Duck Lake*) along the northern shore of Nixon's Harbour. Seven feet can get in over the bar at high tide where you'll find plenty of secure water inside. One skipper took his 67' sailboat drawing 7'4" into the creek on the high tide and had plenty of water where he tied to the mangroves, although he did bump a few times on the way in. The channel is easy to see with its flagged markers on each side; follow the flags in and anchor in any of the three pockets.

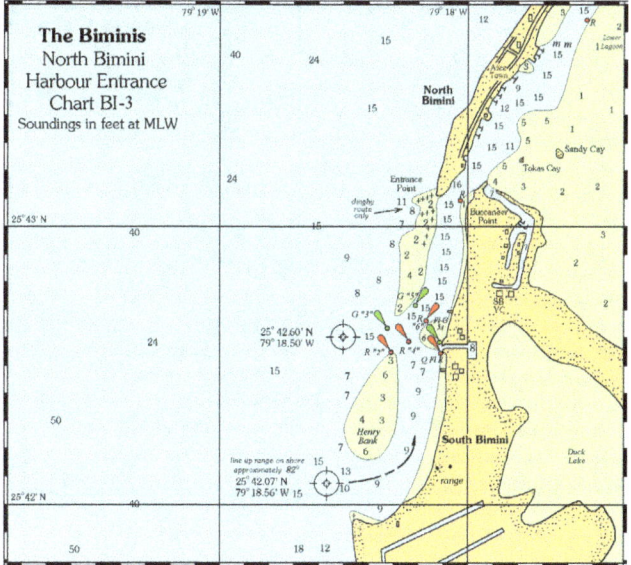

As shown on Chart BI-5 (see below), a waypoint at 25° 41.23 N, 79° 18.50 W places you approximately ¼ mile west of the entrance to *Nixon's Harbour*. From this position steer 90°-95° and pass south of Round Rock. Once past Round Rock line up on the staked entrance channel into the creek and follow the stakes into the deeper water inside.

On the west side of South Bimini lies the entrance to the *Port Royal* canals. Five feet can make it over the bar with spots of 7'-10' inside. Be sure to tie up in vacant areas between houses.

A final note, if the reader is wondering why the *Bimini Sands Marina* on South Bimini is not considered a hurricane hole, it is due to the entrance channel that would funnel any westerly seas directly into the marina making a possibly untenable situation.

The Berry Islands

There are only three places to consider in the Berry Islands, and two of them were hit hard by powerful Hurricane Andrew. *Chub Cay Marina* is a possibility if you don't mind a slip or perhaps tying off between pilings. The marina was devastated by Andrew and quite a few boats destroyed. Something to remember when it's decision making time.

Another possibility would be to work your way into *Little Harbour*. There is a winding channel into the inner anchorage where you can tuck into a narrow channel just north of the Darville's dock in 7'-11' of water with mangroves to the east and a shallow bar and a small cay to the west. *Little Harbour* is open to the north but there is a large shallow bank with 1'-3' over it just north of the mangroves.

By far the best place to get a slip in a hurricane is at Great Harbour Cay, in *Bullock's Harbour* at *Great Harbour Cay Marina*. Check with the dockmaster prior to arrival to make sure there is room at the marina as the holding in the harbour is poor.

Berry Islands: *Great Harbour Cay Marina*

Without a doubt, the best hurricane dockage is to be found at *Great Harbour Cay Marina* in the northern Berry Islands on the island of Great Harbour Cay. In 2016, Hurricane Matthew hit close by and 8 boats and the marina fared well in the 120 mph winds clocked at the docks. The marina manager rode out the hurricane with the boaters in a condo that the marina supplied for the boater's safety!

Great Harbour Cay Marina is entirely landlocked and is situated in a tight "L" shape and well-protected. The marina's dockmaster monitors VHF ch. 16, 68, and 14 and the marina itself has 65 slips which will accommodate boats up to 130' LOA with drafts of 9'.

The route to the entrance to *Great Harbour Cay Marina* begins at the western tip of Little Stirrup Cay and bends its way southeastward across the banks in 9'-12' of water until you arrive at Great Harbour Cay and the entrance to *Bullock's Harbour* and *Great*

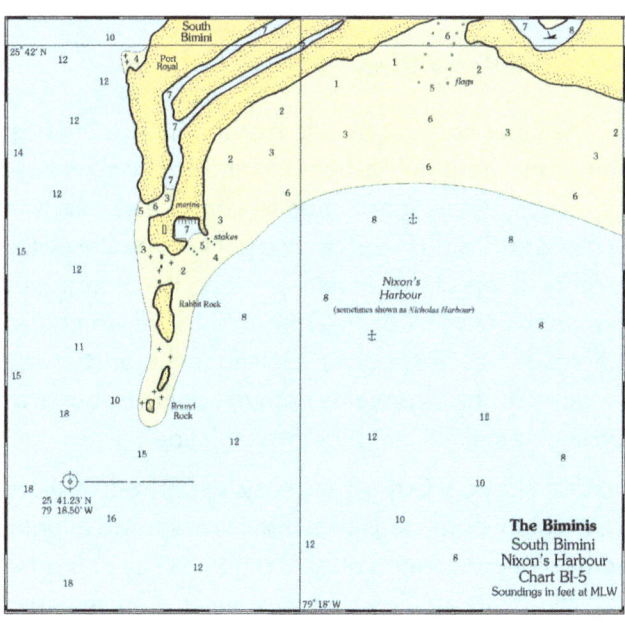

Bullock's Harbour, Great Harbour Cay and marina

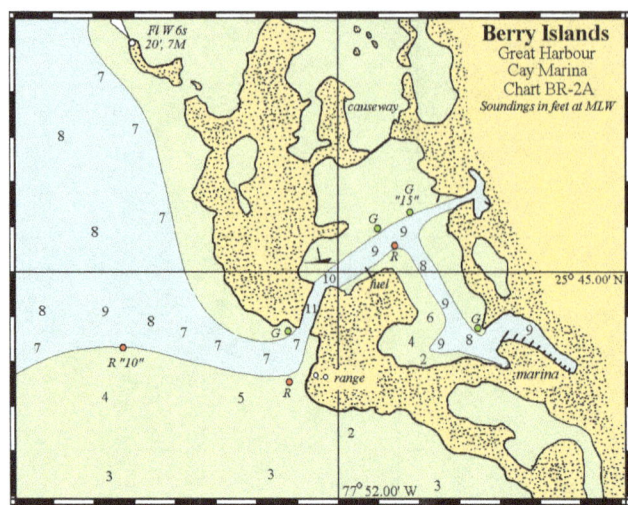

The entrance to *Bullock's Harbour*

Harbour Cay Marina (see photos above). From a waypoint northwest of Little Stirrup Cay at 25° 49.64'N, 77° 57.41' W, take up a course of 174° for the tripod marker that sits on the horizon about 3.61 miles south of you. Once you clear the shallows along the southwestern tip of Little Stirrup Cay you will have deep water on both sides of your course all the way to the marker; if you find yourself in water that is shallowing you are probably too far east, the deeper water lies to the west of you, in fact, to the west the deep water extends all the way to Bimini.

A waypoint at 25° 46.11' N, 77° 56.59' W places you approximately 200 yards east of the tripod marker and you may pass it to either side. Once abeam of the tripod marker take up a course of 111°. This will put the conspicuous Batelco tower at *Bullock's Harbour* just off your port bow. The rest of the markers are single poles and can be taken on either side and you will have no less than 9' to the final marker. Simply steering approximately 110° and looking for the markers is usually the best way to enter. From R10, look just off your port bow ashore and you will see a house on the ridge with a 110° range. The range is designed to bring you in to the last red marker, a red diamond, where you will turn to port and line up on the entrance cut to *Bullock's Harbour* (see Chart BR-2A above).

There will be a green marker just outside the cut to lead you in. At this point you will travel through a narrow cut blasted through the rolling hills to a lake inside and *Great Harbour Cay Marina*. The cut is approximately 80' wide and has 8' at low water (see photo in the previous column).

Once inside follow the marked channel as it winds around to the marina, you will have at least 7' here at low water. The marina is tucked into the southeastern corner and is surrounded by a condo development.

Berry Islands: *Little Harbour*

The inner harbour at Little Harbour Cay is the best hurricane hole in the Berry Islands for anchorage purposes. The entrance channel can take vessels with a 5½' draft inside to anchor in a pocket of water a little over 6' deep at low water, or along the mangroves just north of *Flo's Conch Café* in 7'-11' of water (see photos-top of next page). Getting in the anchorage is tricky as the channel is not well defined, but a 5' draft can make it in just before high tide.

Little Harbour Cay is fairly easy to spot from offshore as it is high, bold, and has a distinctive spread of palm trees along its central ridge. To gain the anchorages at Little Harbour Cay you must enter from seaward

Little Harbour Cay as seen from the south

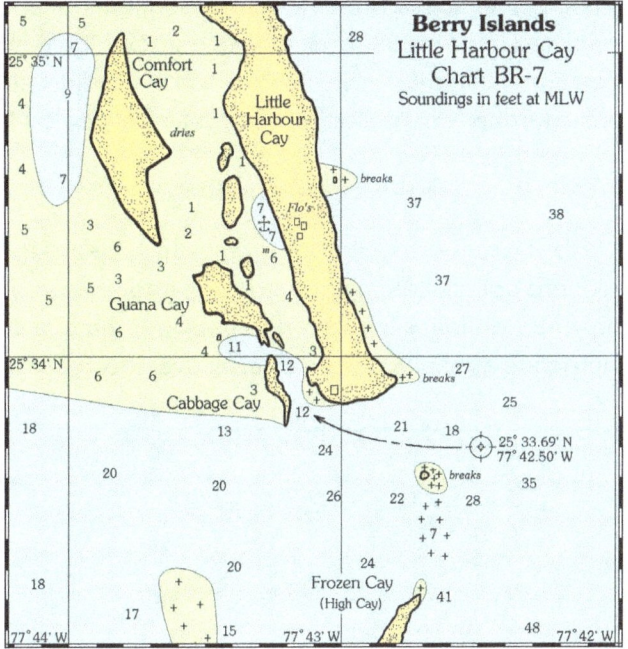

The hurricane hole at Little Harbour Cay

The entrance channel begins just north of Cabbage Cay to the east of the small, unnamed rock off Guana Cay and proceeds in towards the dock below *Flo's*. It takes a small dogleg left and then right just before the deeper water opens up near the dock and flows northward for awhile along the western shore of Little Harbour Cay. Use several anchors here as there is no swinging room. If you have trouble finding the entrance channel call Chester or his nephew Joel on VHF ch. 68 (hail *Little Harbour* or *Flo's Conch Café*).

Berry Islands: *Chub Cay Marina*

The only protection to be found on Chub Cay is at the *Chub Cay Marina*. Be warned that Hurricane Hugo decimated a lot of boats here and it took quite a while for everybody to recover and rebuild.

To find your way into the marina a waypoint at 25°23.90' N 77° 55.08' W, places you approximately one nautical mile south of the entrance and on the 35° range as shown on Chart BR-13 (next page). The range consists of two small daymarks that are lit red at night. If approaching at night look for the bright lights of the power station ashore lying well to the west of the *Batelco* tower with its flashing red light. The range is just a little west of the brightly-lit power station showing one red light over the other. Chub Point Light should show white when you're on station to take up the range.

between the southern tip of Little Harbour Cay and the northern tip of Frozen Cay. The only obstruction is a large rock that is awash in the center of the cut but it is very visible as almost any sea breaks upon it.

A waypoint at 25° 33.69' N, 77° 42.50' W, will put your vessel approximately ½ mile east of the entrance as shown on Chart BR-7 (above). Steer between the partially submerged rock and the southern tip of Little Harbour Cay. The rock can be passed on either side, but the deeper water is on the Little Harbour side.

Vessels with less than a 6' draft who wish to enter the inner harbor below *Flo's Conch Café* should only do so on a rising, almost high tide with good visibility. The ability to read the water is a necessity here if you wish to keep from going aground.

Take up an approximate course of 35° on the range and remember that the deeper water will be to your starboard. If you must stray, by all means, stray east as the principal danger, and one that must definitely be reckoned with, is *Mama Rhoda Reef* lying east and southeast of Mama Rhoda Rock. You can keep

THE CENTRAL BAHAMAS • 51

Chub Cay Marina

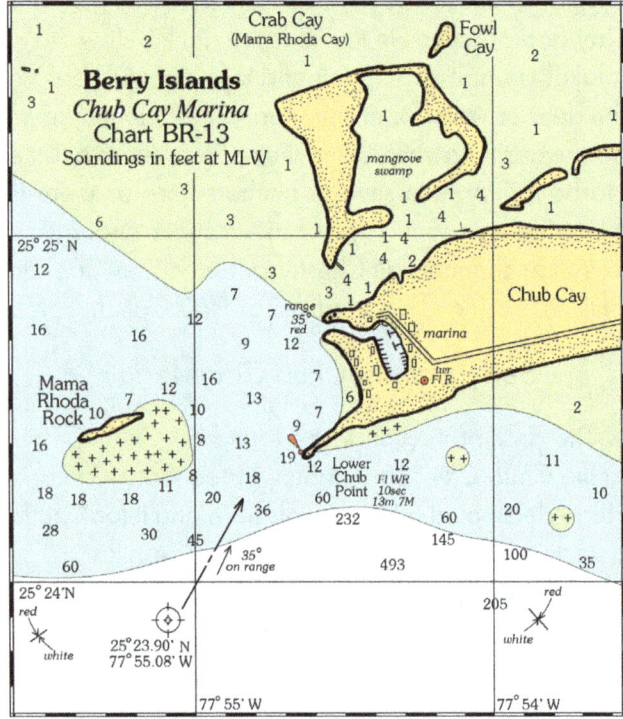

Chub Point Light close to starboard as there is deep water within 50 yards of the light.

The light is sectored showing red in the dangerous sectors. Of course, this light is like all lights in The Bahamas, which is to say it is unreliable at best and subject to change.

Andros: *Kamalame Cay Marina*

Over the last decade, a new marina has opened up on Andros just north of *Staniard Creek* as shown on Chart AN-5 (right). At the southwest tip of Kamalame Cay (Long Bay Cay). Here, a gentleman has built a resort called *Kamalame Cove* aimed at attracting a very wealthy clientele, primarily for the folks who've bought a home on the island. The cove the marina is located on offers a bit of protection if your draft is shallow enough to permit entry and if you cannot get somewhere else with greater protection.

From a position northeast of the northeastern tip of *Staniard Creek* as shown on Chart AN-5, you will have to eyeball your way in as you round the point. The channel has not been dredged so the marina is restricted to drafts of about 4' with a good high tide. You can anchor just inside the point in the basin west of the northern tip of Staniard Cay see photo below).

Andros: *Fresh Creek*

Fresh Creek has a strong current which means the scoured bottom holding ground is not good and your only real option is to get a slip. In 2001, the eye of Hurricane Michelle passed directly over *Fresh Creek*

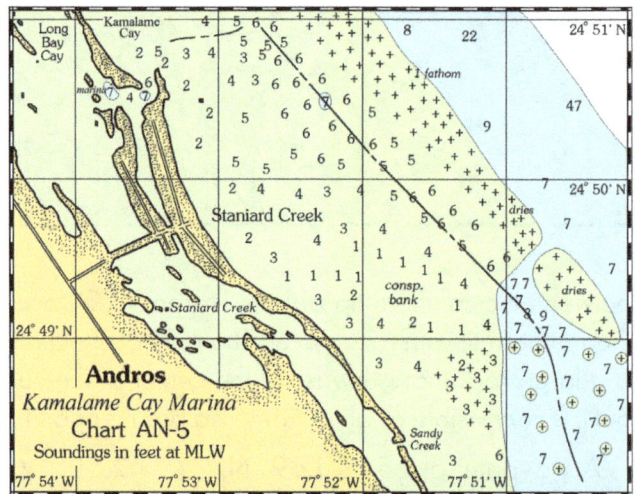

Kamalame Cove Marina

52 • THE CAPTAIN'S GUIDE TO HURRICANE HOLES

with 100 mph sustained winds and no damages were reported at the marina at *Fresh Creek*. The old timers will tell you of a hurricane that long ago raised the level of the creek over 10' and some vessels that were tied off in the creeks for protection were swept inland as much as 5 miles where some still sit today.

A waypoint at 24° 44.25' N, 77° 45.65' W, will place you approximately ¼ mile ENE of the well-marked entrance channel into *Fresh Creek* as shown on Chart AN-7 and in greater detail on Chart AN-7A (both in next column with photos).

At the seaward end of the entrance channel, you will see two steel pilings just southeast of the small rock lying south of Long Cay. Keep these pilings to starboard. Next you will see an amber-lighted *AUTEC* tower, keep it also to starboard passing south of it. Head straight in to the entrance channel from here avoiding the shallow sandbank to starboard between the *AUTEC* tower and shore. You'll see the wreck of the *Lady Gloria* on the southern side of the channel at the edge of the land, and just past her, also on the southern side of the channel just inside the mouth of *Fresh Creek* is a flashing green light.

You might wish to steer around the darker patches you see as you work your way down the entrance channel. There are two shallow bars at the entrance to *Fresh Creek* that carry around 6' at low water. One lies approximately 200 yards east of the entrance and the *Lady Gloria*, and the second lies only about 100 yards east of the entrance. When approaching the entrance keep halfway between the center and the northern shore of the entrance, this is where the deeper water is and keeps you off the rocky shoal north of the southern jetty.

Once inside, favor the southern side of mid-channel, only the southern side of the entrance channel was dredged, the northern side is shallow. On the southern side of the channel are the docks of *Lighthouse Marina*. Most folks anchor between the marina and the bridge but the bottom is scoured and poor holding. The deepest water and the best holding is in the northwestern side of the harbour towards the bridge, but the entire anchorage area is littered with old engine blocks and other debris so use caution when setting your anchors. Two anchors are recommended here as there is little swinging room

Fresh Creek, Andros

Lighthouse Marina, Fresh Creek, Andros

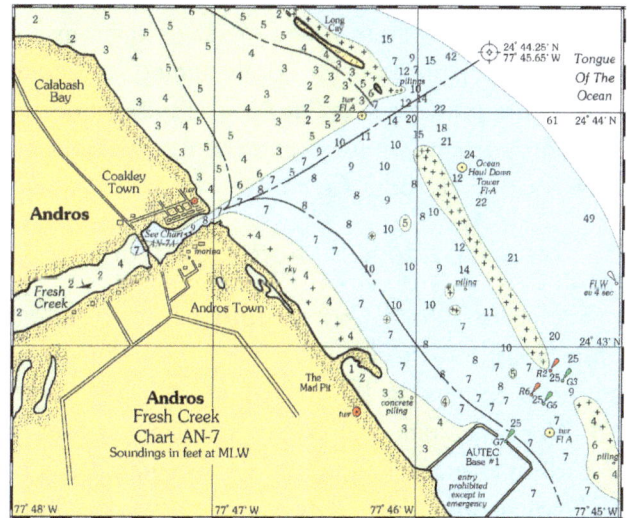

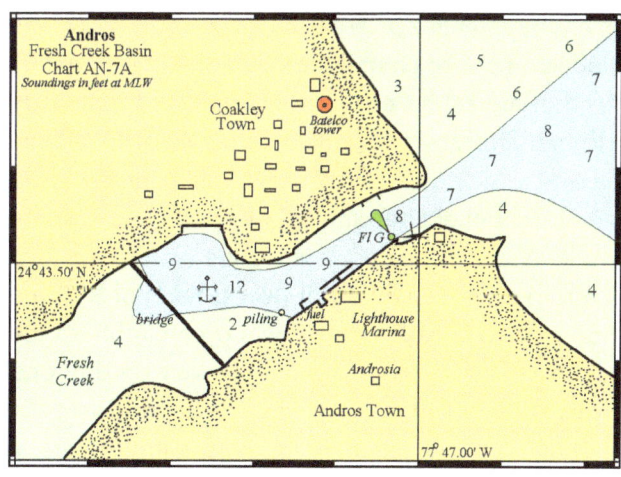

THE CENTRAL BAHAMAS • 53

if it's crowded and there's a lot of current here. It is best to get a slip if one is available. The marina can accommodate vessels to 100' LOA with drafts to 9'

South of the entrance channel to *Fresh Creek* is the breakwater of the *AUTEC* base as shown on the chart. *AUTEC* wants all boaters to understand that all *AUTEC* bases are U.S. Navy installations and entrance is prohibited except in real emergencies. A hurricane does qualify as an emergency, but I have never met someone who tied up their vessel in the *AUTEC* base. *AUTEC Base #1, Fresh Creek,* monitors VHF ch. 16 and answers to *Snapper Base*.

New Providence: *Nassau Harbour*

Here, in the capitol of The Bahamas, *Nassau Harbour* has fair to good holding (and a LOT of debris on the bottom) along with a long east-west fetch so we cannot recommend anchoring in the harbour. There are two hurricane chains crossing the harbour whose approximate locations are shown on some charts for Nassau and in an emergency, you might be able to hook up to one of these. There are several marinas if you wish a slip, including the aptly named *Hurricane Hole Marina* located on Paradise Island. There are also yards around where you can haul out (*Bayshore, Brown's,* and *Harbour Central*).

New Providence: *Lyford Cay Marina*

Exclusive *Lyford Cay Marina* offers excellent protection if you are able to get a slip. The well-marked entrance channel to *Lyford Cay Marina* (see Chart NP-7A next column with photo) will allow vessels of 9½' draft to enter the marina. Vessels with drafts of 11' may tie up inside solely at their own risk.

A waypoint at 25° 04.00' N, 77° 30.94' W, will place you approximately ¼ mile north of the light (Fl W) that marks the entrance channel to *Lyford Cay Marina*. Start steering south staying west of the light. Be sure to stay a little west of the light (if entering the marina channel keep the light to port) to avoid the shallows and breaking reefs to the east. There is a range ashore consisting of twin green lights at the marina, and a higher set of twin green lights on the hill above the marina. These lights stay on day and night. As you begin your passage south down the channel you will notice that there are three pairs of

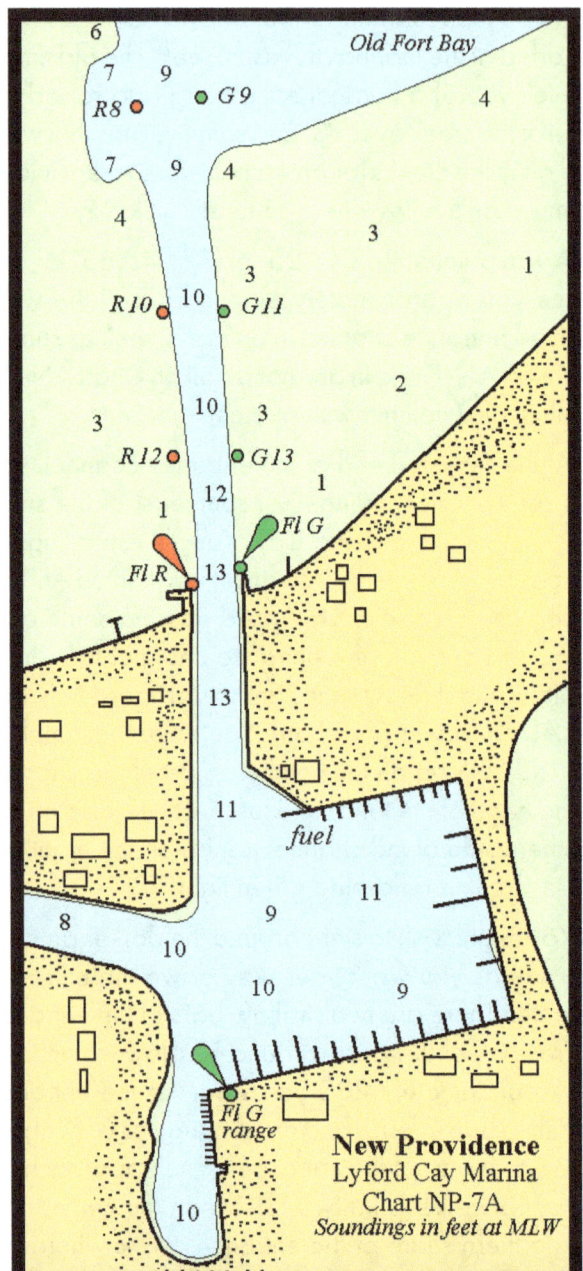

Lyford Cay Marina, New Providence

red and green daymarks, keep between these and finally you will come to the entrance jetties which have red and green lights on their respective sides. Just inside, before the marina, white floodlights shine across the opening.

New Providence: *Albany Marina*

Located in *South West Bay* (as shown on Chart NP-8 below) is the marked, dredged entrance channel (16') to the upscale *Albany Marina*. The marina boasts a 15-acre basin that can accommodate 71 mega-yachts to 300' LOA with a draft of 16'.

A waypoint at 24° 59.00' N, 77° 30.75' W, will place you approximately 1 nm SSW of the entrance to the marina. From the waypoint you can pick up the marked channel leading into the marina. There is no anchoring here, nor is there a haul out, but a protected slip might be available. *Albany Marina* monitors VHF ch. 16 and 74. You can also reach them by phone at 242-676-6020.

New Providence: *Coral Harbour*

Coral Harbour, besides being the base for the *Royal Bahamas Defence Force*, is a series of canals fronted by some very nice homes. The base sits to starboard upon entry and was once a large and elegant marina. If you know someone with a home in *Coral Harbour*, you would probably fare well in a Cat 1 or Cat 2 storm tied up to their seawall.

As you approach the entrance to *Coral Harbour* as shown on Chart NP-8 (below) you will notice the large light tower lying about a mile offshore. This is shown on current DMA charts as being abandoned. It is far from abandoned. It flashes white approximately every 20 seconds to direct the *Royal Bahamas Defence Force* vessels to their base. The best approach into Coral Harbour is to steer toward the light to avoid the shallows just offshore between Clifton Point and Coral Harbour as shown on the chart. Pass on either side of the light and take up a course of approximately 23° for the entrance.

A waypoint at 24° 57.85' N, 77° 29.00' W, will place you approximately 1 nm SSW of the entrance channel. The entrance channel is very conspicuous, lying between two rock jetties just north of the abandoned concrete high rise. The easternmost jetty has a red flashing light, remember red, right, returning.

New Providence: *Palm Cay Marina*

On the eastern shore of New Providence is a wonderful, protected marina, *Palm Cay Marina*. No haul outs, no anchoring, but good protection, for a Cat 1 or Cat 2 storm. The marina has 192 slips that can accommodate vessels to 110' LOA, with drafts to 8', and beams to 36'.

A waypoint at 25° 00.14' N, 77° 15.68' W, will place you approximately ¼ mile southeast of the first set of markers that define the entrance channel to *Palm Cay Marina* as shown on Chart NP-10 (below). The markers here flash red and green at night. Line up between the first set of markers and steer towards the entrance to the complex on a course of 330°. Use caution as the current can push the unwary out of the channel.

Ashore you will see some large light pink condos that look white from offshore. As you get closer you will see the channel dogleg to the west where you will enter the channel into the marina. Make sure that you contact the marina prior to your arrival as the marina is protected by a sea-gate, a chain that is raised and lowered to keep out unwelcome boats.

New Providence: *Rose Island, Salt Pond*

On the southwestern shore of Rose Island is the entrance to a very good hurricane hole shown as *Salt Pond* on Chart NP-11 (top of next page). It is a circular harbour with a small island in the center. The channel is easily 50'-60' wide and 7'-9' deep. Anchor and tie off between the shore and the island. Get there early as everyone in Nassau and the northern Exumas will have the same idea.

Vessels heading to *Salt Pond* from Nassau should head east to Porgee Rock and head northward between Athol Island and Porgee Rock towards the western tip of Rose Island. Once you arrive at the western tip of Rose Island follow the shoreline eastward to enter the channel into *Salt Pond* (see photo below).

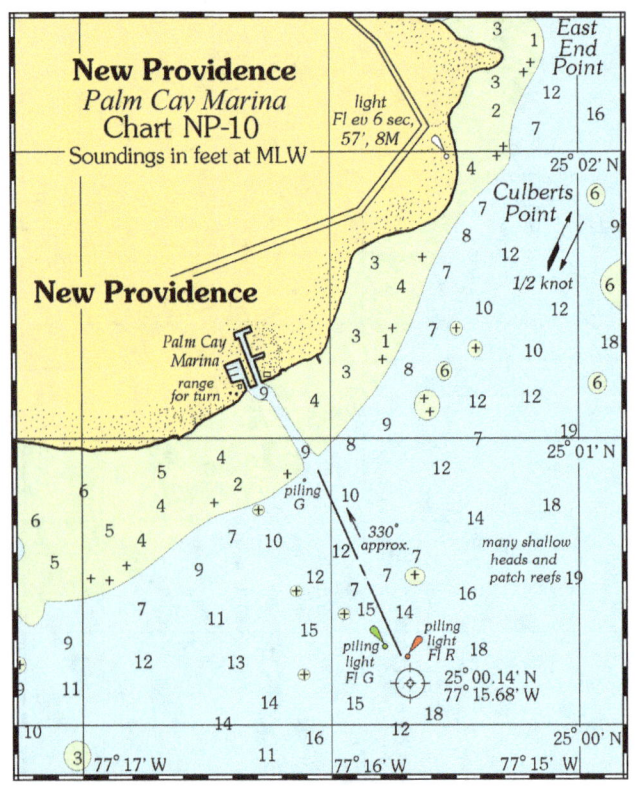

Salt Pond, Rose Island

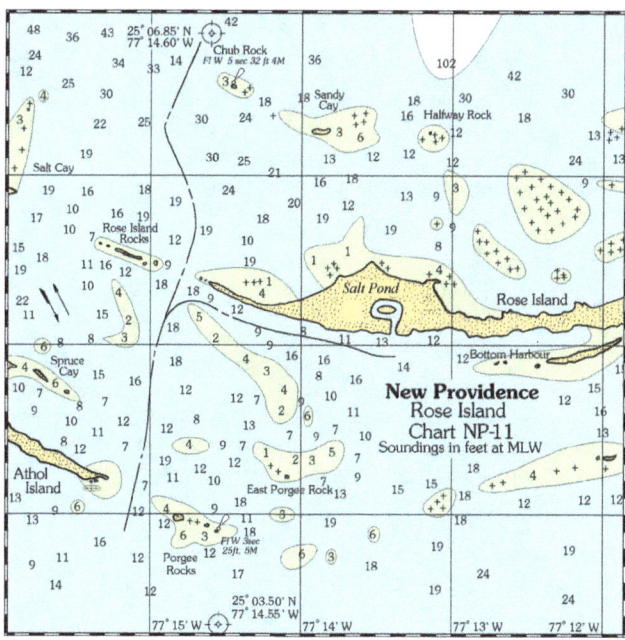

Davis Harbour Marina, Eleuthera

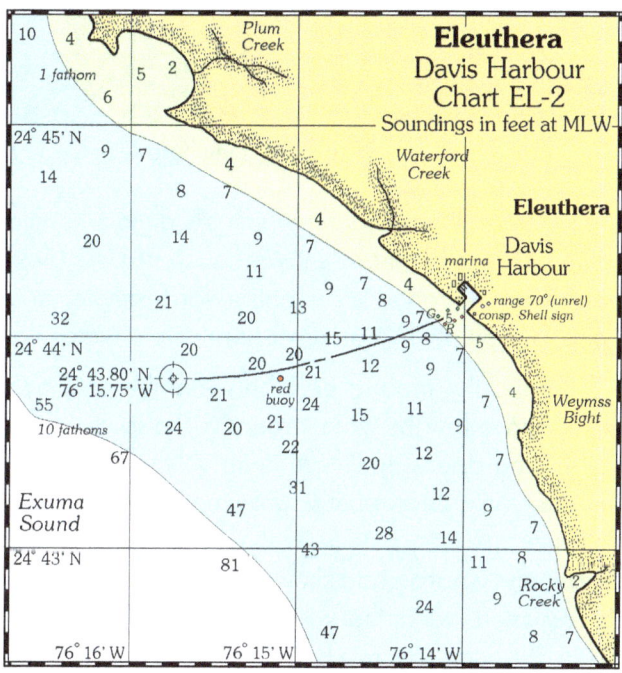

If you are approaching from offshore, a waypoint at 25° 06.85' N, 77° 14.60' W, will place you approximately 200 yards north of Chub Rock. From the waypoint head a little west of south as shown on the chart to pass between Chub Rocks and the western tip of Rose Island. Once past the tip turn to port and follow the shoreline until the entrance to *Salt Pond* opens.

Eleuthera: *Davis Harbour Marina*

As shown on Chart EL-2 (next column with photo), a waypoint at 24° 43.80' N, 76° 15.75' W, will place you approximately ¾ mile west of the red buoy (sometimes it's difficult to see) that marks the entrance channel to *Davis Harbour Marina*, a small but well-protected marina on the southwestern coast of Eleuthera.

Take the red buoy to starboard when entering the marina and head in on a bearing of 70° on the range consisting of a very visible white cross in the rear and a much less visible short white pole in the foreground. A good landmark is the cross and the Shell sign. Take up an approximate course of 70° on the white cross and you will soon make out the three pairs of channel markers in front of you. Take the red ones on the right and head in making a dogleg to port to enter the marina proper after the last pair of poles. The controlling depth for this channel is 5' at low tide. If you are in doubt as to the state of the tide call the dockmaster on VHF ch. 16. *Davis Harbour Marina* (http://cottonbayclub.com/the-marina/) can accommodate 24 boats up to 80' LOA with drafts up to 8'

Eleuthera: *Powell Point*

In the vicinity of Powell Point, Eleuthera, you have three choices for protection. A slip at *Cape Eleuthera Marina*, tying off in No Name Harbour, or tying off in the dredged channels lying west of Cow Rock Point (and just to the east of the marina-see photo on the next page).

A waypoint at 24° 50.23' N, 76° 21.09' W, places you approximately ¼ mile west of the entrance

Powell Point, Eleuthera with *Cape Eleuthera Marina* and *No Name Harbour*

to *Cape Eleuthera Marina* which sits almost ¼ mile south of Powell Point as shown on Chart EL-3 (next page). The entrance is straightforward, simple, and deep, a draft of 12' can enter here at low water.

To enter the marina pass between the entrance jetties and follow the water around to the south where the marina opens up. There is an unlit range at the entrance on a bearing of approximately 90° but the entrance is so easy as to make the range unnecessary. The marina is not a good choice if expecting hurricane force westerly winds that will blow seas into the marina basin. The huge concrete breakwater at the bend in the entrance dogleg has suffered considerable damage and offers testimony to the power of the seas that enter the marina.

Cape Eleuthera Marina suffered the Cat 4 Matthew in 2016 with the eye moving close. However, no damage occurred to the marina slips, villas, or restaurant. In fact, they were up and running as normal on the following day. The owners have no objections to boats calling in for reservations a week preceding any storm.

Just south of the entrance to the marina is *No Name Harbour* (sometimes shown as *Unnamed Harbour*) as shown on the chart and photo. Lately there has been a chain across the entrance, but if it is not there you may enter at your own risk. To enter take Chub Rock

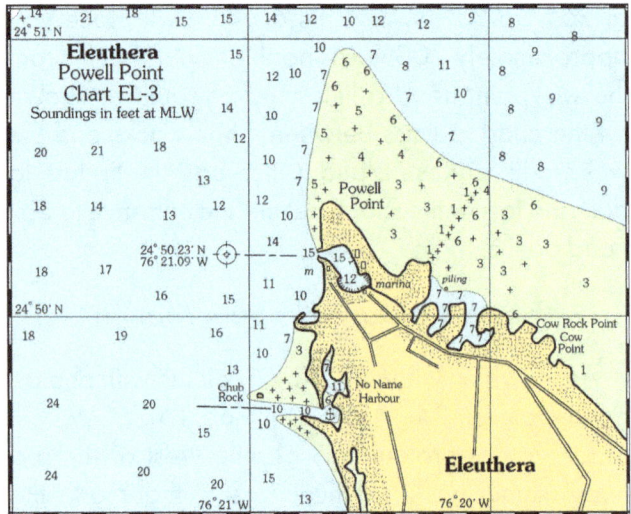

well to port and steer generally eastward staying in the deeper water between the shallow rocky bars to your port and starboard. The entrance is not easily made out from Chub Rock, you'll probably have to pick it out with the binoculars. Once inside you can anchor well up any of the fingers, some of which have 13'-25' in them with only 6'-7' between the deeper parts. *No Name Harbour* offers excellent protection as long as you can anchor and tie off well.

The only problem with *No Name Harbour* is the fact that it has been dredged. This created sheer rock walls and less than ideal holding. Seven feet can enter here at MLW and 6' can work farther up into the small coves that branch off the main harbour. You might consider tying your lines to the trees and setting your anchors ashore here, the holding is not that great being as this is a dredged harbour. The best spot in a hurricane would be in the upper northeastern finger. Skippers will have to tie to trees and probably set an anchor or two ashore on the land to be secure in here and even then, a major storm surge could decimate the place.

On the northeastern shore of the Cape Eleuthera property, east of Powell Point (as shown on Chart EL-3 on the previous page), are some dredged harbours that face north and east. These have 7'-9' of water inside but the entrance is guarded by shallows that restrict entry to vessels of less than 6' and only at high tide. This really is not a recommended hurricane hole as N-E winds could decimate boats inside. *No Name Harbour* and *Cape Eleuthera Marina* offer better protection, however, if there's a place to hide from a storm, you need to know about it and that is why these canals are mentioned.

Eleuthera: *Hatchet Bay*

WARNING! WARNING! WARNING!

Hatchet Bay is often considered a prime hurricane hole because it is so protected by land, but it has a history of damage as the hulls along the shore and bottom will testify. The holding in *Hatchet Bay* is fair in some places, worse in others, and although it may be a fine spot to ride out a norther it is definitely no place to be in a major hurricane as many found out in Hurricane Andrew (you will still see the wrecks along the shore from that catastrophe). In 1999, Hurricane Floyd destroyed over 20 boats anchored in the bay partially due to a 20' storm surge (reported at *Governor's Harbour*) along with 190 mph gusts. A few years later, in 2004, Hurricane Frances brought heavy flooding, damaged vessels, and some structural damage to nearby buildings.

The anchorage on the west side of the bay is grassy with a few rocks while the bottom in front of the old

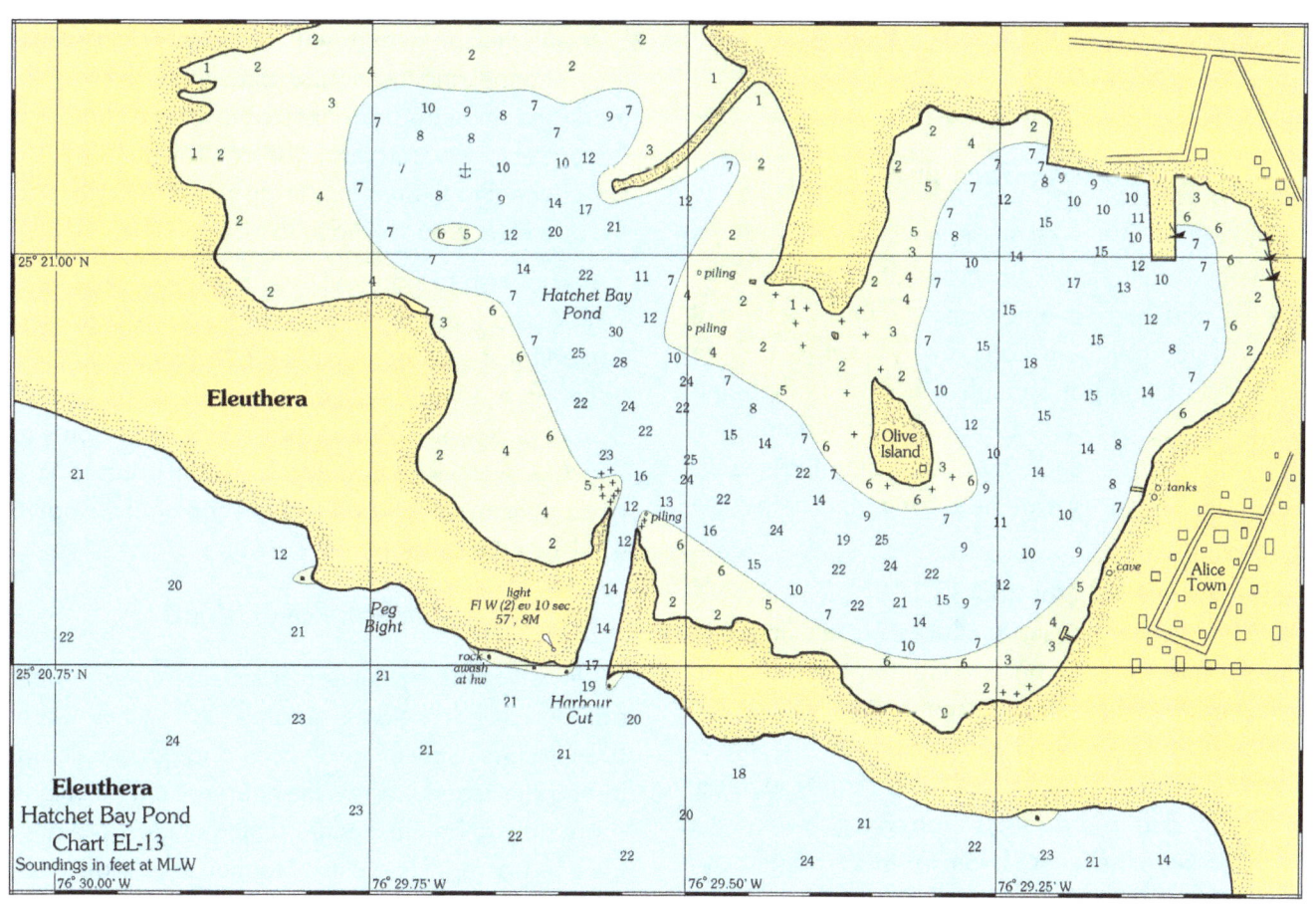

marina is more like mud. Always set your anchors in a sandy patch as the bottom is primarily grass with scattered concrete mooring blocks, engine blocks, and small sunken boats...you MUST anchor in sand if you want to have any kind of a chance for survival of your vessel.

As shown on Chart EL-13 (bottom of previous age), *Hatchet Bay Pond* is the true name of the bay but most call it *Hatchet Bay*. For our purposes, whenever we refer to *Hatchet Bay* we will be talking about *Hatchet Bay Pond*. The entrance to *Hatchet Bay* is sometimes difficult to distinguish from offshore. It is a narrow, 90' wide pass called *Harbour Cut* that was blasted through sheer rock with two small jetties stretching inwards from the *Bight of Eleuthera*.

The light atop the bluff on the western side of the entrance flashes white twice in four seconds and repeats that characteristic every 10 seconds. The light is 57' above the water and is visible for 8 miles. The easiest landmarks for this area are the large white silos, once part of a large plantation, standing like silent sentinels up and down this stretch of coast, and the 265' *Batelco* tower with its flashing red light. A waypoint at 25° 20.50' N, 76° 29.70' W, will place you approximately ¼ mile south of the pass into *Hatchet Bay*. From this position had straight in and anchor wherever you choose.

Eleuthera: *Spanish Wells*

At Spanish Wells you will find *Muddy Hole* off the creek between Russell Island and St. George's Cay. It is the local hurricane hole and 4' can enter here at MLW if you get there early. Every boat (and there are a lot of them) at Spanish Wells will be heading there also so get there EARLY and be warned that you will be surrounded by a lot of unattended vessels. Alternatives are a slip at the marina, and a haul out if there is room.

From a waypoint at 25° 32.10' N, 76° 45.50' W, which places you just southwest of the pilings that mark the western entrance to the harbour as shown on Chart EL-19 (see top of next column). Here you will see two large steel I beams about 8'-10' out of the water. The westernmost one is slightly leaning, probably due to a boat collision. Pass between the two and keep the inner I beam to starboard and you

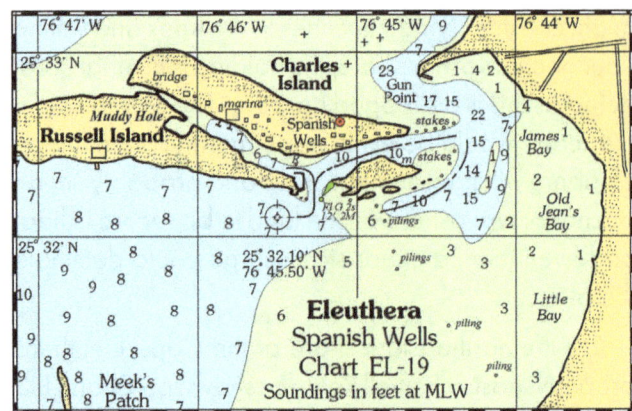

will enter the channel to the harbour. The channel is also marked by a green light at the entrance that flashes green, every 2 seconds, stands 12' high and is visible for 2 nautical miles.

Just a few hundred yards up the channel at the end of the jetty on your port side, another channel branches off to the west (to port) leading to *Spanish Wells Yacht Haven Marina* and *Muddy Hole*. The passage lies between the tip of the shoal bank to port and the piling that you must keep to starboard. Keep the shoal bank to port as you approach the marina and look for the large boat shed, you can't miss it. Between the spot where you turned to port from the main channel and the marina, almost where the stake marks the small boat channel to starboard, there is a 5' spot at low water, this is the controlling depth for this channel to the marina, to go past the marina, to *Muddy Hole*, you will have to play the tide.

If you wish to haul out, you better be very, very EARLY as a lot of folks will have the same idea and you can rest assured that quite a few local folks will have the haul out facilities booked solid a year in advance. *Spanish Wells Marine* can haul boats to 30' LOA, while *R&B Boatyard* can haul boats to 100 tons on their railway and their lift can haul a boat to 45' LOA with a beam of 29'.

Eleuthera: *Royal Island*

The few holes in Eleuthera all suffered considerable damage from Hurricane Andrew. And while Royal Island offers excellent protection and good holding, during Hurricane Andrew the fleet washed up on one shore only to be washed up on the other shore after the eye passed. Co-author Stephen J. Pavlidis once

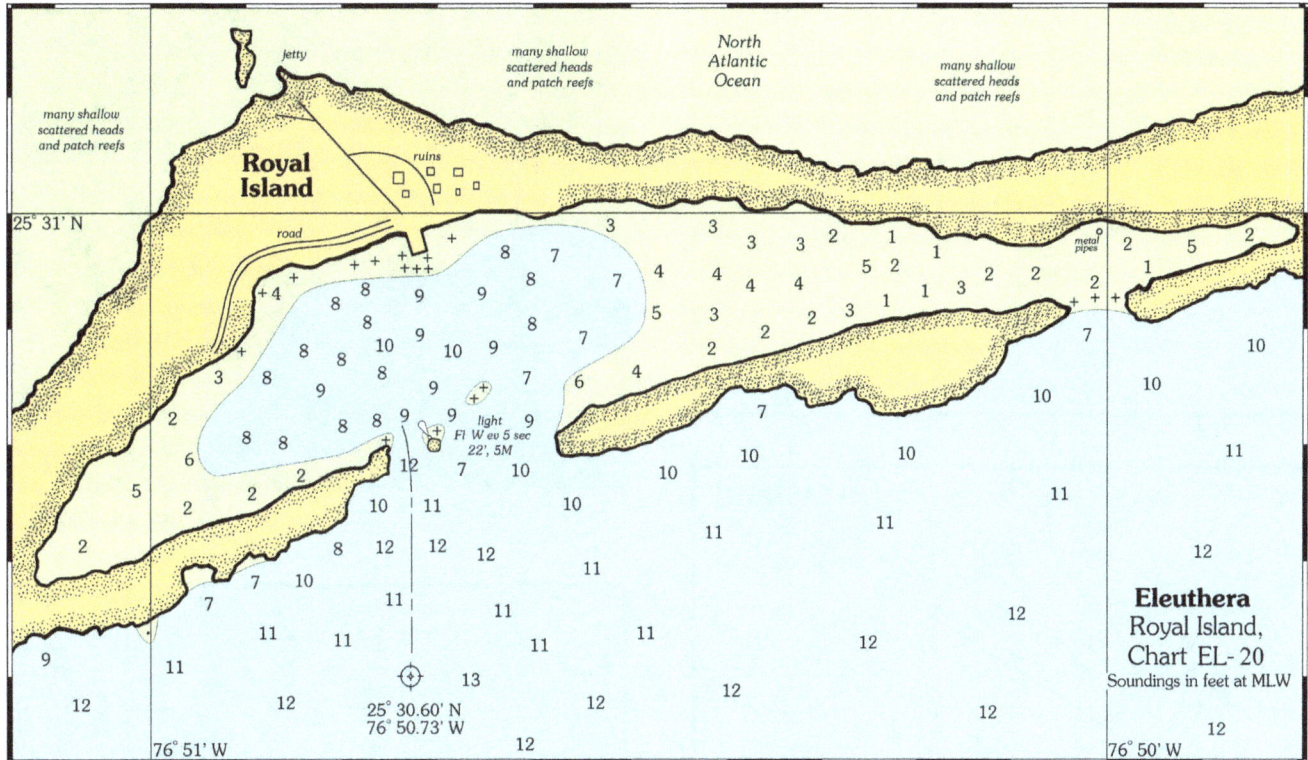

rode out a storm here with 6 hours of 70-105mph winds laying to two CQRs with no problems. That was the night someone was washed off the *Glass Window Bridge* and a policeman went in to save them. The policeman saved the victim but was never seen again.

As shown on Chart EL-20 above, there are actually three openings to the anchorage at Royal Island. The entrance to the northeast is navigable only by very small boats or dinghies. Even then use caution as there is only about a foot of water in places compounded by a small rocky bar just inside. The best landmarks are the two hills, the highest points on Royal Island, the entrance to *Royal Island Harbour* lies about ½ mile north of these hills. A waypoint at 25° 30.60' N, 76° 50.73' W will place you approximately ¼ mile south of the entrance to the anchorage at Royal Island as shown on Chart EL-20 (above).

In the middle of this pass sits a large rock that once was home to the *Royal Island Light* which may or may not be working when you need it. The entrance to the anchorage lies through the narrow cut to the southwest of the light. Keep the rock to starboard upon entering the anchorage. The cut that lies just to the northeast of the rock has a large submerged rock lying approximately 50 yards inside and just barely under the surface. It is possible to pass on either side of the submerged rock but it is so hard to see and the passage on the other side of the light is so easy it is best to ignore this route no matter how inviting it appears.

The holding is pretty good here (except in a Cat-5!) So, don't be afraid to anchor in the white mud/sand/grass bottom. The skippers and crews of the boats that rode out Hurricane Andrew here would probably tell you a different story.

Cat Island: *Springfield Bay*

The one place on the southern coast of Cat Island that might offer decent protection to the cruising boater is at the *Flamingo Hills Resort and Marina* (http://flamingohills.com/marina.htm) at the eastern end of *Springfield Bay* (see Chart CT-1A on the top of the next page).

The full-service marina (see Chart CT-1B on the top of the next page), which is still in the early stages of construction and is not yet open, as of the Summer of 2017 boasts a 12' dredged entrance channel leading to a protected basin with 102 slips that can

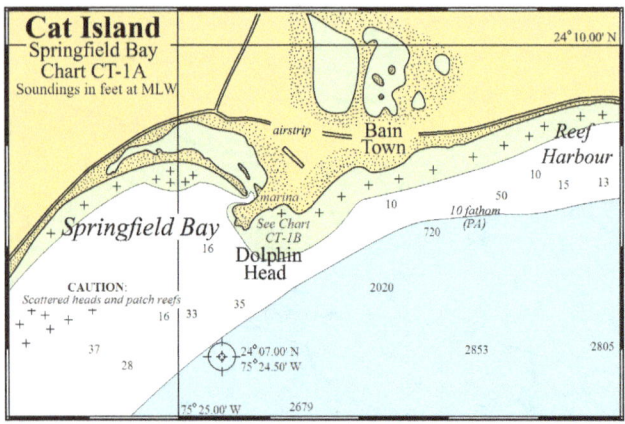

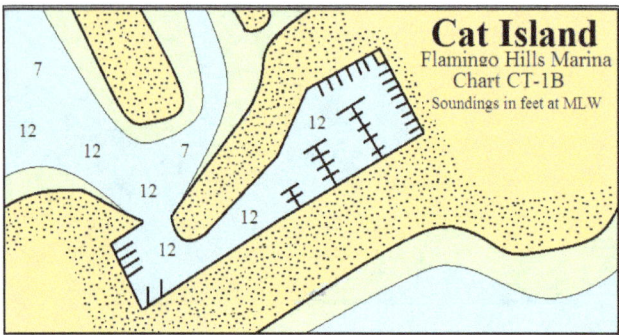

Hawk's Nest Creek, Hawk's Nest Marina, Cat Island

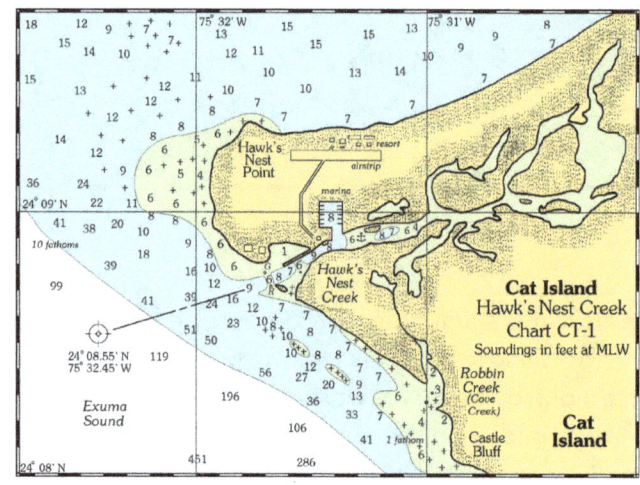

accommodate vessels up to 180' LOA with drafts to 12', even at the marina's fuel dock.

We have included the data we have on the marina and shall chart and detail the entrance channel as soon as it has been dredged and the marina open for business, but for now, all we have is what we show on the charts, so use extreme caution and make sure you have contacted them before seeking protection there.

Cat Island: *Hawk's Nest Creek*

At the very toe of the "foot" of Cat Island, is Hawk's Nest Point and *Hawk's Nest Creek* where you can anchor in the creek or get a slip in the protected marina. The creek will take 6' over the bar at low water, but it shallows quickly once past the marina, down to 5' at MLW very quickly. As you approach McQueen's the water even gets too shallow for most dinghies. Unless you have a shallow draft vessel and can get up Orange Creek, or if you can get into *Bennett's Harbour* along the western shore of Cat Island, your only choice may be *Hawk's Nest Creek* on the southwestern tip of Cat Island.

A waypoint at 24° 08.55' N, 75° 32.45' W, will place you approximately ¾ mile west of the entrance to *Hawk's Nest Creek* as shown on Chart CT-1 (with photo at top of next column). Contact the marina on VHF ch. 16 if you have any questions about entering their channel. The outer edge of the channel lies just north of the off-lying rock and is marked by a red and a green floating buoy. Pass between the two, remember red, right, returning, and head up the creek keeping the conspicuous jetty to port. There is a shallow spot at the mouth of the entrance channel with 6' at low water and another spot about 150 yards further in with the same depth. Once past the end of the jetty the marina's fuel dock will be immediately to port. Just past this dock the marina basin opens up to port with 28 slips accommodating drafts up to 7'.

If you wish to anchor in the creek, head up stream past the marina and anchor wherever your draft will allow. A 5' draft can work up stream a good way and shallow draft vessels even further. Use caution when anchoring in *Hawk's Nest Creek*. The bottom is very, very rocky, and it's hard to get an anchor to set in the thin sand. As if that was not enough, the strong current threatens to drag you along with it; there are countless stories of skippers who have dragged their anchors in this creek. The further upstream you go, the better the holding, but not much better. Make sure your anchor is well set, dive on it if you can. For this reason, it is suggested that should you seek shelter here, try to get one of the marina's 28 slips that can accommodate drafts to 7'.

Cat Island: *Bennett's Harbour*

Bennett's Harbour offers good protection but it is small and open to the north. South of *Bennett's Harbour* is Alligator Point where you will find Bennett's Creek (sometimes called *Pigeon Creek*). The mouth of the creek has a 1.5' (at MLW) bar at the entrance, but shallow draft vessels might be able to work their way upstream to the spot where *Bennett's Creek* winds northward and *Pigeon Creek* continues eastward, you will find 4' here.

A waypoint at 24° 33.75' N, 75° 39.00' W, will place you approximately ½ mile west of the entrance to *Bennett's Harbour* as shown on Chart CT-6A (with photo at bottom of previous column). From this position steer generally east and round the spit of land and its off-lying shoal and turn to starboard to anchor in the small harbour. Bear in mind that *Bennett's Harbour* gets quite choppy when strong winds from the northwest to north push in the seas. There's also a lot of current here, two anchors are a must. *Bennett's Harbour* has room for two to three vessels of moderate draft, no more than 5' or so.

Cat Island: *Orange Creek*

The northernmost settlement on Cat Island is *Orange Creek* (see photo next page), whose creek has waters that are in some places up to 7' deep but the bar at the entrance restricts entry boats with drafts of less than 3' at high water. The creek has a few scattered wrecks along with some fishing boats anchored inside.

A waypoint at 24° 36.80' N, 75° 41.20' W, will place you approximately ½ mile southwest of Arthur's Town. From this waypoint, vessels can work their way northwest to the mouth of *Orange Creek* as shown on Chart CT-7 (next page).

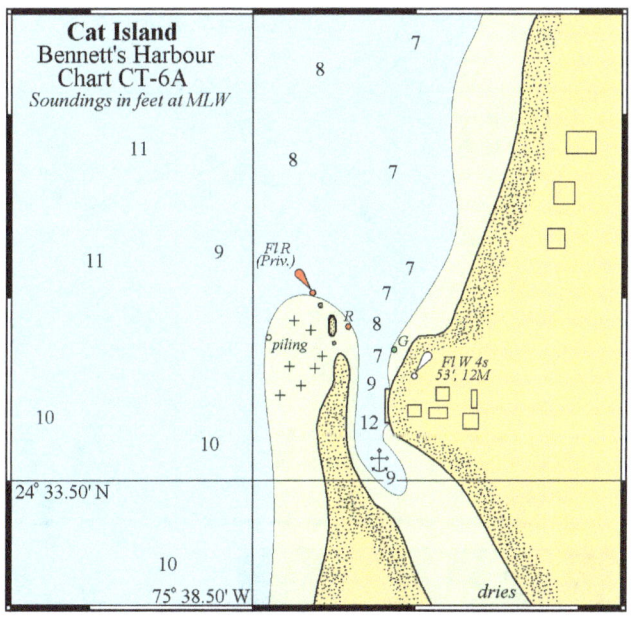

Bennett's Harbour, Cat Island

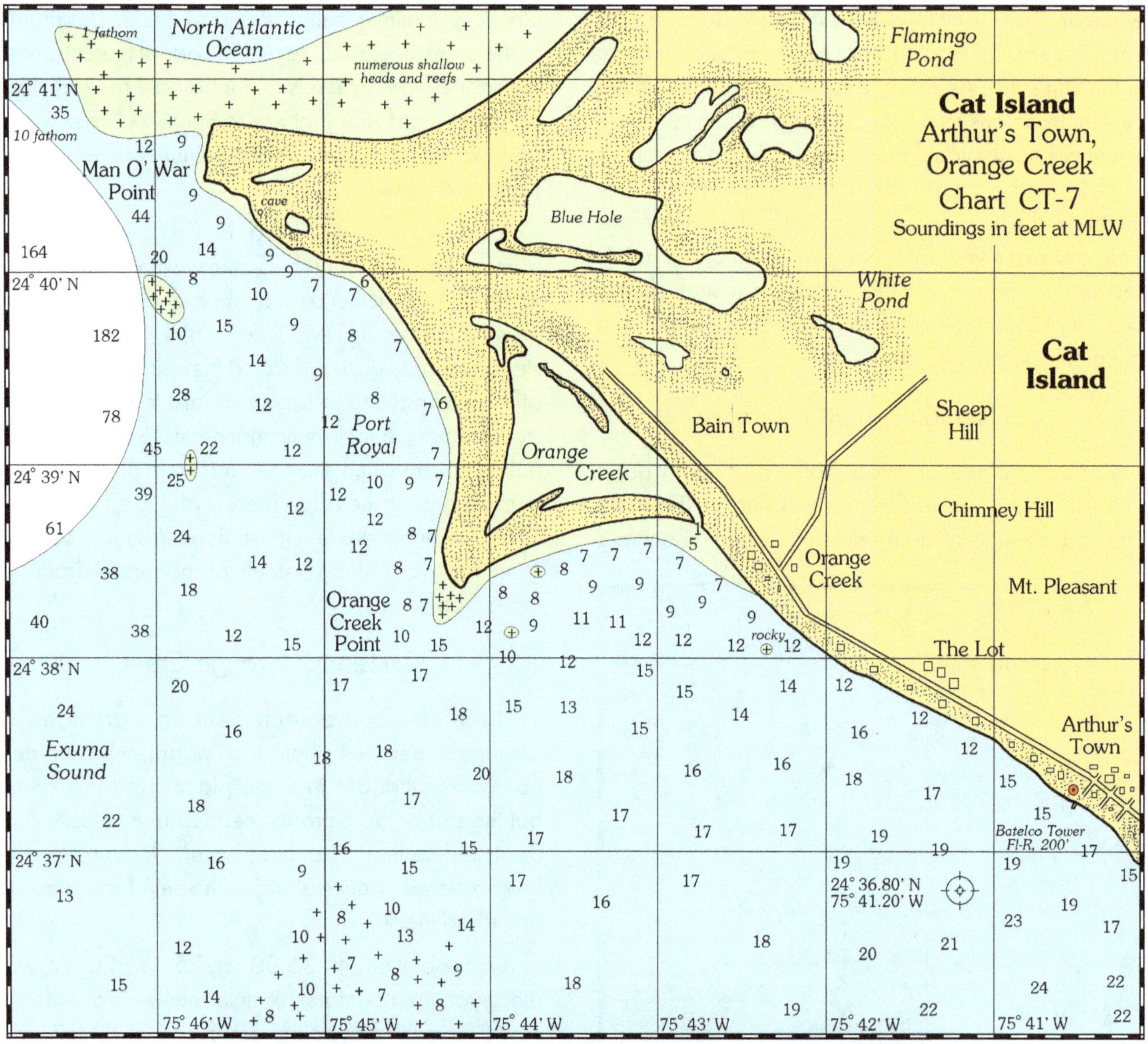

Orange Creek, Arthur's Town, Cat Island

64 • THE CAPTAIN'S GUIDE TO HURRICANE HOLES

Chapter 6

The Exumas

THE EXUMAS ARE KNOWN FOR THEIR PRIME CRUISING GROUNDS WHICH brings lots of boats to this island group making many of the anchorages crowded in season with some boats staying for part or all of the hurricane season. Fortunately, the Exumas are blessed with some of the best hurricane holes in The Bahamas. From Highborne Cay Marina in the north, to the holes at Stocking Island and the anchorages at Red Shanks in Elizabeth Harbour, George Town. You won't be far from protection here whether you want to ride a storm out at anchor, in a marina, or to haul out at George Town or Stella Maris in Long Island. You are also close to Nassau, Eleuthera, and Cat Island where more good protection can be found. There is a new marina under construction at the southern end of Norman's Cay that promises good protection against all but a strong storm surge. Look for its opening in the near future.

Highborne Cay Marina

Highborne Cay Marina is a well-managed marina that is protected by a high hill adjacent to the marina. The marina is very popular and is often full so if you plan to seek shelter here, call early for reservations.

As shown on Chart EX-6A, the entrance to the marina area lies NW of *Highborne Cay Cut*. To enter the marina area from Exuma Sound, a waypoint at 24° 42.10' N, 76° 48.15' W, places you ¼ mile east/southeast of *Highborne Cay Cut*. Head through the cut between Highborne Cay and the off-lying rocks and round up towards the beach and marina keeping an eye out for the small reef that lies just off the small cay to the west of Highborne Cay. Round the tip of the jetty and proceed to the marina docks as shown on Chart EX-6A.

For those arriving from Nassau or the banks, a waypoint at 24° 42.60' N, 76° 52.25' W, will place you well to the southwest of Highborne Rocks.

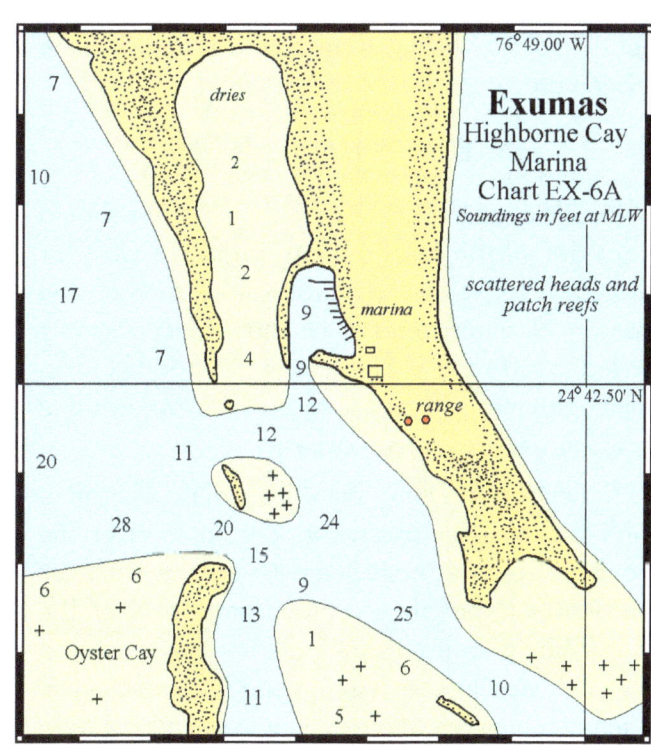

Highborne Cay Marina

From this position take up an approximate heading of 95° magnetic until you approach the entrance into the marina and can pick up the range marks that lie approximately 2½ miles due east. The area directly west of the entrance to the marina has plenty of deep water so strictly following these directions here is not absolutely necessary.

Once past Highborne Rocks you can pilot by eye to the entrance to the marina that lies just north of the small rock that has an orange pole on its northern end. Line up the orange range marks situated on Highborne Cay and head in on a bearing of 90°. Once inside, look to port and line up the orange range marks north of the marina and follow them in on an approximate heading of 15° until you can round the jetty into the marina area.

The Pond at Norman's Cay

Heading south from Highborne Cay, one should consider *North Harbour* at Norman's Cay for protection for vessels with drafts of less than 6'. *The Pond*, or Norman's Pond, as the small bay is commonly called, offers excellent protection and good holding although there is a mile long north-south fetch that could make things rough at best.

The entrance is tricky but will handle a 6' draft at high tide. Most people do not attempt to enter *The Pond* because they have heard it is extremely difficult and that many people run aground in the attempt. The truth is most people run aground when they attempt to pass between the outlying cays and Pyfrom's Cay, an often-recommended route. The waters here are very shallow and should not be attempted by any vessels except those with shoal draft, 3' or less. It is safer and easier to head outside into *Exuma Sound* and pass well to the east of these cays and enter the pond from the northeast in *Exuma Sound*.

Once clear of the outlying cays head to a waypoint at to a waypoint at 24° 37.25' N, 76° 47.50' W, (not shown on Chart EX-8 next page), which places you approximately 1/4 nm east of the deeper water leading to the entrance to *The Pond*.

From the waypoint approach the small rocks leading to the entrance into *The Pond* and you will see what appear to be two cannons, one on the southern rock and one on the northern rock pointing at each other. The cut between these two is the entrance to the channel that leads into *The Pond*. There is a small deep water cut to the north of the northernmost cannon but do not use it, as it shoals quickly just inside.

Once inside the cut look to starboard along the shore of Norman's Cay and you will see a small, privately maintained range, a large white pole with a shorter one in front. Line up the range and follow the dark water (4' at low water) until you can parallel the shoreline in 8' of water.

Beware, there is some very shallow dark water that you must leave to port so don't let it confuse you, use your eyes and take it slow and easy through here. As you parallel the shore give the western tip a large berth to avoid the shallows that work out westward from it. Turn north into *The Pond* and find a good spot

The Pond at Norman's Cay

Warderick Wells: *South Warderick Wells*

Heading south from Norman's Cay in the Exumas you soon come upon *Exuma Park*. Dr. Evans Cottman, in his book "Out Island Doctor," writes about surviving a hurricane in the southernmost creek on Shroud Cay many, many years ago, however today that creek is perhaps too narrow and too shallow for most cruisers. Several cruisers have ridden out hurricanes in the north mooring field at Warderick Wells but in reality, that was one heck of a gamble. But the mooring field and anchorage at South Warderick Wells, between Warderick Wells and Hog Cay, offers better protection but you must avoid damaging the ancient Stromatolites located in the center of the bay.

The easiest entrance is from *Exuma Sound*, where a waypoint of 24° 22.97' N, 76° 36.73W will put you about ½ mile northeast of this anchorage and well southeast of the off-lying rock in 21' of water. The

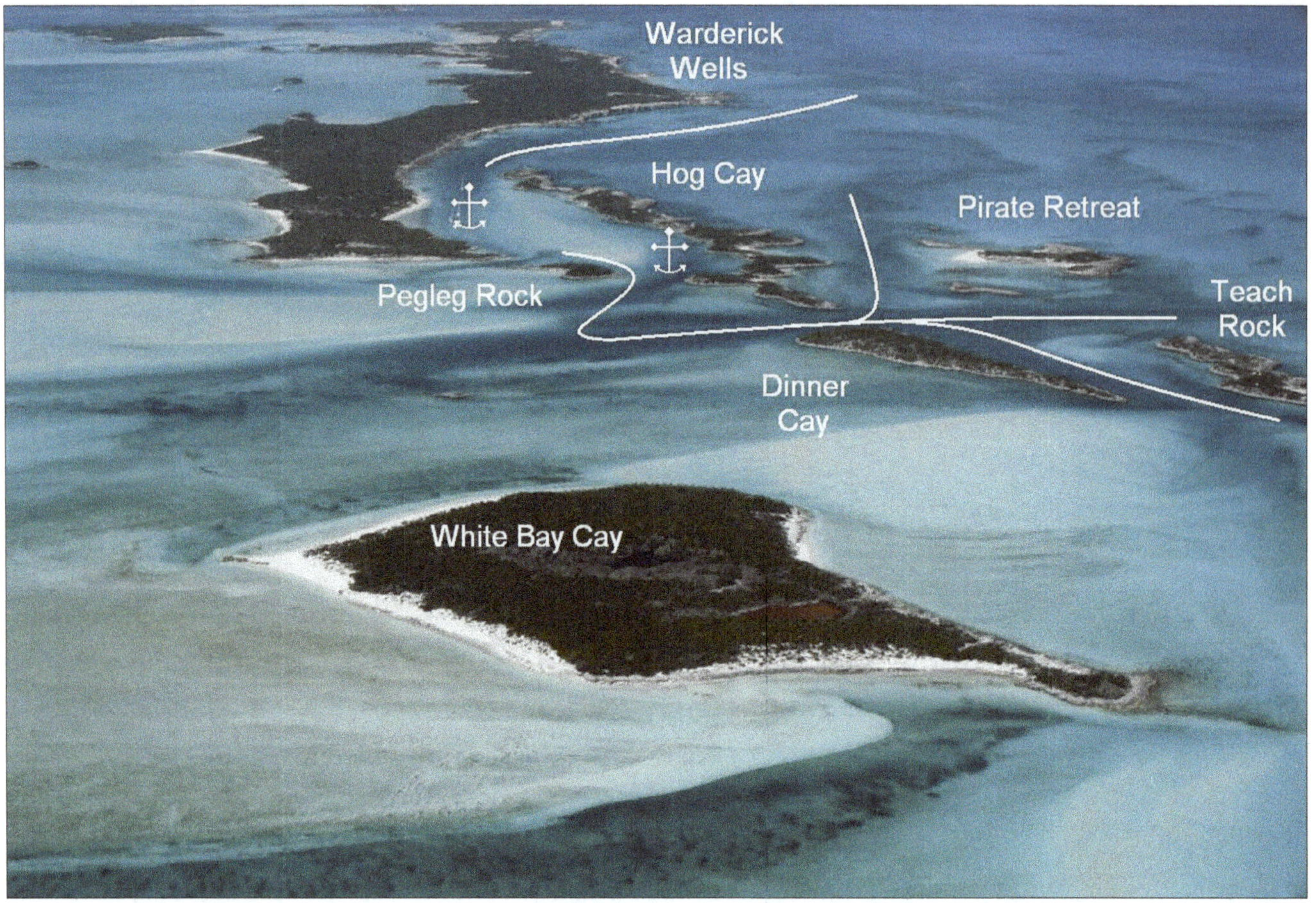

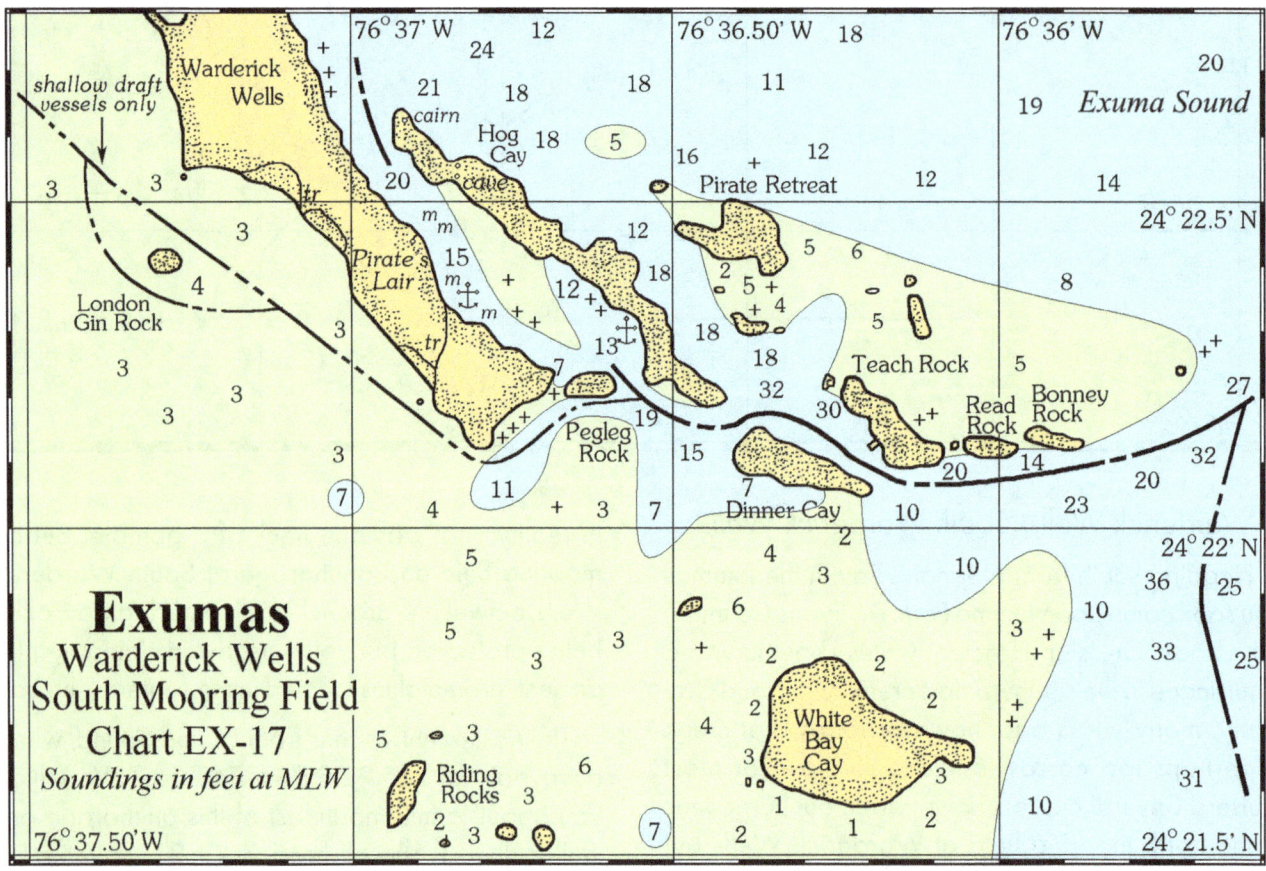

South Warderick Wells anchorage/mooring field

68 • THE CAPTAIN'S GUIDE TO HURRICANE HOLES

Compass Cay Marina

entrance to the anchorage at South Warderick Wells is not immediately seen, but if you steer for the small beach (barely visible from seaward) on a course of 215 degrees, the entrance will soon become evident. As you get closer you will see a small cairn on the north end of Hog Cay. Keep this to port as the mooring field opens up in front of you as shown on Chart EX-17 on the previous page. Head in and keep a lookout for the stray head on your starboard side. The mooring field has a sandbar in the center of it, keep between the sandbar and the eastern shore of Warderick Wells. Anchor on the Warderick Wells side of the bay, or head to the southeast corner of the bay and anchor behind Hog Cay as shown on the chart.

Compass Cay Marina

Compass Cay Marina did just fine in Hurricane Matthew in 2016. The catch here is that if you wish

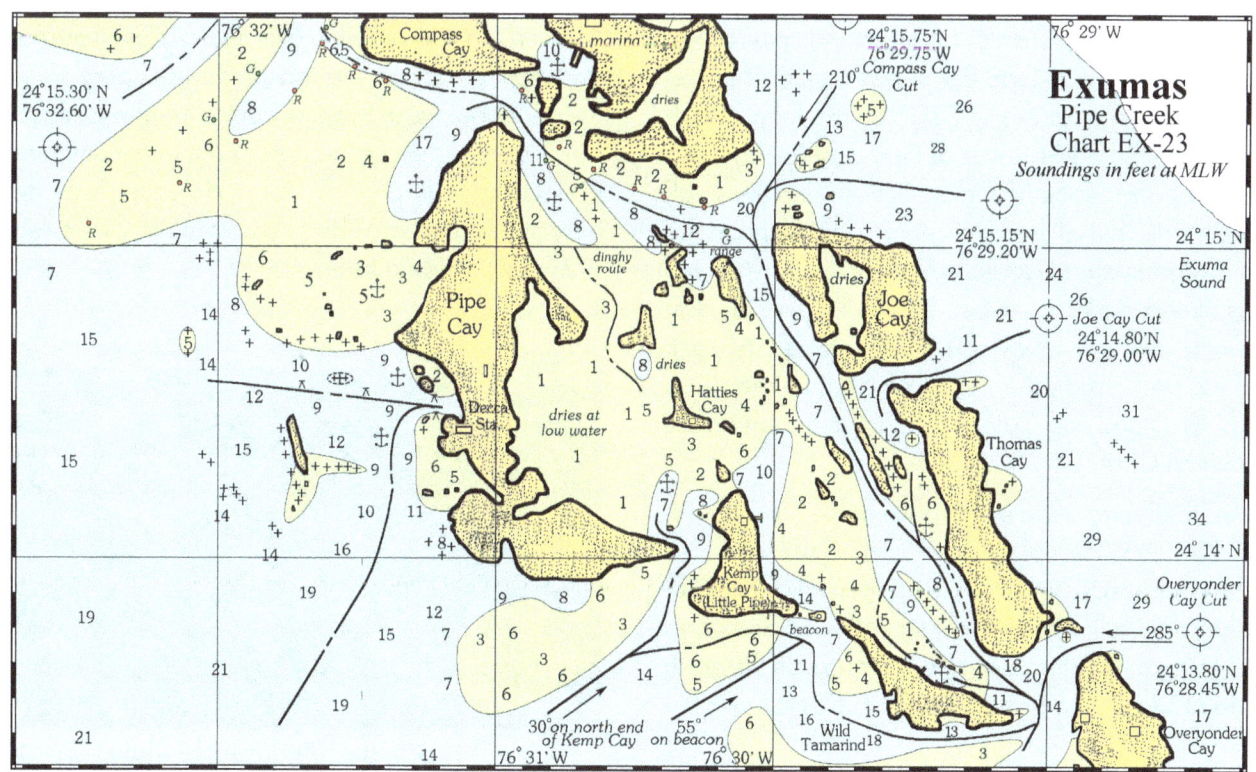

to stay at the marina, you will be required to pay a damage deposit of $5,000 and if any damage occurs to the marina infrastructure due to your boat, you will be required to pay for the damages before you can leave.

There are two ways to access *Compass Cay Marina*, from offshore in *Exuma Sound* and from the banks. Boats with drafts of less than 6' (at low water and 8' at high water) may take the dredged channel round the southwestern tip of Compass Cay (7' at MLW) and make for the entrance to *Compass Cay Marina* as shown on Chart EX-23 (previous page).

If you are heading to *Compass Cay Marina* from Conch Cut, follow close to the northwest tip of Compass Cay in 10' of water and head just west of south, paralleling the shore in 7'-10'. Take care to avoid the rocky bar that juts out westward from Compass Cay about ¼ mile south of the northwest tip. The bar is easily seen in good visibility. Roughly ¼ mile past the rocky bar the deep channel is split by a shallow sandbank. Vessels heading to *Compass Cay Marina* must bear to port to pick up the marked channel leading into the marina (see Chart EX-23 on the previous page).

Skippers wishing to visit *Compass Cay Marina* from *Exuma Sound* may enter at *Compass Cay Cut*. There are two ways to enter this cut. First, a waypoint at 24° 15.75' N, 76° 29.75' W, will place you approximately ½ mile to the north/northeast of the cut. Take up a heading of 210° on the long beach that lies on the eastern side of Pipe Cay (easily seen from sea) and proceed into the cut keeping the small group of rocks that lie in the center of the cut and the shallow rocky shoal just north of them to port. You will pass over a dark brown bar, but it will have 12'-15' of water over it at low water. Pass between the rock that lies well inside the center of the cut (keep it to starboard) and Joe Cay and follow the markers as shown on Chart EX-23.

Another entry is via a waypoint at 24° 15.15' N, 76° 29.20' W, will place you ½ mile east/southeast of the *Compass Cay Cut*. Take up a heading of 285° on the rock well inside the center of the cut and head in keeping the scattered heads and rocks off the north shore of Joe Cay to port and the group of rocks in the center of the cut to starboard. Then follow the entry instructions as per the preceding paragraph.

The old *Sampson Cay Marina*, now closed

Sampson Cay

One of the most protected marinas in The Exumas is located in a tiny cove on the eastern shore of Sampson Cay (see photo above).

Unfortunately, the cay is now private and access to this wonderful hurricane hole is by invitation only.

Staniel Cay

Some cruisers choose to anchor *Between The Majors* (sometimes called Little Major's Spot) as shown on Chart EX-25 on the top of the next page and in the photo below, in fact, co-author Stephen J. Pavlidis has himself done just that but cannot recommend it in anything over a Cat 1 storm. However, we offer it here just in case somebody wants to avail themselves of the protection offered.

The anchorage *Between The Majors*

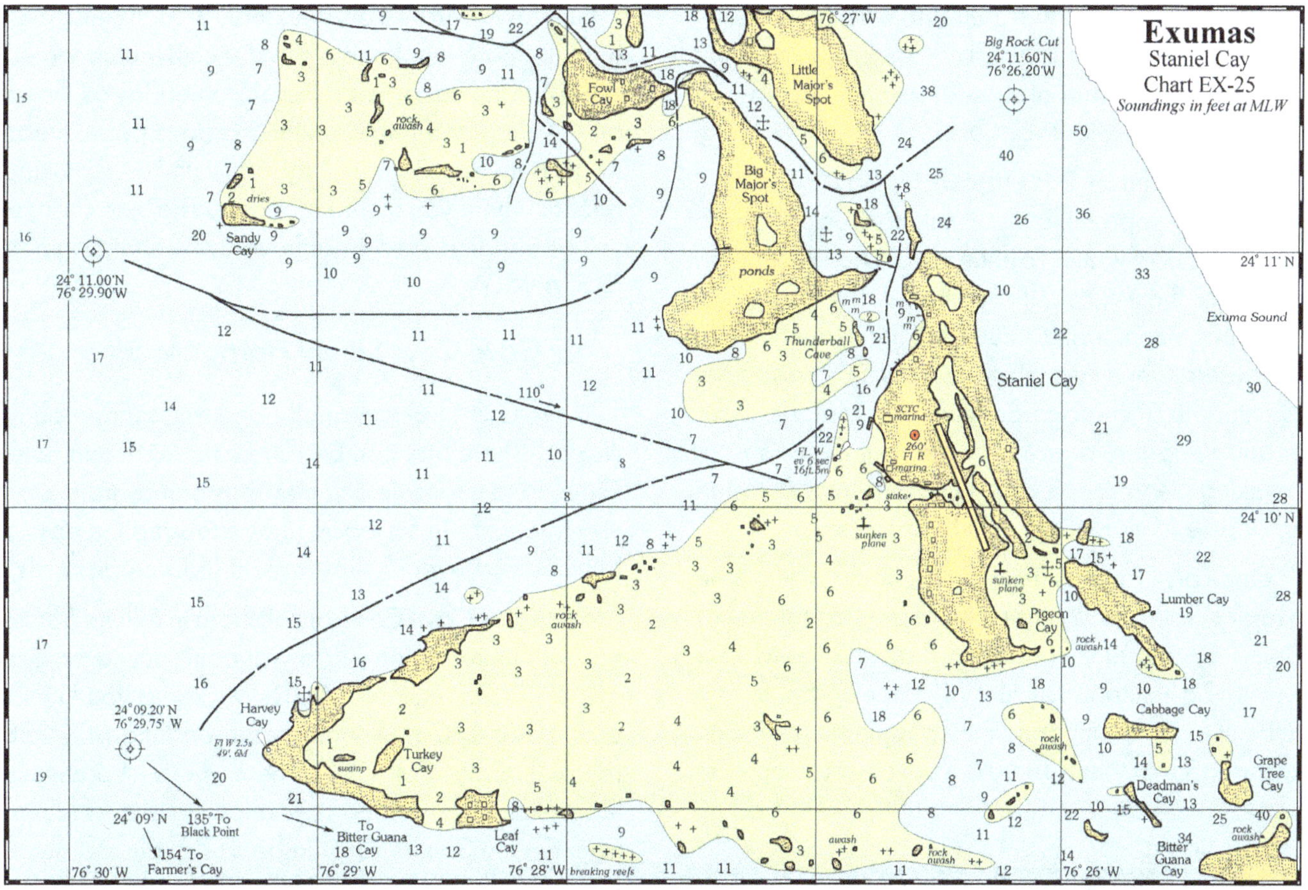

Staniel Cay, the creek leads north from Pigeon Cay and Lumber Cay

The anchorage *Between The Majors* is protected by cays that rise to over 30' and the holding is excellent. If approaching from offshore in *Exuma Sound*, you will have to use *Big Rock Cut*.

A waypoint at 24° 11.60' N, 76° 26.20' W, places you approximately ½ mile to the northeast of the cut in 50' of water. To enter the cut, head southwest and favor the northern half of the cut. There are some rocks in the center of the cut with only 8' over them at low water. Once through the cut you may head north to enter the anchorage *Between The Major's*. Head in and keep the large rocky bar that is awash and breaking (very conspicuous) lying off the southwestern tip of Little Major's Spot well to starboard.

Once past it turn to the NNW, and parallel the Big Major's Spot shore and pick out a place to anchor. Do not anchor at the north end *Between The Majors*. There is a white coral reef (5' over it) that is very difficult to make out and many people have hit it not knowing it was there. There is a private mooring here.

Shallow draft vessels (less than 3') can work their way up the creek (follow the distinctive blue water) just inside the easternmost leg of Staniel Cay as shown on the chart and photo on the previous page. Head north between Pigeon Cay and Lumber Cay and anchor where you are comfortable. The blue channel will be winding, and it will have a few shallow spots of 3' at MLW.

Cave Cay: *Safety Harbor Marina*

Cave Cay boasts a small pond and marina on its western shore that can be used as a hurricane hole. The marina is for sale and they may or may not permit you to anchor in the basin (they dredged the basin), but they do have 35 slips for rent (NO credit cards).

To access Cave Cay from offshore in *Exuma Sound* you will need to head to a waypoint at a waypoint at 23° 54.25' N, 76° 15.20' W, places you ¼ mile east of *Cave Cay Cut* as shown on Chart EX-28B (below). *Cave Cay Cut* is one of the best passes in this stretch of the Exuma cays allowing access to and from *Exuma Sound* from the banks. Wide and deep, it is not as dangerous as Galliot Cut.

Enter the cut between the south end of Cave Cay and Moon Cay that lies to the north of Musha Cay. The cut is wide and deep and the strength of the current makes for rough conditions when it opposes a strong wind. Follow the channel and steer to starboard once past the SW tip of Cave Cay and the entrance to the marina and basin will open up on your right.

Safety Harbor Marina at Cave Cay

The Pond at Rudder Cut Cay

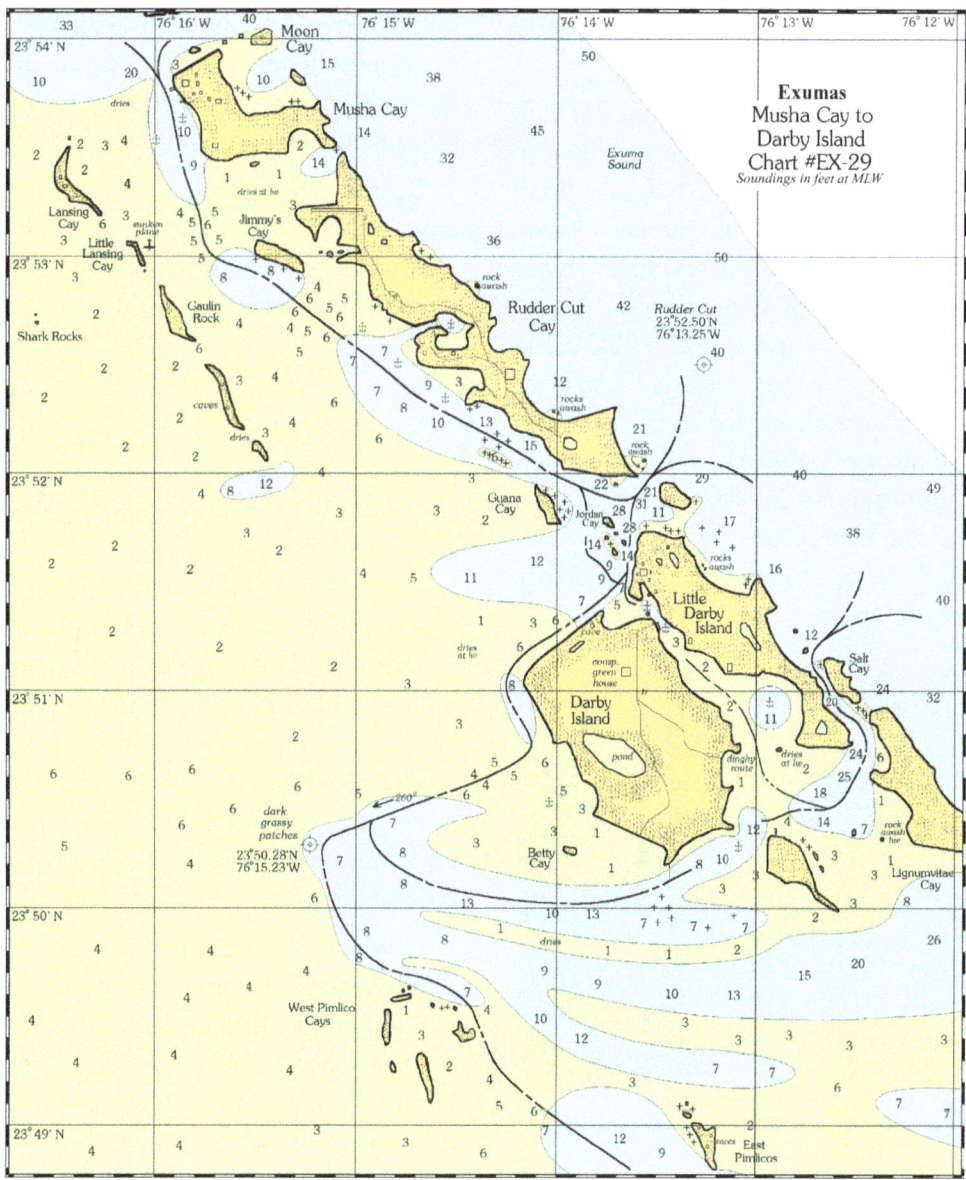

Rudder Cut Cay: *The Pond*

Just a bit south of Cave Cay sits Rudder Cut Cay and its protected little cove usually just called *The Pond*. Many experienced captains like the pond at Rudder Cut Cay as a refuge but I see the eastern shore as being very low. A strong hurricane with a large storm surge and high tide might make this anchorage a death trap. However, we offer it here in the hope that someone can use it and there is not a great storm surge.

As shown in the photo (on the previous page), and on Chart EX-29 (above), *The Pond* is protected except for the eastern shore (at the bottom of the photo).

The entrance to *The Pond* has shallow bars both north and south of the entrance. The channel will accommodate a 6' draft vessel at low water however there is a 5' bar just at the entrance to the lake at low water. Once inside there is room for 8 boats in 6'-8' of water but beware as there is a lot of debris on the bottom.

If approaching Rudder Cut Cay on the banks side from Musha Cay, pass between the very obvious sandbank between Musha Cay and Rudder Cut Cay and the not so obvious sandbank to its west. Favor the Musha Cay side of this route staying about 20-30 yards west of the sandbank. This route will accommodate a 5' draft at low water except for a low spot of 4.8' just off Jimmy's Cay. Once past Jimmy's Cay keep an eye out for the entrance channel

THE EXUMA CAYS • 73

leading into *The Pond*, follow it in and anchor where you choose.

Vessels in *Exuma Sound* may enter at Rudder Cut (Chart EX-29 above). The cut can be absolutely deadly in strong onshore winds and an outgoing tide. Even in light onshore winds and swell coupled with an outgoing tide you will see small breakers forming over ½ mile offshore. The current flows like a river for quite a way into Exuma Sound. A waypoint at 23° 52.50′ N, 76° 13.25′ W, places you approximately ¼ mile to the NNE of the cut. From this point line up on the house on Little Darby Island on a heading of 185° and enter the cut passing between the southern tip of Rudder Cut Cay and the small unnamed cay to the north of Little Darby Island. An approach from this heading instead of coming in on a more easterly heading will help you avoid the main flow of the breakers until you are almost in the cut itself. Keep a sharp lookout for the rock that is awash at high tide about 50 yards off the southeastern tip of Rudder Cut Cay.

Do not attempt to pass between the unnamed cay and Little Darby Island, this cut is foul with rocks and heads. Once inside Rudder Cut you may turn to starboard towards *The Pond*.

Great Exuma: *Emerald Bay Marina*

In early October of 2016, *Emerald Bay Marina* (see photo in previous column) sustained a nearly direct hit by Hurricane Matthew causing scattered damage but proving for the second year in a row that the marina is a good hurricane hole. The damage amounted to one boat suffering minor damage and there were no injuries to staff or guests. In 2015, Hurricane Joaquin hit the marina causing scattered damage to dock hardware such as electrical and water pedestals as well as some shrubbery on the marina's grounds. The marina has no haul out facilities but does have 150 slips that can accommodate vessels to 240′ LOA, with a beam of up to 50′, and draft to 14′.

A waypoint at 23° 38.50′ N, 75° 54.50′ W, will place you approximately ½ mile east/northeast of the marked entrance to *Emerald Bay Marina*, shown in detail in Chart EX-41A (bottom of previous column). From the waypoint (not shown on the chart) head SW and enter the marina proper between the red and green markers and the lit jetties.

Great Exuma: *Stocking Island*

The George Town area is home to what may be the finest holes in The Bahamas. Holes #0, #2, and #3 at Stocking Island are excellent hurricane holes in every sense offering protection from wind and wave. The only problem here is that these holes will be crowded, and Hole #3 is usually full of stored boats with absentee owners.

The Holes at Stocking Island (Chart EX-45 at top of the next page) are packed in the winter when boaters often cram themselves in, lying to two or more anchors to avoid swinging into their nearby neighbors. The entrance is called Hole #1 and is open to the west and southwest and lies just off *Volleyball Beach*.

Emerald Bay Marina, Great Exuma

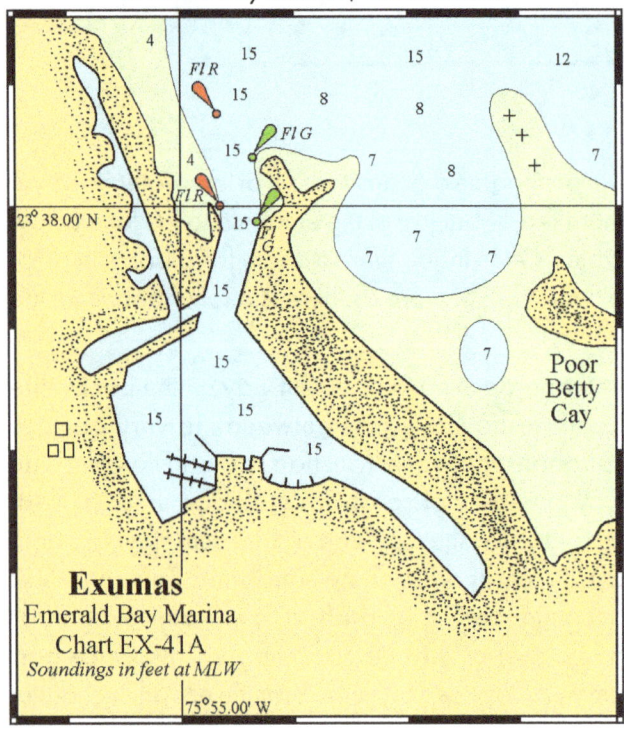

The Holes at Stocking Island

inside Hole #2, on the southeastern tip of the hole on the western shore of Stocking Island is the dredged harbour for *St. Francis Resort and Marina*.

You can proceed into Hole #3 in the same manner, keeping to the port side in the channel to avoid a reef to starboard. The third hole is deeper and is usually full of unattended stored boats managed by *Little Toot* (VHF ch. 16). The *Kevalli House Marina* located here has 9 slips for vessels to 70' LOA with drafts to 6.5'. Check with the marina, they may have a slip available.

Great Exuma: *Crab Cay and Red Shanks*

If you cannot get hauled out at Master Harbour Marine, and of course you don't want to use the *Exuma Yacht Club Marina* in Kidd Cove as it is too open and has a poor hurricane record, you can find some very good hurricane protection can be found in the *Red Shanks* anchorage area if you can get in close to the mangroves, and inside the western arm of Crab Cay as shown on Chart EX-48 at the top of the next page.

To enter *Red Shanks* from George Town you have two routes from which to choose. If you round Rolle Cay (Chart EX-48) and follow the eastern shore of Crab Cay you will have to weave your way through some small but shallow patch reefs near the southeastern end of Crab Cay. The water will shallow to 5'-6' at low tide but with good light you can easily pilot your way through.

The best route is to pass through the reefs that are marked by buoys just south of Elizabeth Island and take up a course of 120° on the conspicuous house on Man-O-War Cay. When the southeastern tip of Crab Cay begins to come abeam, look for the large patch reef that lies well before the shallows west of Man-O-War Cay. Pass to either side of the reef although the better water is to the east of it. Steer approximately southwest until you can round up into the *Red Shanks* anchorage keeping John Devine Cay to port.

There are actually three areas to anchor here. The first one lies in the deep water just south of Crab Cay and northwest of Red Shanks Cay. The other two anchorages lie more to the west along the Crab Cay shore. The entrances to both should only be attempted at high tide and with good visibility. To enter the first inner hole, head toward the southern shore of Crab

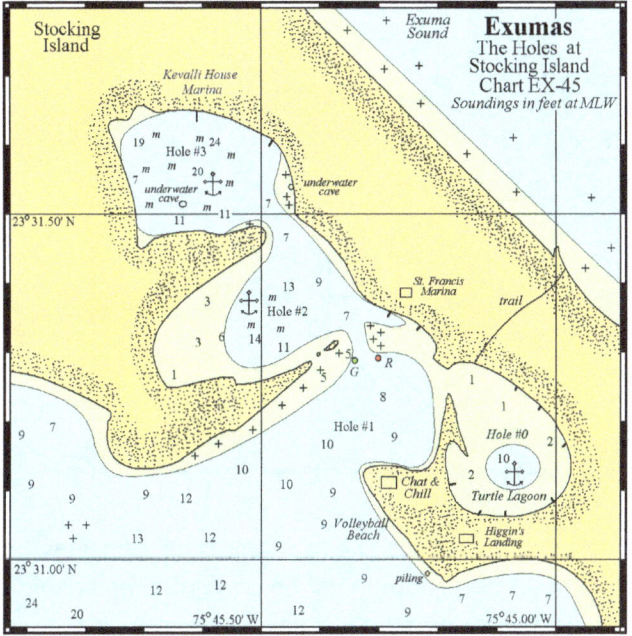

The entrance from Hole #1 into Hole #0 is for shallow draft vessels only. The narrow, twisting, shallow channel (marked by stakes) can only be maneuvered at high tide by the shallowest of drafts. For this reason Hole, #0 is often referred to as *Multihull Hole* and sometimes *Turtle Lagoon*. There is deep water, 6'-10' at low tide, if you can make it inside.

A draft of 6' can be taken from Hole #1 into Hole #2 at low tide. Enter the cut between the red and green markers as shown on Chart EX-45, and watch out for the curving reef to starboard, keeping a bit more to the port side of the channel. Keep an eye out for the occasional submerged rock to port before the entrance. As of this writing the reef was marked by a privately maintained line of small white buoys. Just

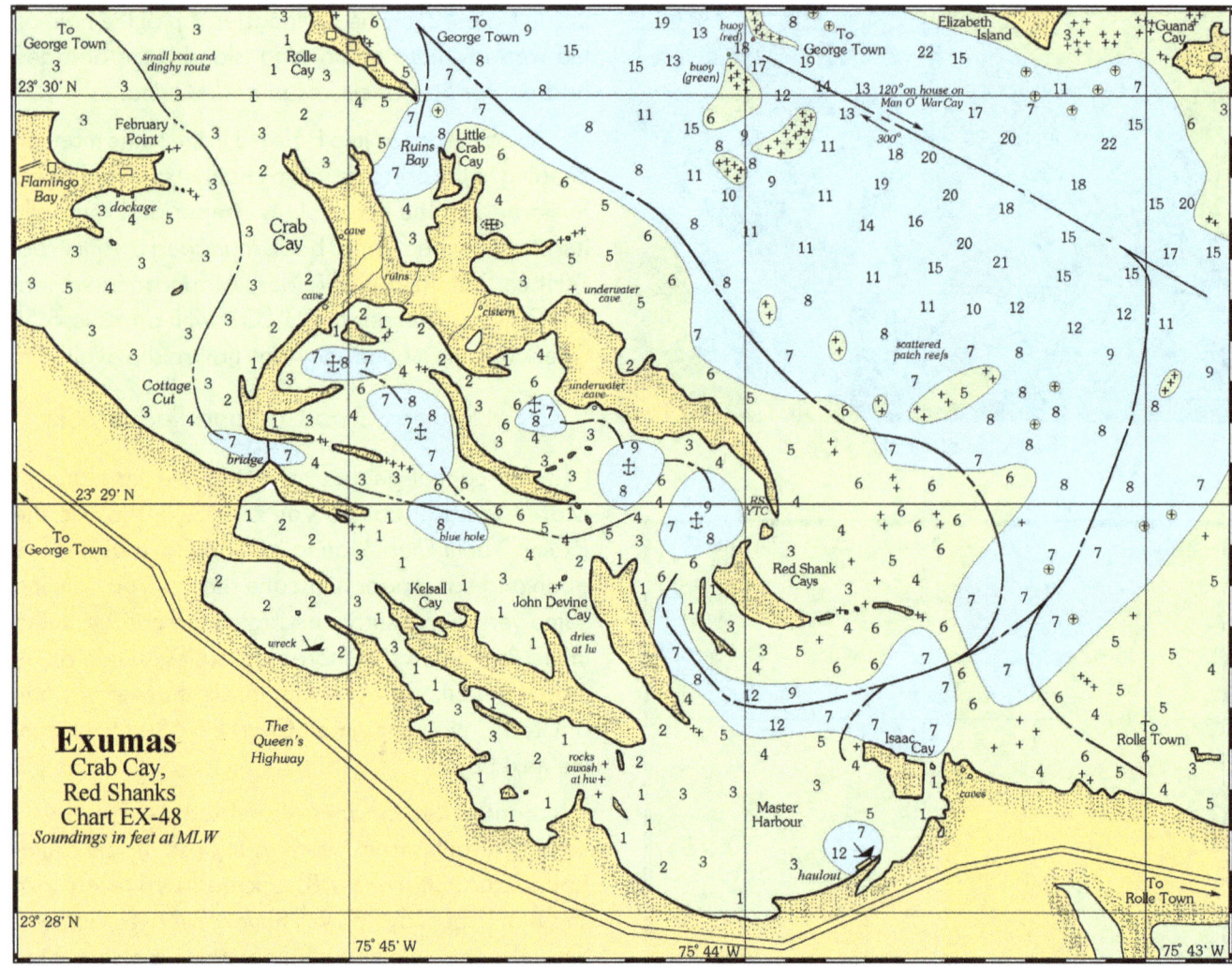

Cay to clear the shallows that work northward from John Devine Cay and parallel the shore into the hole. You can take 6' through here at high tide. The entrance to the second hole is slightly more difficult as you must pass between two sandbars in a channel that will take 6' at high water. The deeper water lies slightly closer to the Crab Cay shore and you must steer around the point to avoid a shallow bar. The bottom here is generally sand and if you take it slow and easy you should not run into trouble.

Another spot with good protection lies to the west of Red Shanks, inside the western arm of Crab Cay. This anchorage is good for vessels with drafts slightly exceeding 6' but to enter you must play the tides. There is 7'-9' at low water in places here and the holding is good. This anchorage is usually occupied by shoal-draft multihulls due to the shallows in the approach channel from *Red Shanks*. The entrance lies between the first two anchorage basins in *Red Shanks*. The entrance is over the shallow sandbank between the two holes and should only be attempted at high tide or just before high tide. Steer between John Devine Cay and the small rock that lies to its northwest, favor the small rock but not too closely. Round the narrow unnamed cay to starboard in 4'-6' at low water and pass into the deeper depths (7'-9') that lie to its west. Watch out for the rocky bar that runs eastward from the small cay that lies to the south of the anchorage area as shown on Chart EX-48. Do not anchor in the deeper water labeled "Blue Hole," it is exactly that.

76 • THE CAPTAIN'S GUIDE TO HURRICANE HOLES

Chapter 7

The Southern Bahamas

IF YOU ARE CRUISING THE SOUTHERN BAHAMAS, FROM THE CROOKED/ACKLINS District to Mayaguana or Inagua you will not find a truly safe hole. Although a large sailboat managed to ride out Hurricane Klaus lying between Samana and Propeller Cay it is not advisable to attempt to test your luck in that anchorage. Options include either heading north to better protection at George Town or continuing on to The Turks and Caicos for protection If you have enough time you could try to make Luperón in the Dominican Republic, which is as good a hole as any in the Caribbean, but NEVER try to race a hurricane! With that in mind we will show you what protection there is to be found in the Southern Bahamas, at Long Island and Ragged Island in the Jumentos.

Long Island: *Joe Sound*

Just south of Long Island's northern tip at Cape Santa Maria is *Joe Sound*, a good anchorage in almost any conditions, and even fair hurricane protection as long as there is not a huge storm surge. The land to the west is low and a storm surge of 9'-15' such as Hurricane Lili brought in 1996, could make this anchorage untenable. Actually, the western shore of Great Exuma was struck with a 15' surge during Hurricane Lili as the eye passed over the channel between Cat Island and Long Island heading NE and carrying 92mph sustained winds.

The entrance to *Joe Sound* is through a very narrow cut that is hard to see except from straight on. A waypoint at 23° 36.85' N, 75° 21.55' W will place you approximately ¾ mile southwest of the cut as shown on Chart LI-1 (see next page).

The entrance lies about ¼ mile south of the very distinctive angular white houses. Never attempt this cut at night unless you are very familiar with it. As I mentioned earlier the cut is extremely narrow although a friend of mine, the owner of a 50' trimaran with a 30', beam claims he goes in and out at high tide. Enter the cut (7' at high tide) being careful to avoid the rocks on the southern side of cut. Once in the channel, slightly favoring the southern side, follow the slight curve until you pass into the deeper water (7'-9') on the other side. When you are through the cut you may turn to port and anchor wherever your

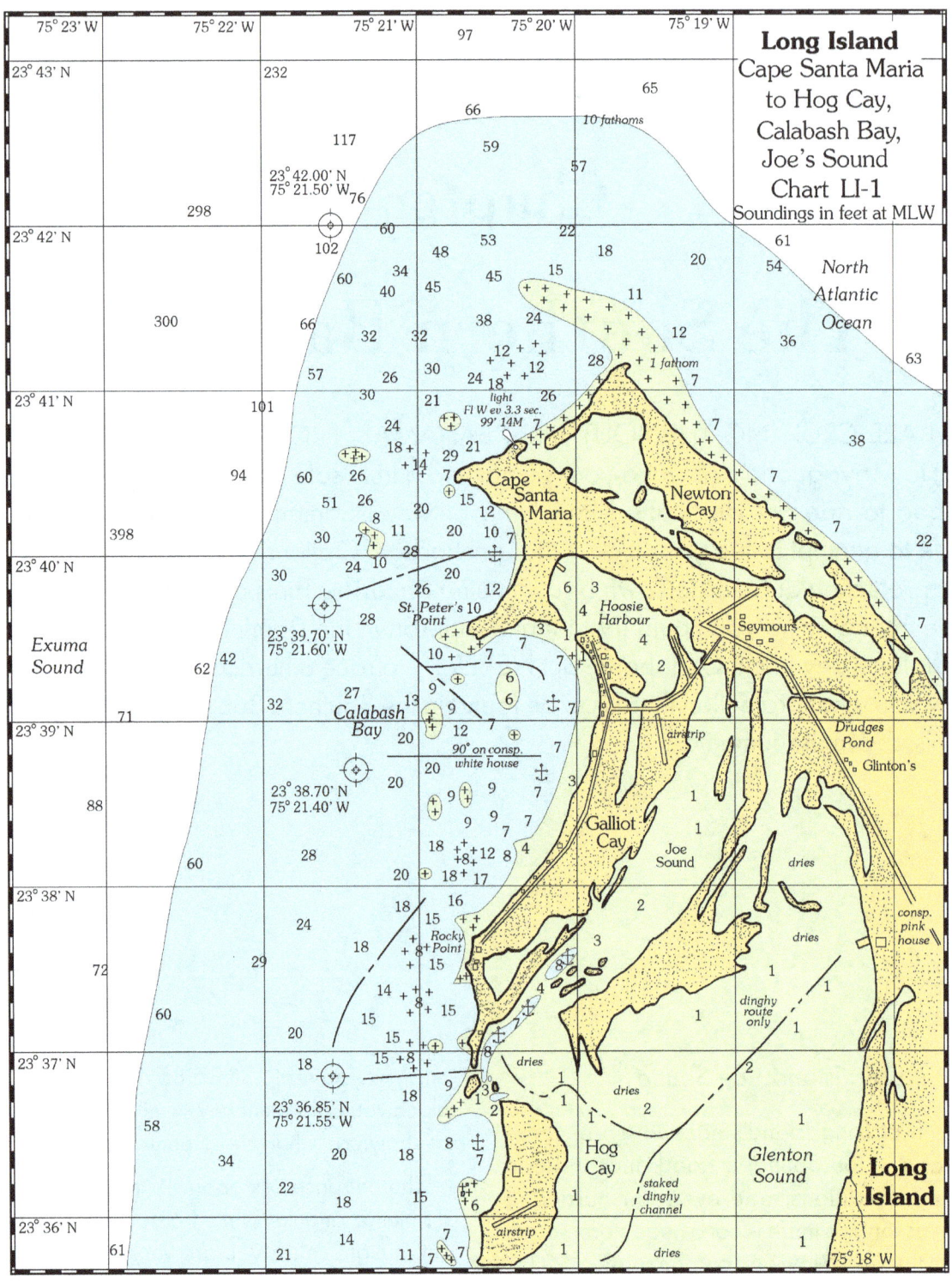

draft will allow. Shallow draft vessels can go further up into *Joe Sound*, really nothing more than a creek. The sides dry at low water with about 6'-9' in the deep-water passage.

About ½ mile north of the entrance, after crossing a 5' bar (at MLW) and between two groups of mangroves, vessels will find 8'-9' at low water before the creek shoals. As many as a dozen boats can lie in *Joe Sound* safe from the fury of the seas but open to the wind.

Long Island: *Stella Maris*

Stella Maris, Star of the Sea, is a huge resort complex lying about halfway between Dove Cay and

78 • THE CAPTAIN'S GUIDE TO HURRICANE HOLES

Simms along Long Island's western shore and should be your first choice for shelter. The resort is home to a full-service marina and boatyard with 8 slips that can accommodate a 20' beam and 8 alongside ties for vessels to 100'. *Stella Maris* offers great protection for those who want to get a slip, haul out (marine railway, 80' LOA, 6' draft, and will accommodate catamarans), or tie off in the canals to the north of the marina.

The marina cannot be seen from the normal routes of boats passing around Dove Cay as shown on Chart LI-2 (below). If bound north or south along the western shore, a waypoint at 23° 33.03' N, 75° 19.90' W will place you approximately ½ mile southwest of Dove Cay and approximately 3½ miles west of the entrance to the marina. From this waypoint steer 85°-90° and you will see a large orange buoy; just past it is a row of stakes that lead into the marina. Except for the very first one, all the stakes will be topped with an orange float, but they do not show up well on radar. Take these stakes close to port (within 3 yards) and you will have at least 5' at MLW (7' at high tide) the entire way to the marina on an approximate course of 90° - 95°. At the last stake you will turn to port to enter the marina complex.

Be sure you call the marina on VHF prior to your arrival and to secure accommodations. As you enter the marina proper the fuel dock (gas and diesel) will be to starboard and the slips directly in front and to port of you. Well to port is a narrow "S" shaped canal that works its way around the marina and is an excellent hurricane hole where several local boats rode out Hurricane Lilly in 1996 with no damage. The canal is 6'-7' deep at low water and there is 7'-8' at the docks.

Long Island: *Dollar Harbour*

Along the western shore of Long Island, a string of cays stretches from the Jumentos and Nuevitas Rocks eastward to Long Island itself. At the eastern end of these cays, just off the mainland of Long Island, can be found a hurricane hole at *Dollar Harbour* (the entrance has a 5' controlling depth at MLW, see Chart LI-5 on the next page).

The only way to reach the area for vessels of 3' draft or more is to approach from the west at Nuevitas Rocks or from the south or southeast along the southeastern shore of Long Island. Vessels with less than a 3' draft can, by playing the tide, gain access to the *Dollar Harbour* area directly from Salt Pond. From

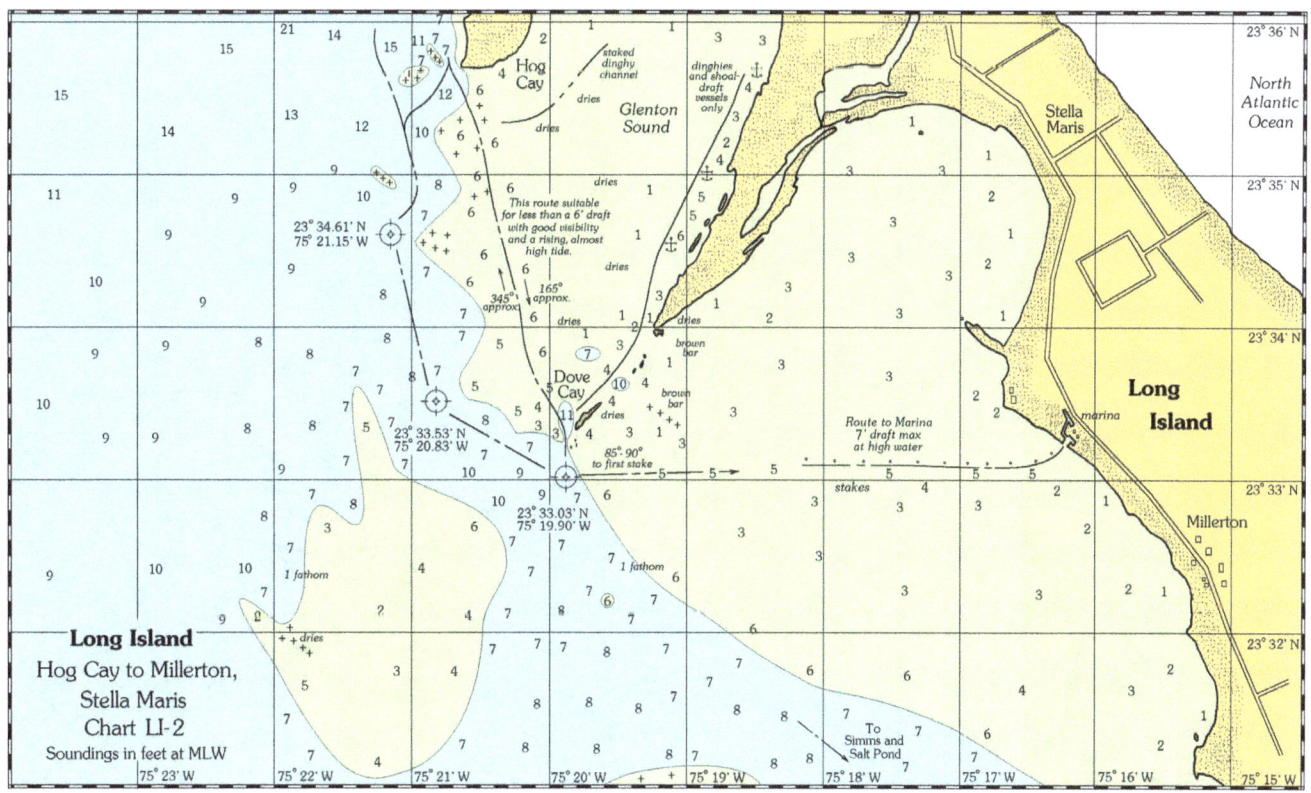

Salt Pond you may head directly for Upper Channel Cay. Pass west of Upper Channel Cay and work your way around its western tip in 8' of water keeping the conspicuous brown bar to starboard. The water will shallow to 2' for a stretch of half a mile or more until you will pick up the conspicuous deep blue water, (7' and more) that winds its way to the small cut between Conch Cay and Wells Point. Most of the surrounding waters dry at low tide and every now and then you'll see a mangrove bush that has taken root.

You can anchor in *Conch Cay Harbour* or between Conch Cay and Wells Point, but there is a lot of current there. To make *Dollar Harbour*, pass between Conch Cay and Wells Point and head westward paralleling the shoreline of Conch Cay and Dollar Cay staying about 50-100 yards off in 2'-3' of water at low tide. Once you arrive at the deep blue water between Conch Cay and Dollar Cay turn to starboard and anchor wherever you can get the best protection from the forecast wind direction.

Deeper draft vessels, those vessels with drafts of over 3', must approach *Dollar Harbour* from the south. A waypoint at 23° 10.00' N, 75° 15.55' W, will place you approximately ¾ mile south of the entrance to *Dollar Harbour* as shown on the chart.

From this position steer towards the eastern tip of Sandy Cay. You will be able to see a channel of slightly deeper, slightly bluer water between the sandbanks to your port and starboard. This channel of blue water is what you will want to enter but first you must pass over a bar with almost 5' over it at low water. After that you must steer around a few large patch reefs that are easily seen. Never attempt this passage in poor visibility, at night, or in a strong onshore swell as a slight miscalculation may not be forgiving to your keel. Once past the bar work your way to the area between Sandy Cay and Dollar Cay. There is a very obvious sand bar in the middle of the two cays and you can pass it on either side in 5' at low water to gain the deeper water just beyond. Once inside you

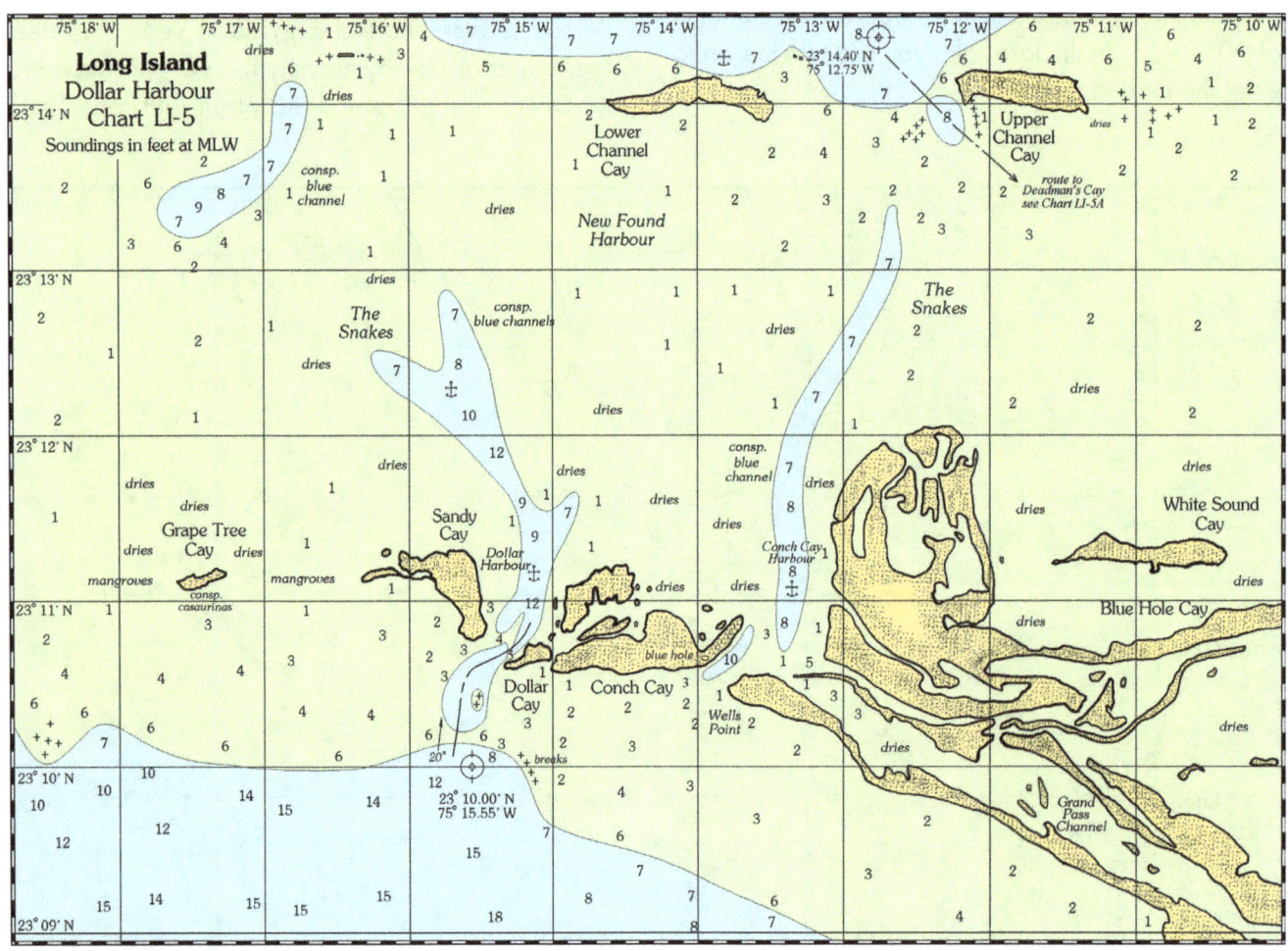

will have excellent holding in a sandy bottom but be prepared for current.

Long Island: *Deadman's Cays*

The larger communities such as Deadman's Cay, Buckleys, Mangrove Bush, and Cartwright are located in an area of shallow water well south of Salt Pond only suitable for shoal draft vessels, dinghies, and small outboard powered boats seeking shelter from a hurricane.

Shallow draft vessels can head southeastward from Upper Channel Cay towards Lower Deadman's Cay to anchor off Snapper Creek Cay as shown on Chart LI-5A (see below). From a waypoint at 23° 14.40' N, 75° 12.75' W, head southeastward (approximately 138° T) passing south of Upper Channel Cay as you make your way for a waypoint at 23° 11.75' N, 75° 10.35' W, which brings you to a position northwest of Snapper Creek Cay where you can pick up the channel of deeper water (15' deep in some places) that leads around the cay's western tip and carries on along the southern shore of Snapper Creek Cay.

Piloting by eye is essential on this shallow water route as the controlling depth is only 3' at MLW, you'll have to play the tides very carefully here (the tend to average approximately 3.5 hours after Nassau tides). Don't try to anchor off the southwestern tip of Snapper Creek Cay in the area shown as The Boilers as the holding is poor and the bottom is scoured and hard packed.

Long Island: *Little Harbour*

On the eastern shore of Long Island, we cannot recommend Clarence Town and its marina as a hurricane hole, the area is far too open to wind and sea. However, about 10 miles southeast of Clarence Town, *Little Harbour* offers good protection in all wind directions. Although strong easterly seas can make the bay quite rough, even untenable in the worst of conditions. We only offer this anchorage as an option if needed, and NOT as a preferred hurricane hole.

The entrance to *Little Harbour* as shown on Chart LI-7 (see below) is easy to enter even with 6' following seas. To enter take the southernmost of the two openings, between the unnamed cay and the mainland of Long Island favoring the northern side of the entrance between the two. There were once two rock cairns ashore that you would pass between, nowadays they've been torn down and a couple of white stakes take their place, one on the small cay to the north, and two stakes on the mainland to the south of the entrance. Do not attempt this entrance in strong easterly weather or with a heavy ground swell running,

A waypoint at 22° 58.65' N, 74° 50.30 W, will place you approximately ½ mile east of the narrow, inconspicuous entrance. The best water at the entrance lies about ¾ of the way north across the entrance from the mainland tip. Line up the opening on a heading of 270° and you will have 11' through here at low

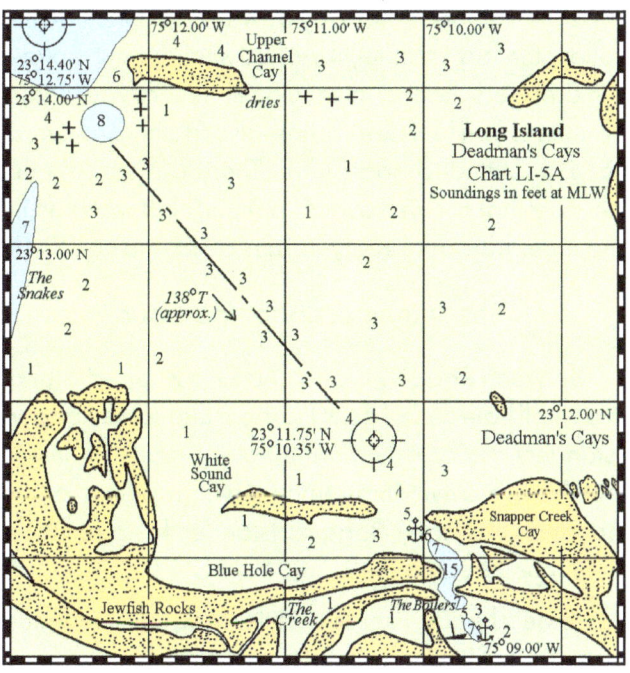

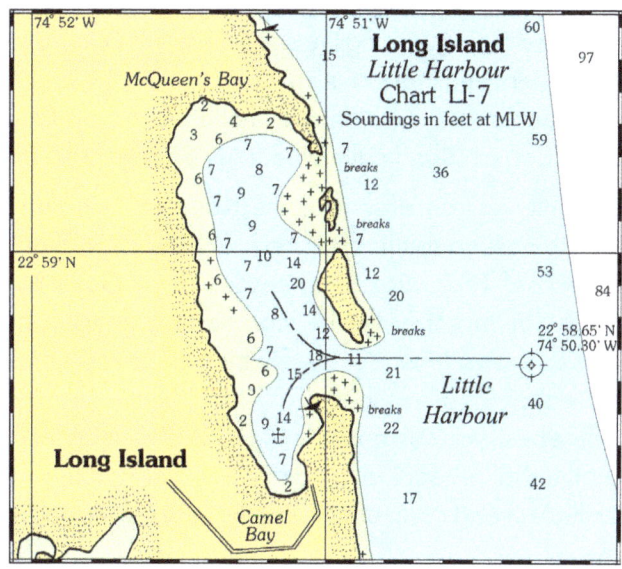

water. If a sea is running you will notice it breaking south of you about halfway across the opening and also around the small cay that lies to the north of the entrance. Use caution when entering as some rocky ledges and heads line the sides of the entrance channel.

In east through south to west winds the best anchorage is in *Camel Bay* at the south end of *Little Harbour*. Once inside the harbour, turn to port and tuck in wherever your draft will allow. In northerly winds you can tuck in at the north end of the harbour at *McQueen's Bay* with its good holding sandy bottom. Never anchor in *McQueen's Bay* in heavy easterly winds and seas, move to *Camel Bay* instead.

The Jumentos: *Ragged Island*

There are only two possibilities in the Jumentos chain, and both are in the vicinity of Ragged Island, one is the harbor at Duncan Town, and the other the local hurricane hole at *Boat Harbour*.

At Duncan Town, a boat with a draft of less than 5' can work its way up the mangrove lined channel to anchor just off the town. Here you will find 4'-6' at high water with mangroves and cliffs surrounding you. This would be a nice hurricane hole if it were just a couple of feet deeper.

South of Hog Cay lies *Ragged Island Harbour* and the entrance to Duncan Town as shown on Chart JU-14 (next column). If you are approaching from offshore we do not recommend the cut between Hog Cay and Ragged Island unless you are familiar with this area or are adept at reading the waters. There are three reefs guarding the entrance to this cut, two at the eastern end that make up what is called the *Outer Bar Reef*, and one inside directly between Hog Cay and Ragged Island called the *Inner Bar Reef*.

If entering from offshore come in on a southwesterly course passing southeast of the rock beacon on *Black Rock Reef*. This course takes you inside the *Outer Bar Reef*. Southwest of the beacon lies the dangerous and extensive *Inner Bar Reef*, which is easily seen as it breaks with almost any sea. Although the northern side is best, you can pass to either side of it. Watch out for the shallow spot, 5' at low water, lying southwest of the reef and north of Gun Point. There is a new light

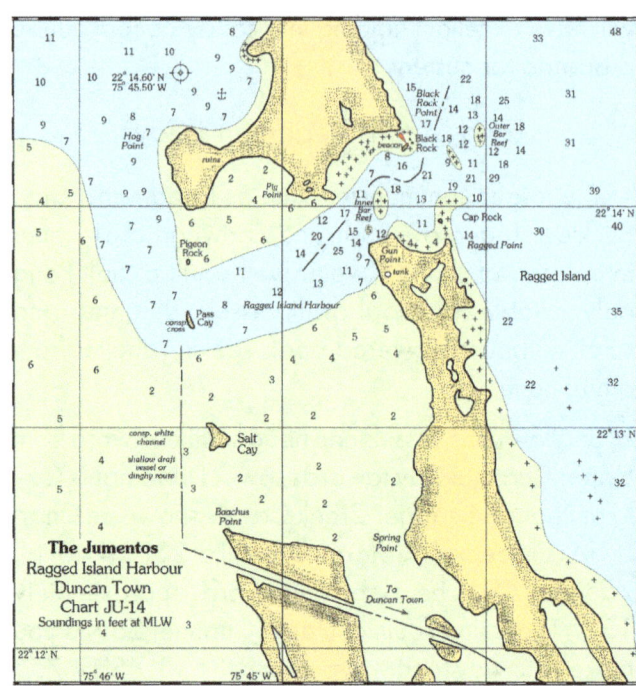

on Black Rock, sometime shown as Bulva Rock (see Chart JU-14).

If you are traveling south on the western side of the Jumentos, head for the southern end of Hog Cay past Pigeon Rock, Pass Cay, (with its conspicuous large cross), and enter the shallow bank just west of Salt Cay, sometimes called Pigeon Cay. As you enter the shallow banks you will begin to see a white channel between two grassy shoals. This is the channel to follow in the harbor at Duncan Town.

You may see the remains of the old channel heading off to the east about halfway down the channel. Do not follow the old channel, it will just shallow out. Follow the dredged channel into the basin secure your vessel wherever you feel best. There will certainly be several local unattended boats here. There are spots in the channel that are well over 6' deep at high water while the inner harbour barely carries 5'-6', mostly 3'-4'.

The Jumentos: *Boat Harbour*

Between Ragged Island and Little Ragged Island, is a small hole called *Boat Harbour* that some Ragged Islanders use as a hurricane hole. There is 9' inside but there is a winding channel with a 3' bar at the entrance. Ask any Ragged Islander for directions, they'll be happy to help.

Vessels headed south to Little Ragged Island must pass west of Ragged Island to clear the shallows lying just off the western shore. As shown on Chart JU-15 (see below), pass Wilson Point and turn to the NE to enter *Southside Bay*, and enter the shallow route to the deeper water of *Boat Harbour* (called that because Lockhart Cay was once the home to a boat-building facility). The route has a controlling depth of 3' at MLW, but there is 7'-12' inside the hole if you can get in there.

Crooked/Acklins District

The only protection in the Crooked/Acklins District will be found in the maze of creeks between *French Wells* and *Turtle Sound* for boats with drafts of 3' or less, or by going through *The Going Through* towards the *Bight of Acklins*. Here you will find a maze of shallow creeks leading to numerous small mangrove lined holes, perfect little hidey-holes for the shallow draft cruiser (less than a 3' draft) seeking shelter.

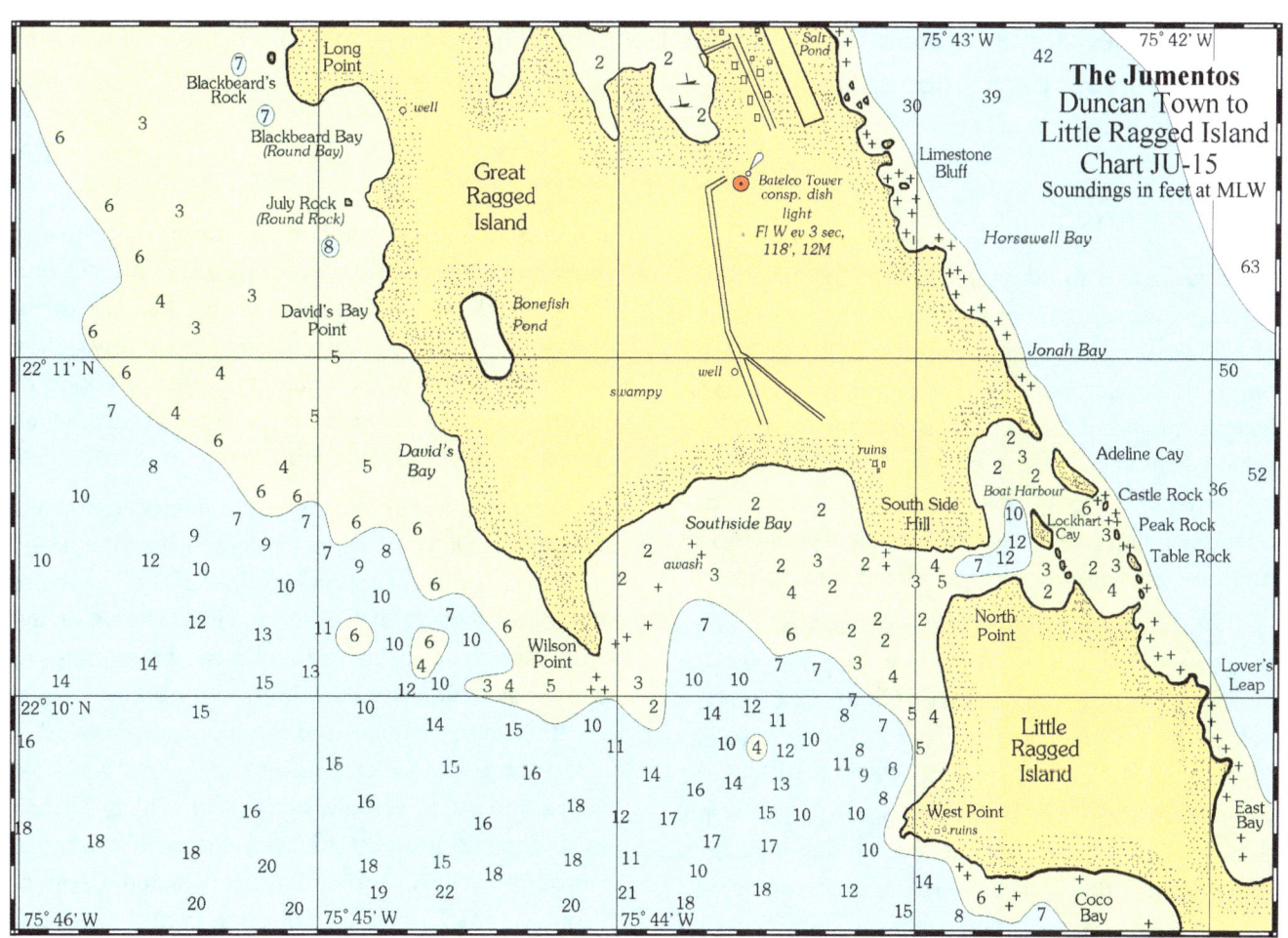

Chapter 8

The Turks and Caicos

THE CAICOS ISLANDS LIE ON THE NORTHERN EDGE OF THE HUGE CAICOS BANK west of the Turks Islands and separated from their sister cays by the ocean deep Turks Island Passage. In the Caicos Islands, Providenciales (Provo) offers a well-protected marina (Turtle Cove) on the northern shore, while along the southern and eastern shores several dredged canals offer an opportunity to get well inland, hopefully away from any damaging seas though you will still be affected by a storm surge. Shallow draft boats, those with drafts of less than 3', could work their way into some of the creeks between North Caicos, Middle Caicos, and East Caicos if needed. South Caicos' *Cockburn Harbour* is excellent in most conditions but it is unsuitable as a hurricane shelter. Bear in mind that when anchoring in any of the dredged canals around Provo, that the bottom will likely be poor holding; you'll have to set some of your anchors on shore here.

Provo: *Turtle Cove Marina*

Turtle Cove Marina is protected 360 degrees from wind and waves, whatever problems occur are born of a lack of appropriate pilings, a 5'-6' approach depth, no floating docks, and a lot of other boats. Side tying to the marina docks is best with several anchors setting your vessel away from the dock. Use a lot of chafe gear. Try to find old automobile tires to place over any pilings near the boat. Use all of your fenders and allow for the storm surge. Check into a hotel.

There are two routes to *Turtle Cove Marina* from outside the reef, but the best one is the primary route through *Sellar's Cut* as shown on Chart TCI-C10 (see next page). To enter through *Sellar's Cut*, a waypoint at 21° 48.40'N, 72° 12.40'W, will place you approximately ¼ mile north of the green daymark that defines the entrance channel through *Sellar's Cut* as shown on Chart TCI-C10. There are supposed to be a matching set of red and green markers here, but wind and waves seem to remove one or the other every year. Remember that as you approach the cut and find yourself wondering where the other marker is. Always remember that any floating aid to navigation mentioned in this book is subject to disappearing or moving between the time we go to print and the time you arrive at your destination.

From the waypoint at *Sellars Cut* head generally southward until you can pass between the outer green daymark (red right returning), in 13'-22' of water and follow the rest of the markers in as shown on the chart. At one time you had a choice of two routes to take when inside the reef, the more northerly route has been discontinued and the markers removed in favor of the far easier *English Cut*. I have run both routes and much prefer *English Cut*, which, though narrow (a 52' long by 30' wide trimaran can make it through with no problem) has a minimum depth of 7' and is far shorter. Once through *English Cut* keep the green markers to port as you head towards the marina entrance. Watch out for the shallow heads lying just north of the shoreline about 100 yards northeast of the entrance to the pond. The winding entrance channel itself is narrow and has a shallow sandbar at its eastern entrance, on your port side when entering, and a rocky bar across the channel on your starboard side. Once inside, give the final inner turn on your port side a wide berth before rounding to port into the pond itself.

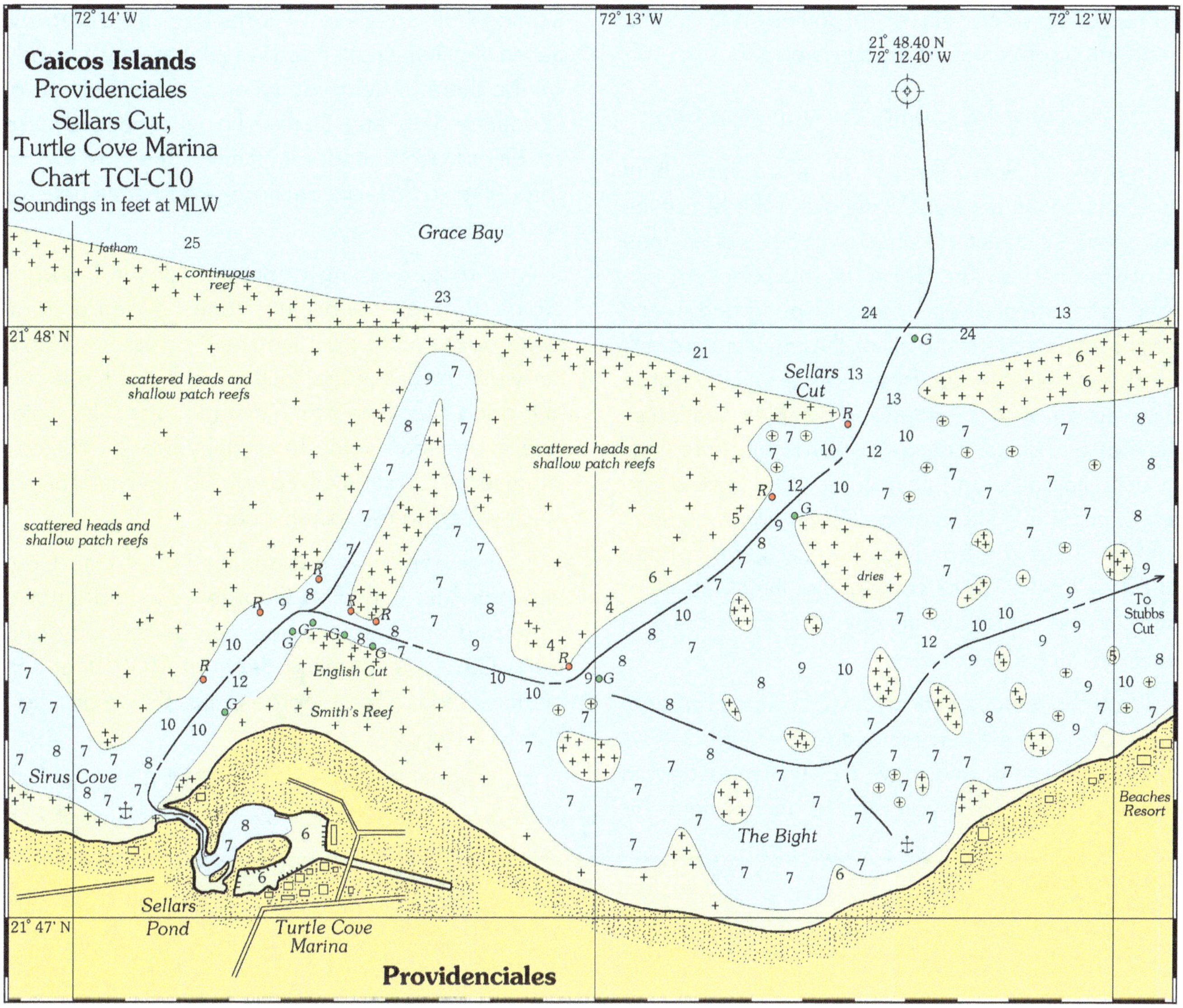

If unsure about how to enter *Sellar's Cut* and the entrance to *Turtle Cove Marina*, call the marina on VHF ch. 16 and the dockmaster will be happy to have someone come out to lead you in.

One last note about *Turtle Cove Marina*. During Hurricane Frances in 2004, folks were walking around

Turtle Cove Marina

THE TURKS AND CAICOS • 85

in knee-deep water in the central portion of the marina. That's an approximate storm surge of 6'-8'.

Provo: *Leeward Going Through-Pine Cay*

In Leeward Going Through, there is a small canal leading into the Leeward community. The bar at the entrance restricts entry to vessels with drafts of less than 5' at high tide. Leeward Going Through has often been used as a hurricane hole by some skippers and should be considered as such solely on the imposing strength of *Blue Haven Marina* if you seek a slip. The large *Blue Haven Marina and Resort* complex dominates the area and offers floating slips with significant pilings to accommodate up to two 200' LOA vessels and withstand a 15' storm surge. The marina monitors VHF ch. 16 and 14 and can be reached by phone at 649-946-9910, or you can visit their website at http://www.bluehaventci.com/ or email them at contact@bluehaventci.com.

If you choose to anchor at Leeward Going Through, be aware that there is, a tremendous current that flows through here on 'normal' days, making it questionable in a hurricane. So, if you plan to anchor here, be warned that nearly every charter boat in the area will be leaving their unattended vessels here and nobody will be there to help you when these vessels drag. Co-author Stephen J. Pavlidis has experienced these conditions and cannot recommend Leeward Going Through as a first-rate anchorage hurricane hole for this reason.

A word of warning about the tides in Leeward Going Through during hurricanes. When a storm surge approaches from the south across the banks, the water rushes in the southern sides of these cuts at a good clip. One past hurricane raised the water level in Leeward Going Through by over 6'. You can imagine the current involved with the movement of that much water, so use extreme care.

Further north, there is shelter NE of Pine Cay, between Pine Cay, Fort George Cay, and Stubbs, Cay, and another spot between Stubbs Cay and Dellis Cay (3' controlling depth at MLW) that can be accessed via *Fort George Cut* as shown on Chart TCI-C13 (below).

A waypoint at 21° 53.70'N, 72° 07.90'W, will place you approximately ½ mile northwest of *Fort*

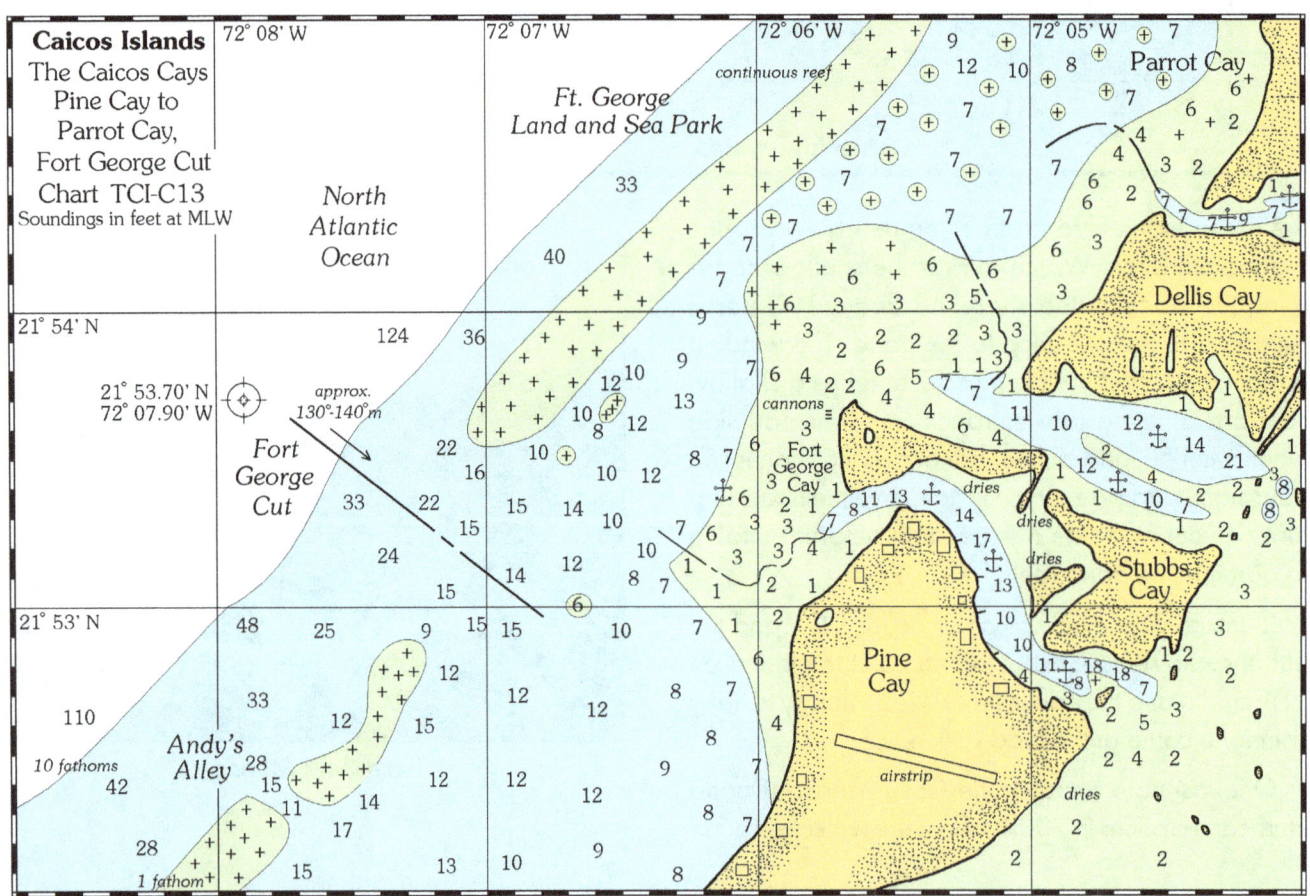

George Cut. Enter the cut on an approximate heading of 130°-140° magnetic. The heading here is not so important as simply staying between the reefs, but not to worry, *Fort George Cut* is wide and deep. Once inside you can head northeastward until you can take up an approximate southeast heading towards the point on Pine Cay as shown on Chart TCI-13. There is really no way to describe the entrance here other than to say that it curves around generally back towards the northeast until you reach the darker, deeper water between the cays. A high tide, excellent visibility and the ability to read water are what will get you through here. There are several sandbars that you must zigzag between and some shallow grassy patches that give a false impression of being deeper than they look. Once inside you will find a deep anchorage that goes well eastward between Pine Cay and Stubbs Cay.

To access the anchorage between Dellis Cay and Stubbs Cay as shown on Chart TCI-C13, head northeastward from *Fort George Cut* between the reef and Fort George Cay, avoiding the large shallow bar that sits northwest of Fort George Cay and the small shallow reefs between the bar and the outer barrier reef. Line up the southwestern tip of Dellis Cay and head in on it on a heading of 140° magnetic. You will probably have to dodge some shallow spots as all these sandbars change frequently in this stretch of cays.

As you approach the southwestern tip of Dellis Cay you will find that you also have to pass between two shallow yellow-colored sandy bars, 1'-2' at MLW. One works out westward from Dellis Cay, the other also lies east/west just a little northwest of the first bar. You will have to turn to starboard to pass between the two to make it into the deeper water at the entrance to the anchorage. The anchorage itself has two arms, one on the Dellis Cay side, one on the Stubbs Cay side. Both offer good protection.

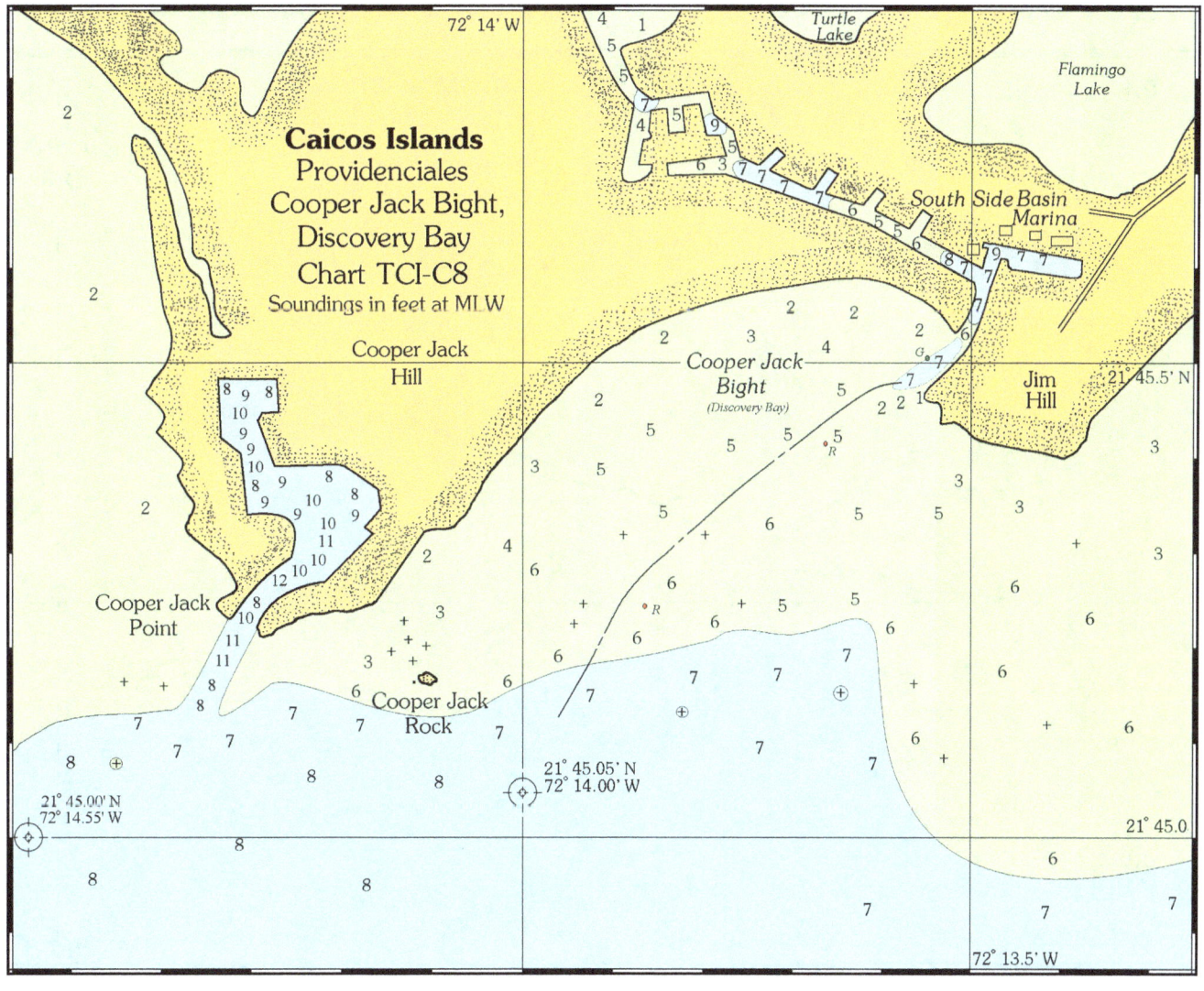

Provo: *Discovery Bay*

More protection lies on the southern shore of Providenciales in *Discovery Bay*. Here you'll find numerous canals that are a favorite of local boaters and are well-protected although a strong storm surge will certainly enter here.

Problems here include numerous derelict boats already in the area, a 5' approach depth (although a 6.5' draft can enter with a high tide), and a lack of places to tie off although there are a few mangroves. The area is fairly low and flat, not offering much blocking of the wind. Theft may be an issue if you leave your boat here unattended. More important is the local who doesn't like boaters entering *Discovery Bay* and swears he will try to scare them off when they enter. The last I heard he was NOT effective.

To reach *Discovery Bay*, head to a waypoint at 21° 45.05' N, 72° 14.00' W, placing you approximately ¼ nautical mile southwest of the entrance channel into the canals and the marina as shown on Chart TCI-C8 (previous page). At this point, you can head generally NE past Cooper Jack Rock until you pick up the markers leading in to the marina. Watch out for scattered heads and shallow bars on this route, the whole southern shore of Providenciales has numerous scattered heads and bars strewn about that are easily seen in good light and avoided. Be sure to take the point and its shallow sandbar well to starboard. Round into the deeper water of the entrance channel (There is a small range set up on shore for this purpose.).

Provo: *Caicos Marina*

East of *Discovery Bay* lies the *Caicos Marina and Shipyard*. The marina offers a long concrete seawall for side-ties as well as a good haul out facility.

A waypoint at 21° 44.80' N, 72° 10.30' W, will place you a little over 1 nautical mile south of the entrance channel leading into the Caicos Marina

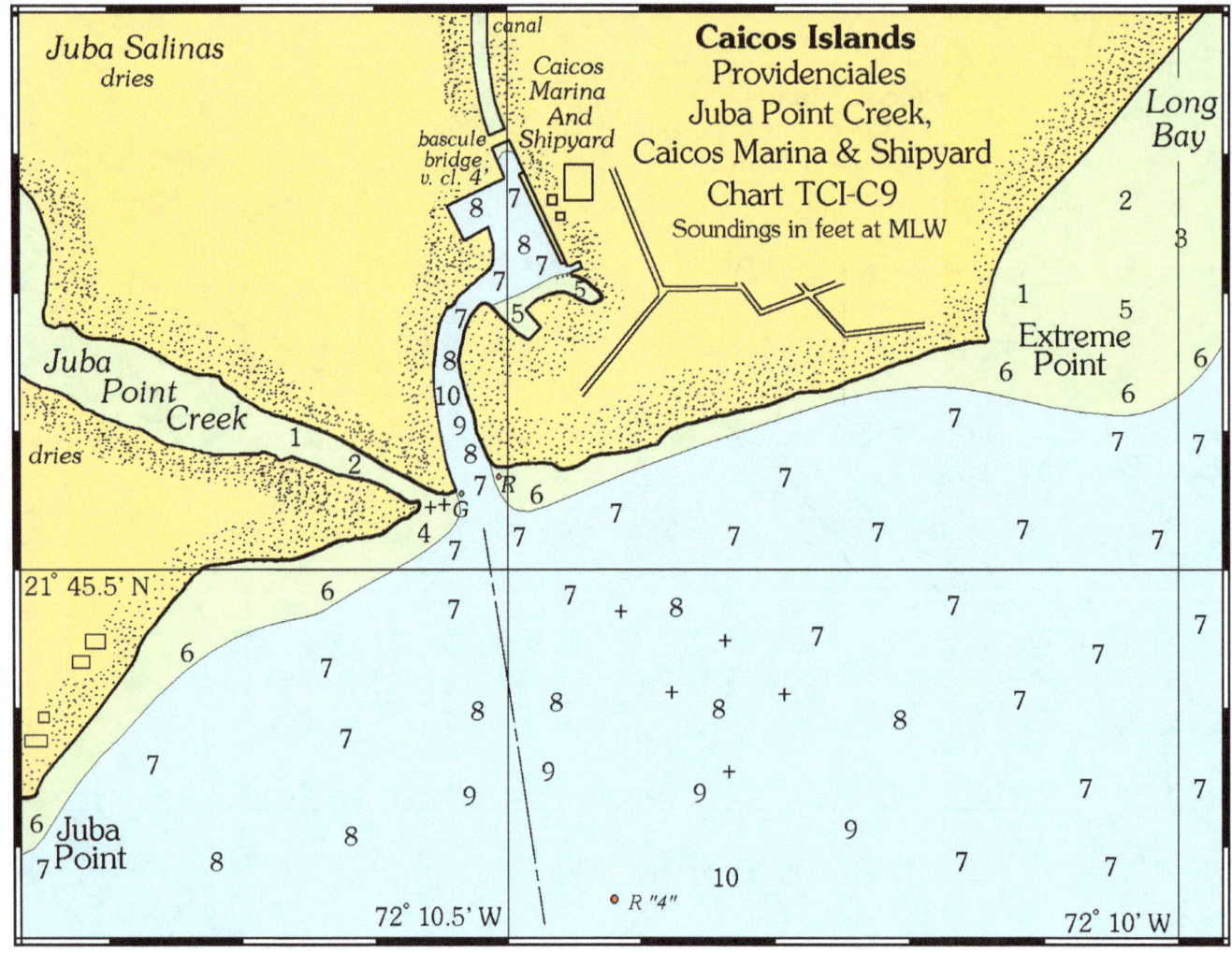

88 • THE CAPTAIN'S GUIDE TO HURRICANE HOLES

and Shipyard. From this waypoint you should see the outer sea buoy (white light atop) and the two red markers between it and the shore at the entrance to the marina. From this position, the entrance channel is approximately 1 nautical mile distant on an approximate course of 345° magnetic, the course actually bends more northward after R "4").

The channel mouth (as you will see on Chart TCI-C9 at the bottom of the previous page), is marked by tall red and green striped pilings (red-right-returning), and as you approach the entrance keep a lookout for any stray heads in the surrounding waters. There is a shallow spot between the channel markers that carries 7' at low water. The channel curves to the east as you approach the marina complex with its huge building and tall cranes that are usually the first sight of the marina from seaward. The two small coves southeast of the marina dock and lift are private and are not part of the marina complex; they would make a fair hurricane hole if needed. A canal heads northward from the marina past a small bridge towards a number of private homes in the Long Bay Hills area, but draft is limited to about 3'-4' at MLW.

West Caicos: *West Caicos Marina*

West Caicos was to be the 'flagship' island destination for the prestigious *Ritz Carlton* hotel chain. Infrastructure was begun in 2008 and resulted in numerous condominiums being constructed, roads being built, and, to the boater's delight, a marina carved out of the west side of the island. With the financial disaster of 2008, the development is deserted other than a few security personnel. The skeleton marina provides a protected anchorage of 360 degrees from wind and waves. There is a slight elevation to the island offering some protection coming from the terrain. There should be no problem with other boats. You will not be allowed to go ashore, although, in the event of a hurricane, I would consider taking up safety in one of the unfinished condominiums.

As shown on Chart TCI-C14A (see below). the entrance to the *West Caicos Marina* lies through a

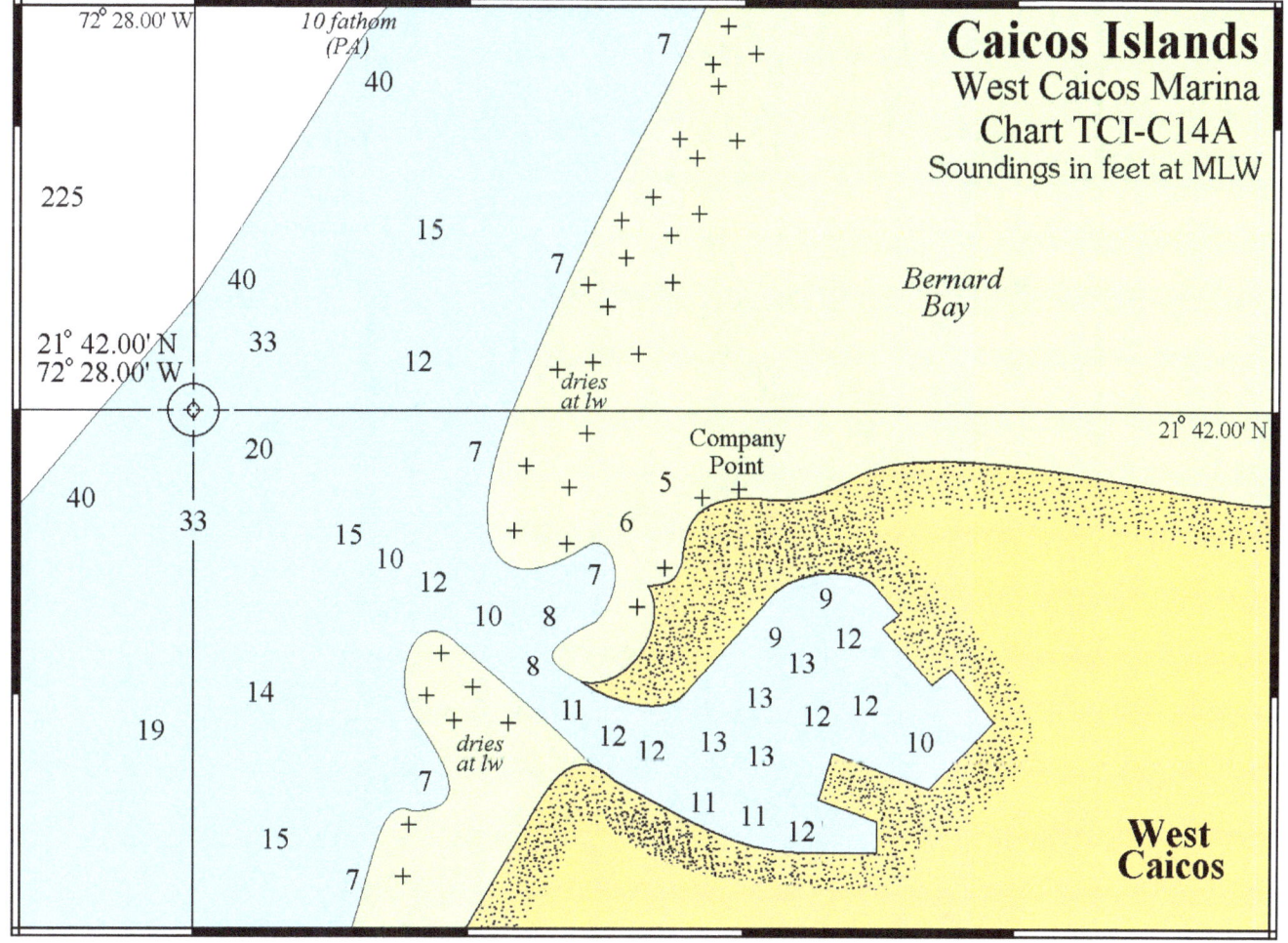

natural break in the reef just west of Company Point. A waypoint at 21° 42.00' N, 72° 28.00' W, will place you approximately ¼ nautical mile northwest of the entrance channel leading into the marina. From the waypoint, steer approximately 125° to pass between the entrance jetties and enter the marina basin (the use of your eyes to steer by is far more important that sticking to a 125-degree course-line). Use extreme caution, there are shallow reefs on both sides of the channel, and if that were not enough to worry about, Provo dive boats use the channel to pass to and from the western shore of West Caicos. The dive boats usually don't slow down for cruising vessels and a dangerous situation could come about if one was headed outward from *Bernard Bay* as you were headed into the marina. Keep your eyes open here and let's avoid any problems.

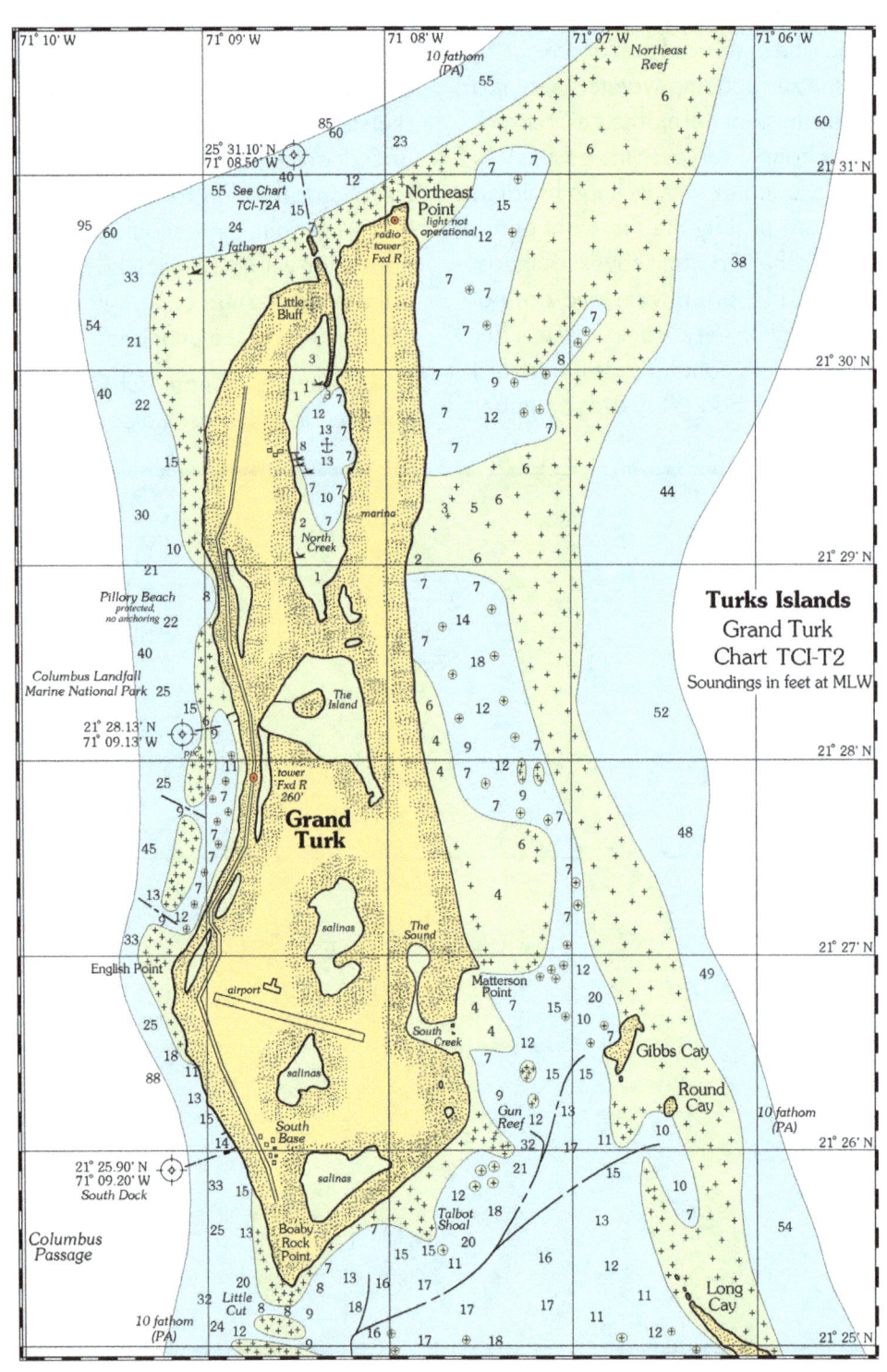

90 • THE CAPTAIN'S GUIDE TO HURRICANE HOLES

Note: if you arrive late to the storm, meaning a day before the storm is to hit, this channel may not be passable as the waves will break across the entrance. Arrive early! Be prepared! You will enter the channel in 12' of water until it opens up into the marina.

There will be 15'-20' inside with good holding bottom. Do not tie to the existing dock. Anchor in the northeast corner of the marina while setting anchors on shore, as well as in the water. Use chafe gear for the lines on shore as they run across the sea wall. Set out as much anchor chain as you have, assuring they will set horizontally, as it is relatively deep. Follow the weather reports to assist in pointing the bow into the wind. Be cautious in tying to coconut trees as they don't have a very deep root system.

Grand Turk: *North Creek*

In the Turks Islands, the only choice for shelter is to round the northern tip of Grand Turk and seek protection inside *North Creek* (as shown on Chart TCI-T2 on the previous page) if conditions allow entry. The entrance channel is limited to about 6½' on a normal high tide but once inside the water deepens to over 12' in places. There is quite a bit of north/south fetch to take into consideration however in this narrow lake.

When approaching the northern shore of Grand Turk from the west or the north your landmarks will be the large red and white checkerboard water tower, the old and unused lighthouse on Northeast Point, and the radio antenna that shows a fixed red light at night.

As you approach the northern shore from the west, the actual entrance will be hard to discern against the background of the shoreline, but it will appear to lie below the white-roofed building that sits about ¼ mile south of the water tower. A waypoint at 21° 31.10'N, 71°08.50'W, will place you approximately ½ mile north of the narrow entrance channel and the jetty as shown on Chart TCI-T2 (and in greater detail on Chart TCI-T2A at the top of the next column). At the time of this writing there were two white PVC stakes near the end of the jetty that lies on the western side of the entrance channel; bear in mind that these are private markers and are subject to moving or removing without notice.

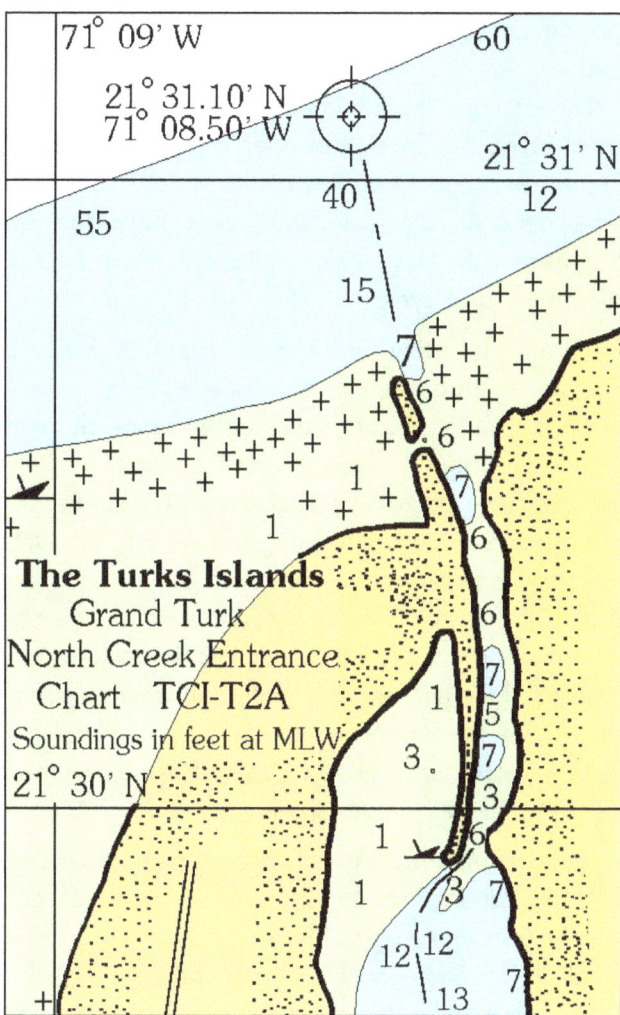

In normal conditions the swells will break across the reefs to the east and the west of the channel with lighter action actually in the channel itself. The prudent mariner will likely pass by the entrance once or twice at a close but safe distance to get a good feel for the way the channel lies. The channel is the only entrance to the lake inside and all the water

The Entrance to *North Creek* as seen from the southeast

goes in and out of this one channel with the tide, so expect a lot of current. You'll have better steerage if you go against the tidal flow, but I have done both and had no difficulties. However, if it is your first time, I recommend going against the flow, or if possible to time your entry for high water, slack, or just as the tide begins to ebb. Never attempt this entry with the sun directly in your eyes.

When you have decided to enter the channel, line up and approach the entrance keeping the jetty close to your starboard side. The entrance channel is narrow, about 75' wide along the end of the jetty but it widens inside. You should be able to discern the deeper, bluer water alongside the jetty and the yellow and brown of the shallow rocky water to the east of the channel. The reefs to the east and west of the entrance channel stretch farther northward than the end of the jetty so keep your eyes open to avoid them. Enter the channel keeping close to the jetty, but not too close; you'll have 6' just inside the entrance and 8' in places through this first section.

The jetty is made up of large rocks, blocks of concrete, dredged materials, and numerous rusty car and truck frames, engine blocks, axles, and one small front-end loader. Any of these can easily damage the inattentive skipper's vessel if too close. The jetty has a small break in it as it reaches the shore, so don't confuse this with the channel you are in. The break is shallow and has the top of a large rock in the center of it at high water. Both sides of the channel shallow slightly at this break so keep to the center of the visible channel if at all possible and you'll have 5' at MLW. Proceed along the entrance channel paralleling the jetty and as you pass between the jetty and the shoreline to the east the channel will widen a bit. Favor the jetty side as the eastern shore is very shallow and the channel still follows close to the jetty.

As the anchorage comes into view you will notice that the jetty has a large rusty crane on it as well as a barge at its southern end that also has a crane on it. Between these two cranes you will find the shallowest spots in the entrance channel. The bottom through here is sand and the current has built up over a dozen or so shallow sand mounds (easily seen in good light) that lie perpendicular to your course and stretching from the jetty to the shallows of the eastern shore. These sand mounds, one or two of which only have about 3'-3½' of water over them at MLW, are not very wide and if you bump you will likely be able to power over them. Sometimes there are shallower spots on one side or the other of these mounds, so you might be able to zigzag your way through. Depths between these mounds are generally 6'-7' at MLW.

As I mentioned earlier, 5½' can enter here but if you don't choose a good tide to enter on you might have to power your way over these humps. I believe that in an emergency, a 6' draft could enter here on an extremely high tide if the skipper didn't mind powering over these humps. The only problem would be that this same skipper would then need another extremely high tide to get back out.

As you approach the end of the jetty by the remains of the huge barge and crane, don't get careless, you have one or two more obstacles to avoid. At the eastern end of the entrance channel, across from the end of the jetty, lies a huge piece of steel bar that once marked the eastern end of the entrance channel. Instead of being exactly vertical, this bar leans over and is awash at high water and only about 6" of it juts above the surface at low tide. This steel bar could do a lot of damage to your boat, so keep an eye out for it and don't stray too close to it. Just as you reach the end of the entrance channel at the barge you will come upon a grassy shoal that is not shown on other charts of the area. The reason for this is that it has just appeared over the last few years. The shoal, easily seen in good light, only has 3' of water over it in most places. If you have a high tide you can pass it on either side, but if the tide is low you can only pass between the shoal and the end of the jetty in 5'-6' at MLW close in to the jetty.

Once around the grassy shoal area you will be in 8'-15' of water and you can anchor wherever your draft allows and you feel comfortable. The anchorage shallows west and northwest of the end of the jetty and south of the dock on the eastern shore of the lake, with 5' at MLW.

Chapter 9

The Dominican Republic

ALONG THE NORTHERN SHORE OF THE DOMINICAN REPUBLIC THE BEST protection is in the tiny harbor at Luperón, probably one of the finest hurricane holes in the entire Caribbean. Although the rivers on the southern shore of the Dominican Republic look inviting and offer good protection, caution must be exercised as torrential rains will cause flooding and very strong currents, not to mention all manner of flotsam and jetsam floating down on you.

Luperón

The entrance to Luperón lies between a large shoal area on the west side of the entrance and another area on the east side. Caution is called for when entering the channel to Luperón, in poor light the shoals are difficult to see. Approaching the entrance slowly in the morning light, you may be able to make out the shoal fairly easily, as the water may still be clear then. Later during the day, as the winds pick up (about 0900 the breeze fills in and blows all day until around sunset); it gets a little harder to discern the shoal from offshore.

There are some hills to both the right and left of the entrance channel, but the very conspicuous light-colored roof of a hotel/resort complex makes a good landmark. Keeping the hotel well to starboard as you head south to the entrance waypoint, the entrance to Luperón lies approximately ½ - ¾ mile to the east of the hotel.

As shown on Chart DR-1 (next page), a waypoint at 19° 55.50′ N, 70° 56.51′ W, will place you approximately 1 nm north of the entrance channel. A good course is to keep the eastern cliff close on your port bow until you can make out the buoy that marks the shoal on your starboard side. As of this writing there are two buoys (bear in mind that this may change at any time) and the westernmost buoy is in shallow water and should be kept WELL to starboard upon entering.

Keep an eye out for the breaking shoal off the eastern shore and take the westernmost buoy to starboard upon entering. The entrance channel leading into the anchorage is often hard to make out until you get fairly close and it opens up before you.

As you approach the entrance, keep close to the eastern shore where the deeper water lies; keep the buoys well to starboard. Inside you'll see *Cano Quitano* to the east. Often just a staging anchorage for those leaving Luperón it offers good protection if your draft allows you to enter and tie up in the mangroves.

As the entrance to the inner harbor opens up to starboard, pass roughly midway between the northern and southern shore watching out for the shoals off each one. As you enter the harbor the deeper water lies along the mangroves on the southeastern shore.

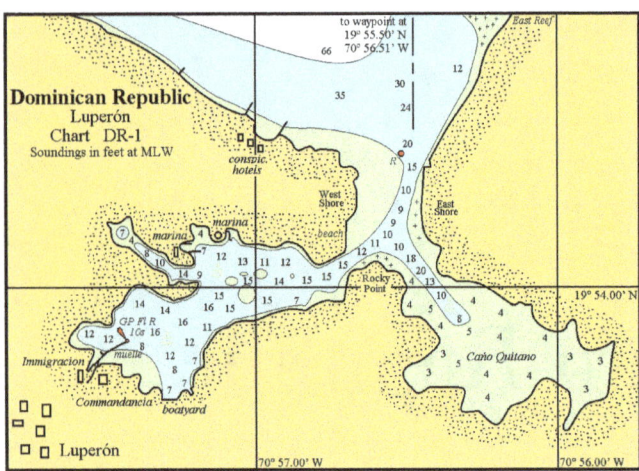

If you pass along those mangroves you'll avoid the large, shallow mud shoals that plague the center of the harbor and change and grow with each passing season. The shoals rise up from 20' depths to lie about 1'-2' under the surface.

Here you can get a slip, haul out, or find a spot in the mangroves that pleases you. The bottom is mud that holds very well once your anchor sets which can take a day or two. The town dock is marked with a red light (GP Fl R, ev 10 sec.) that works well although it is not visible outside the harbor.

Cofresi: *Ocean World Marina*

A few miles east of Luperón lies *Ocean World Marina*, a place you can get a slip (not recommended here during a hurricane) or haul out (35-ton lift). The marina has 104 slips and can accommodate vessels to 250' in length with drafts of just under 12'. But bear in mind that this marina is open to the NW-NE and large storm seas will do a lot of damage here; (even in moderate conditions, swells bring considerable surge into the marina).

As shown on Chart DR-2 (top of next column), a waypoint at 19° 50.05' N, 70° 43.65' W, will place you ½ mile northeast of the entrance channel into *Ocean World Marina* (you will see the yellow sea buoy located at 19° 50.095' N, 70° 43.535' W). The entrance channel offers no problems in normal to moderate weather however, in heavy northerly swells the entrance may break all the way across making entrance dangerous if not impossible. It's a straight shot down the entrance channel keeping between the buoys.

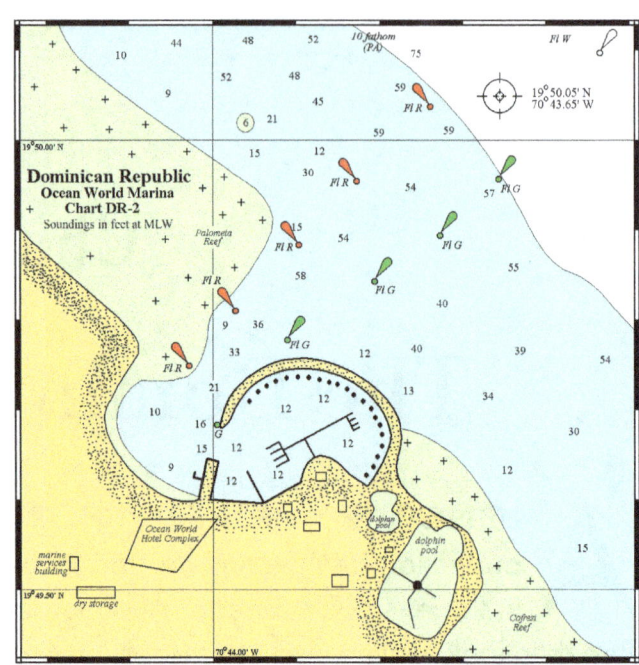

Samaná: *Puerto Bahía Marina*

Lying just to the west of Samaná is the well-protected *Puerto Bahía Marina* offering much better protection than anything in and around Samaná.

As shown on Chart DR-8 (below), a waypoint at 19° 11.40' N, 69° 21.40' W, will place you approximately ¼ nm SW of the marked entrance channel. From this waypoint head between the markers and the jetties and into the marina basin. You'll notice some current upon entering the marina, and once inside there is a bit of surge (and a 3' tidal rise and fall) so secure your vessel accordingly. The

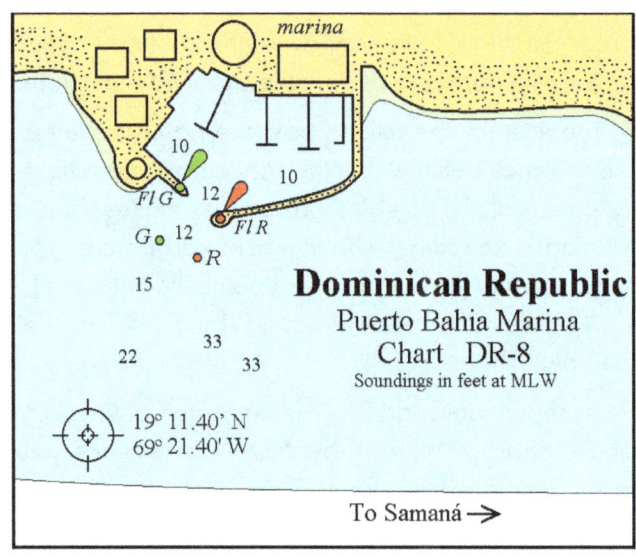

94 • THE CAPTAIN'S GUIDE TO HURRICANE HOLES

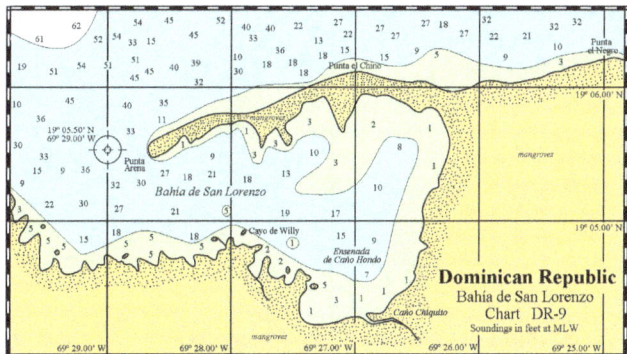

marina can accommodate vessels up to 150' LOA with drafts to 10'.

Bahía de San Lorenzo

Situated on the southern shore of Bahía de Samaná, approximately 12 miles SW of Samaná and *Puerto Bahía Marina*, is the *Bahía de San Lorenzo*, the highlight of *Parque Nacional de los Haitises*, or simply, *Los Haitises*. The area consists of 58 islands and numerous winding, mangrove-lined waterways.

As shown on Chart DR-9 (above), a waypoint at 19° 05.50' N, 69° 29.00' W, will place you approximately ¼ nm W of Punta Arena. From the waypoint, head south and then east, passing south of the Punta Arena peninsula, working your way eastward.

Anchor where your draft allows in the best mangrove protected cove or creek you can find. The bottom is mud but the holding is good.

The Southern Coast: *Barahona*

If you find yourself along the southern shore of Hispaniola and a storm threatens, you have several choices for shelter, none as fine as Luperón, but you still have several choices to anchor, haul out, or get a slip.

If approaching from the west, Barahona is often the first stop for cruisers checking in to the Dominican Republic. It is an active commercial harbor but there is a snug little anchorage in its northeast corner complete with some mangroves.

The entrance channel to Barahona lies between two reefs, *Arrecife Yunca* and *La Piedra Prieta*. As shown on Chart DR-10 (bottom of next column), a waypoint at 18° 13.30' N, 71° 04.00' W, will place you approximately ¼ nm NE of the lit (Fl G 3s) sea

Mangrove Creek, Los Haitises

buoy. From the waypoint take the sea buoy to port; there is plenty of deep water around the sea buoy so you can pass it on either side. As shown on the chart there is a lit range ashore, the bearing when entering is 243° magnetic.

There is a protected anchorage in the small cove just past the *Club Nautico* docks as shown on the chart. Just follow the buoys into the anchorage area north of the town dock; use caution in poor visibility, the markers are red and green pillar buoys and unlit. Anchor in the NE corner of the cove, the holding is fair to good in a mud bottom. On the northern shore of the cove is a power plant and it is well-lit at night. The dock on the western shore is for coal vessels so do not anchor near it.

The Southern Coast: *Puerto de Haina*

Only about 7 nm SW of Santo Domingo is the entrance to Puerto de Haina, a large and busy commercial port situated on the *Rio Haina*. The area is known for crime and the only reason I mention it here is due to the marina and boatyard located just upriver.

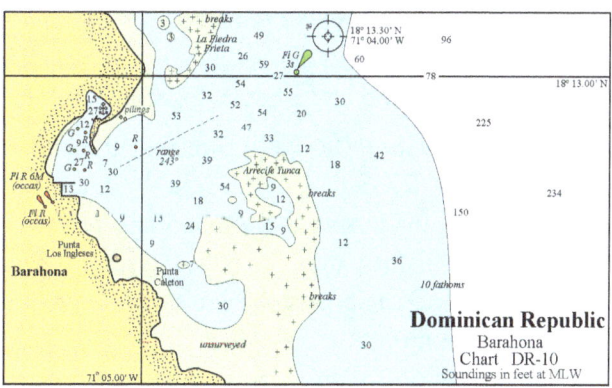

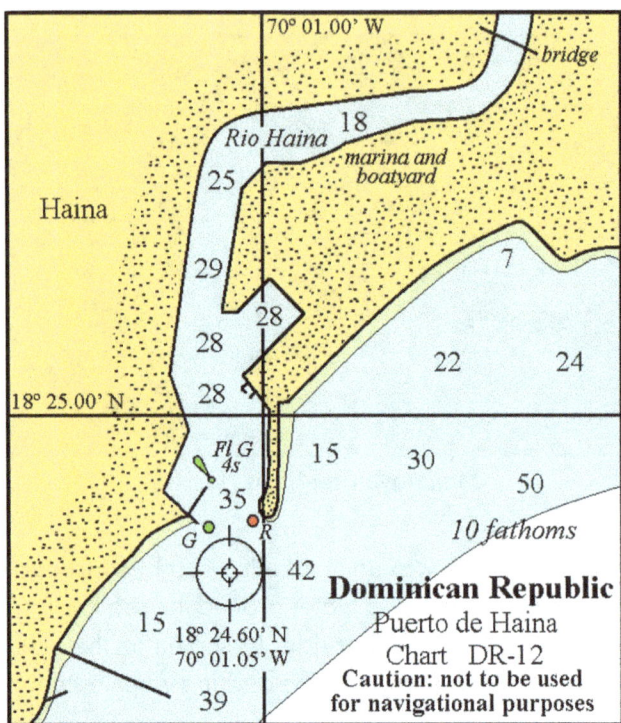
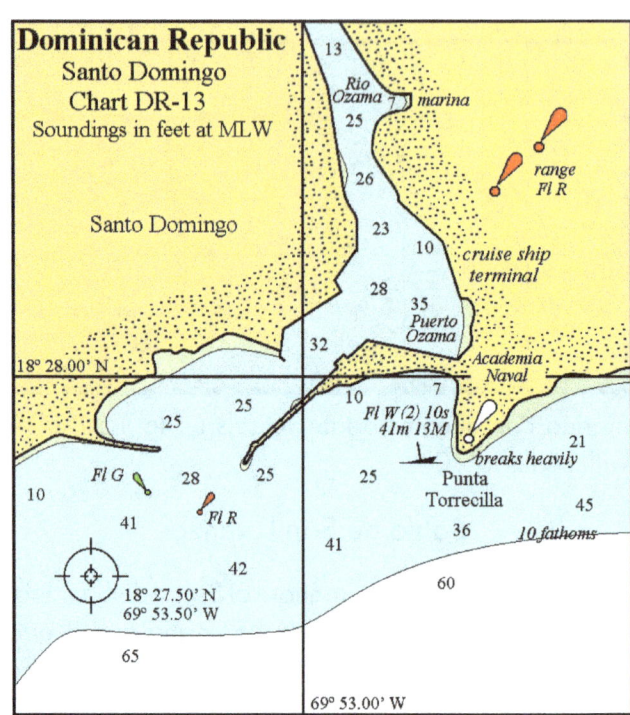

Do not consider this area a recommended anchorage, rather view it as an option in an emergency.

As shown on Chart DR-12 (above), a waypoint at 18° 24.60' N, 70° 01.05' W, will place you approximately ¼ nm S of the entrance channel. From the waypoint pass between the jetties and head up river, staying mid-channel, and after the river turns to the east you will see the Club Nautico marina and boatyard to starboard on the southern shore of the Rio Haina.

The Southern Coast: *Santo Domingo*

Santo Domingo is far down on the list of places to seek refuge, there is no place to anchor, the current can be quite strong, but there is one small marina on the *Rio Ozama*'s eastern shore where you might find shelter if they have room. Warning, all dockage here is Med-moor with buoys for your bow or stern line. This is one of those places that we mention only for the fact that it is there and has, at best, Spartan offerings.

As shown on Chart DR-13 (top of next column), a waypoint at 18° 27.50' N, 69° 53.50' W, will place you approximately ¼ nm S of the marked entrance channel between the jetties. From the waypoint head NE up the *Rio Ozama* past the cruise ship terminal staying midstream until you arrive at the marina. There is a lighted range to lead you in, but it isn't really necessary for the average cruising boat as there is good water unless you get too close to the shore or jetties.

Just past the *Club Nautico* marina is a floating bridge (not shown on the chart) that blocks further inland river access to boats. There is a fair amount of current in the river, even more after strong rains.

The Southern Coast: *Casa de Campo Marina*

Casa de Campo Marina offers protected slips and a yard with a 120-ton *Travelift* for hauling out. Some of the slips have finger piers while others are Med-moor. The marina can accommodate vessels to 250' LOA with drafts from 9'-15' with a tiny 6" tidal rise and fall in normal conditions.

As shown on Chart DR-15 (bottom of previous page), a waypoint at 18° 23.60' N, 68° 54.40' W, will place you approximately ¼ nm SW of the well-marked entrance channel. From the waypoint head in a NNE direction and you will pick up the lit markers to guide you into the marina. You will notice the conspicuous red and white striped light at the end of the jetty to starboard. If you need help, hail the marina on VHF ch. 16 or 68 and they will send out a boat to guide you.

The Southern Coast: *Cap Cana Marina*

On the eastern shore of the Dominican Republic is the large and protected 130-slip *Cap Cana Marina*, part of a huge condominium project. The marina was designed for larger sportfishing vessels, not the average cruising boat, and for that reason fendering can be difficult as many slips are alongside a wall with an overhang at the top. The marina has 130 slips including 81 slips that can handle vessels over 130'

As shown on Chart DR-17 (below), a waypoint at 18° 29.85' N, 68° 22.00' W, will place you approximately ½ nm SE of the marked entrance channel to the marina. Never attempt this entrance in strong easterly seas.

The entrance channel has a least depth of 7' at MLW with a tidal range of approximately 1.5'. Do not confuse the entrance channel for *Cap Cana Marina* with the entrance channel to *Punta Cana Marina* which lies just to the north. *Cap Cana Marina* offers a pilot service (0800-1800 daily) if needed; the marina can be reached on VHF ch. 16 or 72.

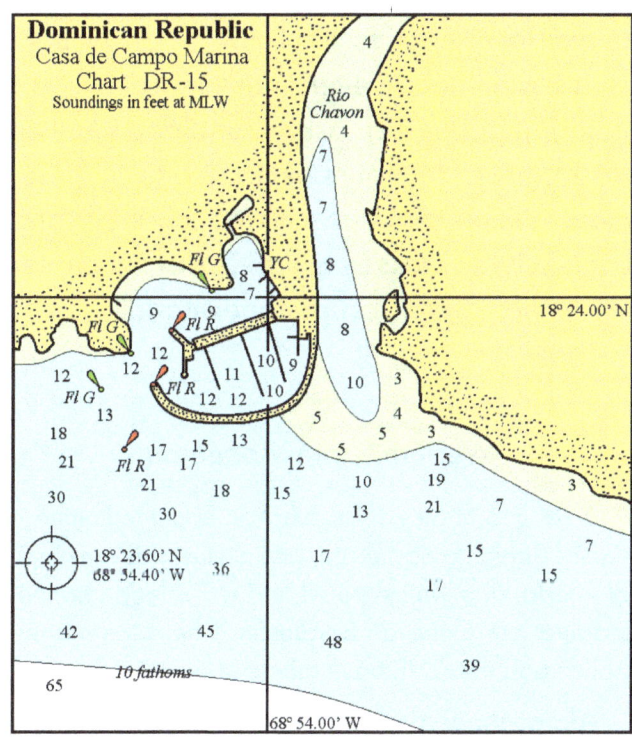

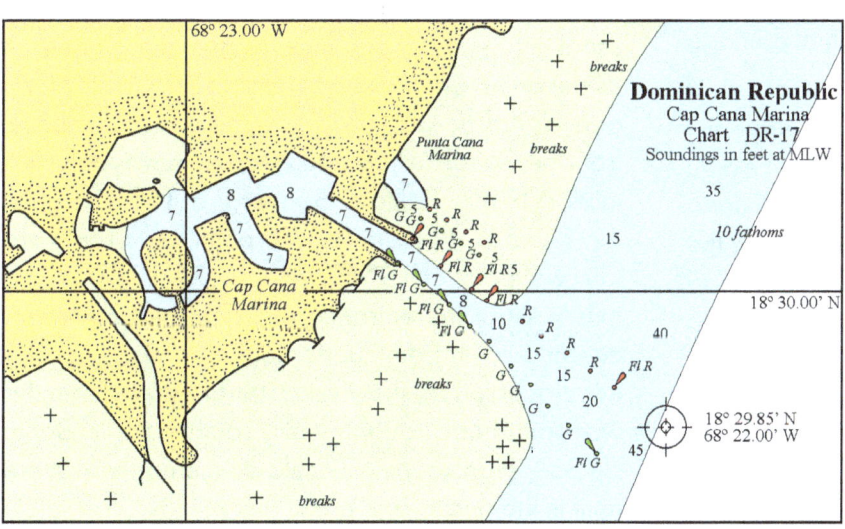

Chapter 10

Puerto Rico

ALTHOUGH PUERTO RICO GETS IN THE WAY OF A LOT OF HURRICANES, THE island has several nice hurricane holes as well as places to haul out for protection. One thing you will notice about Puerto Rico is that in nearly every harbor you will see an anchored or moored boat that shows signs of damage or neglect. Ask around and you will learn that these are usually hurricane damaged vessels, don't let yours become one of these. Hurricane Maria really did a number on Puerto Rico. When Hurricane Maria left the island Puerto Rico had no power anywhere on the island and in two weeks only had 5% of the power back on. Puerto Rico will be suffering a long time from the effects of Hurricane Maria.

Western Shore: *Puerto Real*

On the western shore of Puerto Rico lies the small, protected harbor of Puerto Real (not to be confused with the Fajardo dockside area of the same name) as shown on chart PRW-5 (next page). The mangrove swamp in the northwest corner of the bay is a great hurricane hole, but entry is limited to vessels with drafts of 4' at low water...the entrance will take about 5' at high water. If you wish to haul out or get a slip, *Marina Pescaderia* can help, they suffered minimal damage in Hurricane Irma and Hurricane Maria.

As shown on Chart PRW-5 (next page), a waypoint at 18° 04.10' N, 67° 12.70' W, will place you approximately 1 mile west of the entrance to Puerto Real and well inside *Arrecife Tourmaline*. From the waypoint just off the entrance to Puerto Real, head generally east towards the harbor favoring the Punta La Mela side of the entrance as shown on the chart. A newly marked channel will guide you into the harbor. Once inside, keep an eye out for the shallow spot in the center of the harbor.

Western Shore: *Boqueron*

A second choice for protection is at the southern end of *Bahía de Boqueron*, also on the western coast of Puerto Rico where you'll find a 7' deep channel leading into *Cano de Boqueron*, where the marine police and the DNR have their docks.

A large shoal called Bajo Enmedio lies in the center of the mouth of the *Bahía de Boqueron* and offers a bit of protection to vessels anchored inside. Around this shoal are two easy entrances to *Bahía de Boqueron* as shown on Chart PRW-6 (next page), *Canal Norte* and *Canal de Sur*. If approaching from offshore, from the *Canal de la Mona*, head for a waypoint 4½ miles WNW of *Bahía de Boqueron* at 18° 01.90' N, 67° 17.50' W. From this waypoint head eastward to a waypoint at 18° 01.90' N, 67° 13.00' W, ½ mile west of the entrance to Canal Norte (vessels heading south from Puerto Real may also head for this waypoint) as shown on Chart PRW-6. From this waypoint head roughly east/southeast into *Bahía de Boqueron* keeping ¼ mile off the northern shore until in the anchorage area. This will take you safely

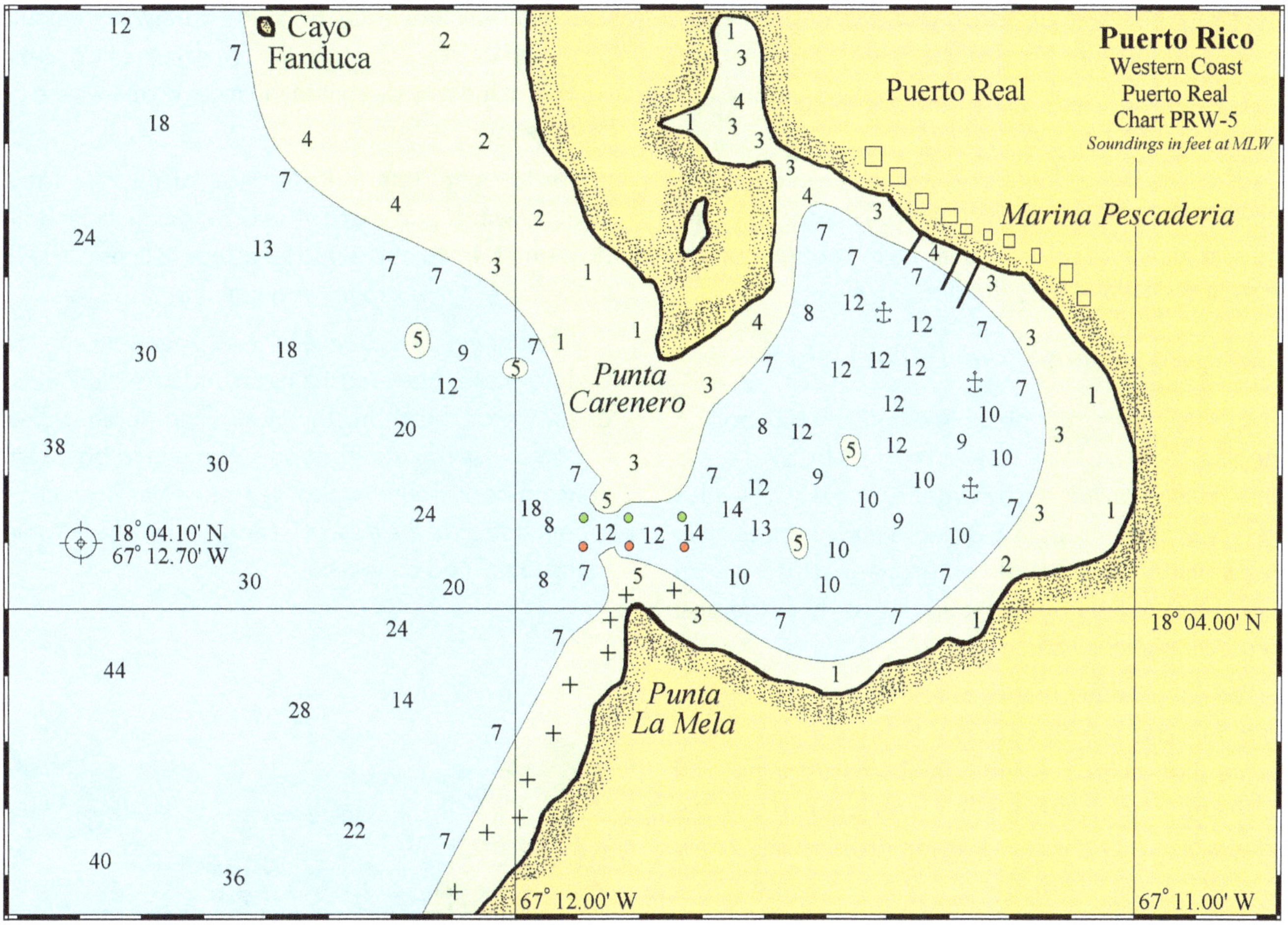

Puerto Rico
Western Coast
Puerto Real
Chart PRW-5
Soundings in feet at MLW

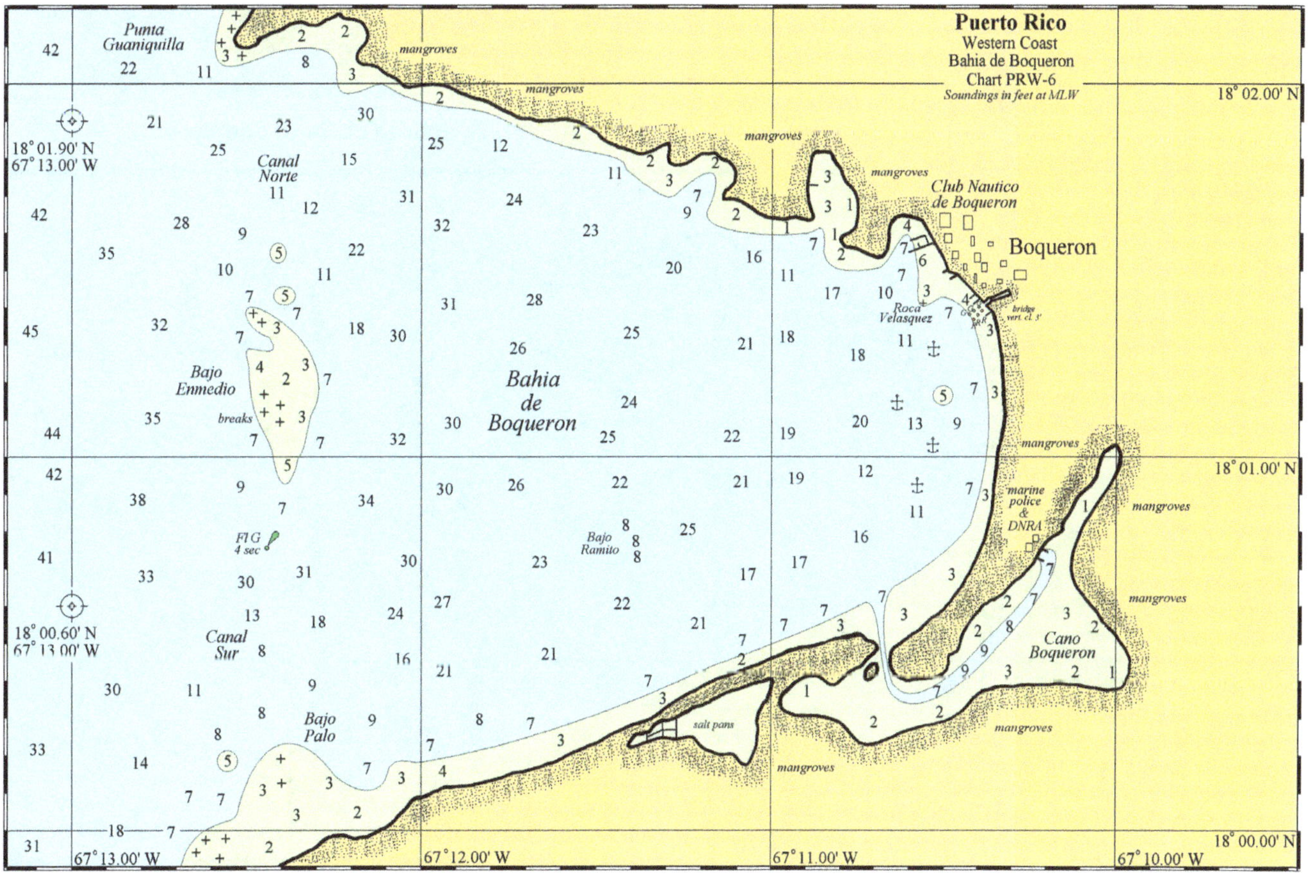

Puerto Rico
Western Coast
Bahia de Boqueron
Chart PRW-6
Soundings in feet at MLW

PUERTO RICO • 99

past *Bajo Enmedio* and once past the shoal, the bay is deeper and safe of hazards.

The hole here is the cove lying in the southeast part of the bay as shown on the chart. Follow the stakes in and keep between them as you round the mangroves and head northeast towards the police docks. You'll have a minimum of 7' the entire way though outside the channel the waters shoal rapidly.

Southern Shore: *La Parguera*

Good shelter, and possibly a haul out (controlling depth 4.5' at MLW) if there is room, is located in the area of La Parguera. La Parguera is best known for its bioluminescent lagoon called *Bahía Fosforescente* lying just to the east of La Parguera as shown on Chart PRS-2 (see below), and this is where the best protection can be found.

Do not attempt to enter here at night, good visibility is required as the shoal to port is not easily seen. On some charts there is a small cay shown just to the west of the entrance to the bay, this is no longer a cay; the mangroves have filled in the gap and the cay is now part of the mainland.

Entering the bay, you will find 10'-13' of water as you avoid the shoals on both sides of the channel (see chart) .The best protection here for vessels of 6' draft is in the creek that leads to the northeast and a smaller creek that leads off to the northwest. Vessels drawing 4' or less can take the creek in the northwest section of the bay deep into the mangroves where 6' of water can be found.

Try to anticipate the direction of the strongest forecast winds and place your bow facing it. Set both bow and stern anchors and tie at least 2 lines per side to the largest mangroves you can find.

Another viable option is to work your way into the 8' deep water between the mainland and Isla Cueva as shown on the chart. The controlling depth is 3' at MLW for this route and storm surge would be a real concern here. There is also deeper water protection for anchoring to the east in La Parguerra where some mangrove lined canals exist.

Hauling out in La Parguera

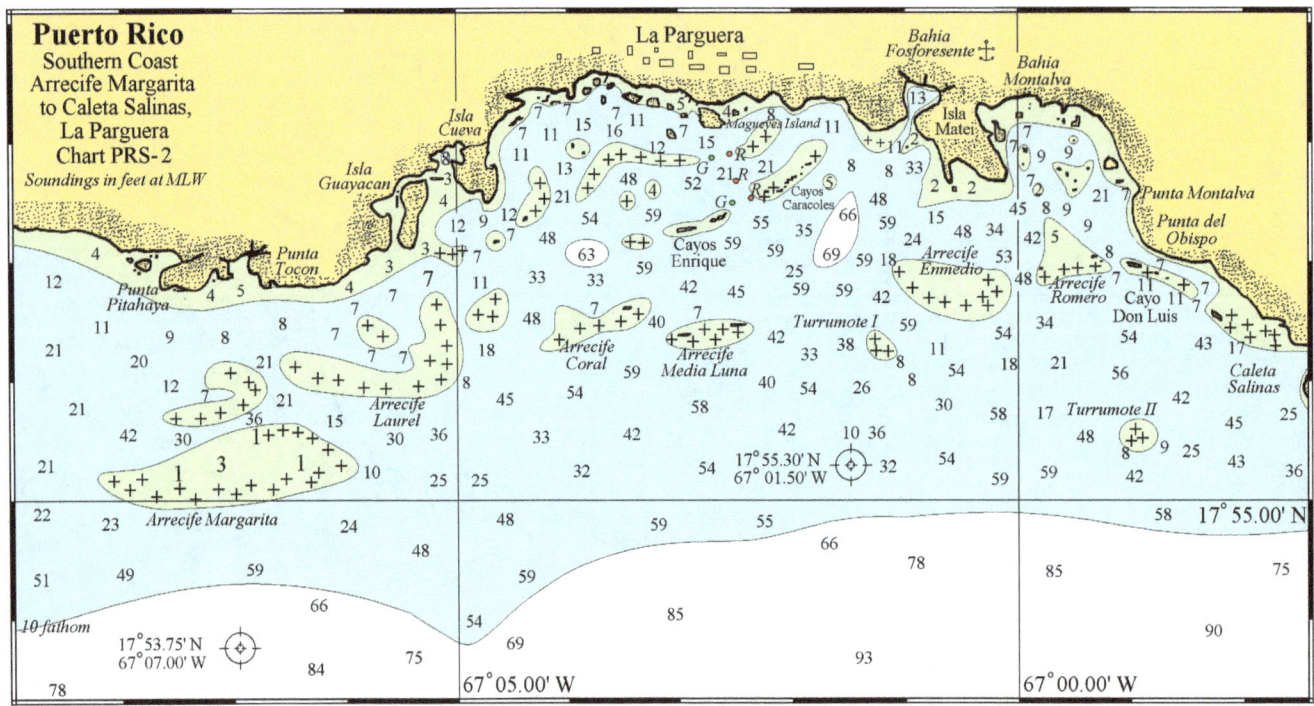

Southern Shore: *Bahía de Guanica*

A short distance east of La Parguera is *Bahía de Guanica*, a large commercial harbor. As shown on Chart PRS-3 (see below), a waypoint at 17° 54.85' N, 66° 54.75' W, will place you ¼ mile southwest of R "2," the first marker leading into the bay. Head just a bit east of north following the buoys as shown (there is a lit range bearing 354° T). Once inside *Bahía de Guanica* head west and anchor in the extreme southwestern corner, just south of Punta Pera, as close in as you can, this will offer the most protection. Another option is to anchor in *Bahía Noroeste* as shown on the chart. In either cove, be sure to check your anchors as most of the bay is poor holding and there is a lot of debris on the bottom.

Southern Shore: *Bahía de Guayanilla*

As shown on Chart PRS-5 (see next page), a waypoint at 17° 56.60' N, 66° 45.81' W, will place you approximately 1½ miles south of the marked entrance channel leading into *Bahía de Guayanilla*.

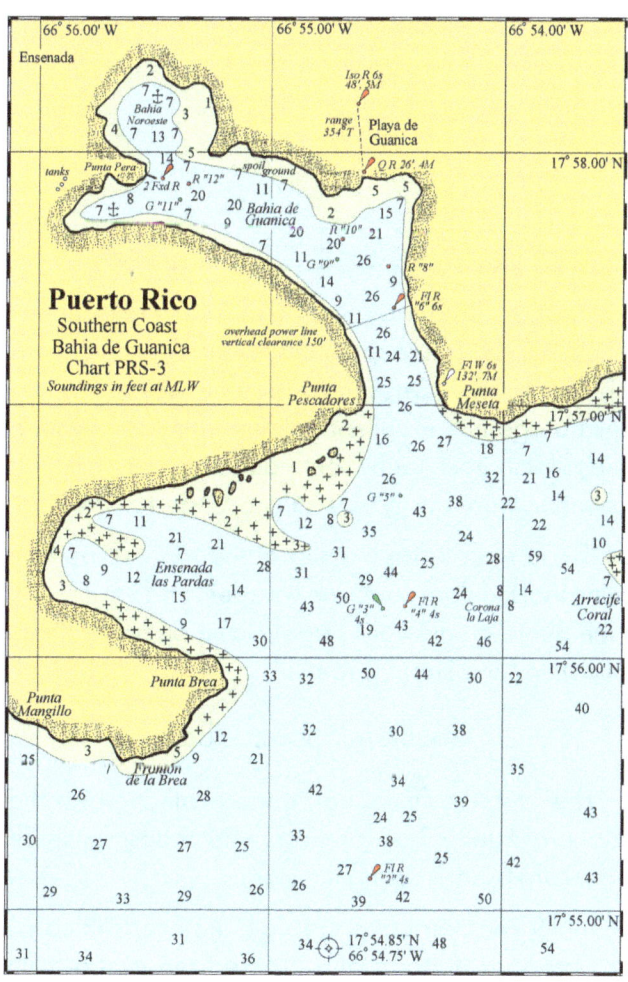

Visible from offshore is a rectangular container lift off Punta Gotay on the western side of Punta Guayanilla. The tank farms east of Punta Gotay, the tanks at Punta Pepillo, and the large stack south of Guayanilla are also very conspicuous

From the waypoint head just a bit west of north until you can pick up G "1" and keep it on your port side. This marks the beginning of the entrance channel and ahead of you you'll notice a lit range (358° T). The channel is wide and deep here and keep the buoys where they are supposed to be (red on right) and pass between Punta Gotay and the shoal to its west whose eastern edge is marked by G "5."

The best protection lies to the east, but the area is highly industrialized, and the holding is poor with some scattered debris. You can work your way into the small cove north/northeast of Punta Gotay by taking R "8" to starboard and following the narrow channel of deeper water southeastward into the cove as shown on the chart. The protection is not very good here, a better spot lies more to the northeast, through the narrow channel into the small cove as shown on the chart. Another option, and this one puts you smack dab in the middle of the industrial area, is to pass the huge well-lit pier and anchor in a tiny cove at its northeastern end.

Another option is to anchor inside Punta Verraco at the western end of the harbor. To get the best protection here you should have a shallow draft vessel and work your way into the cove as far as your draft allows.

Marina de Salinas

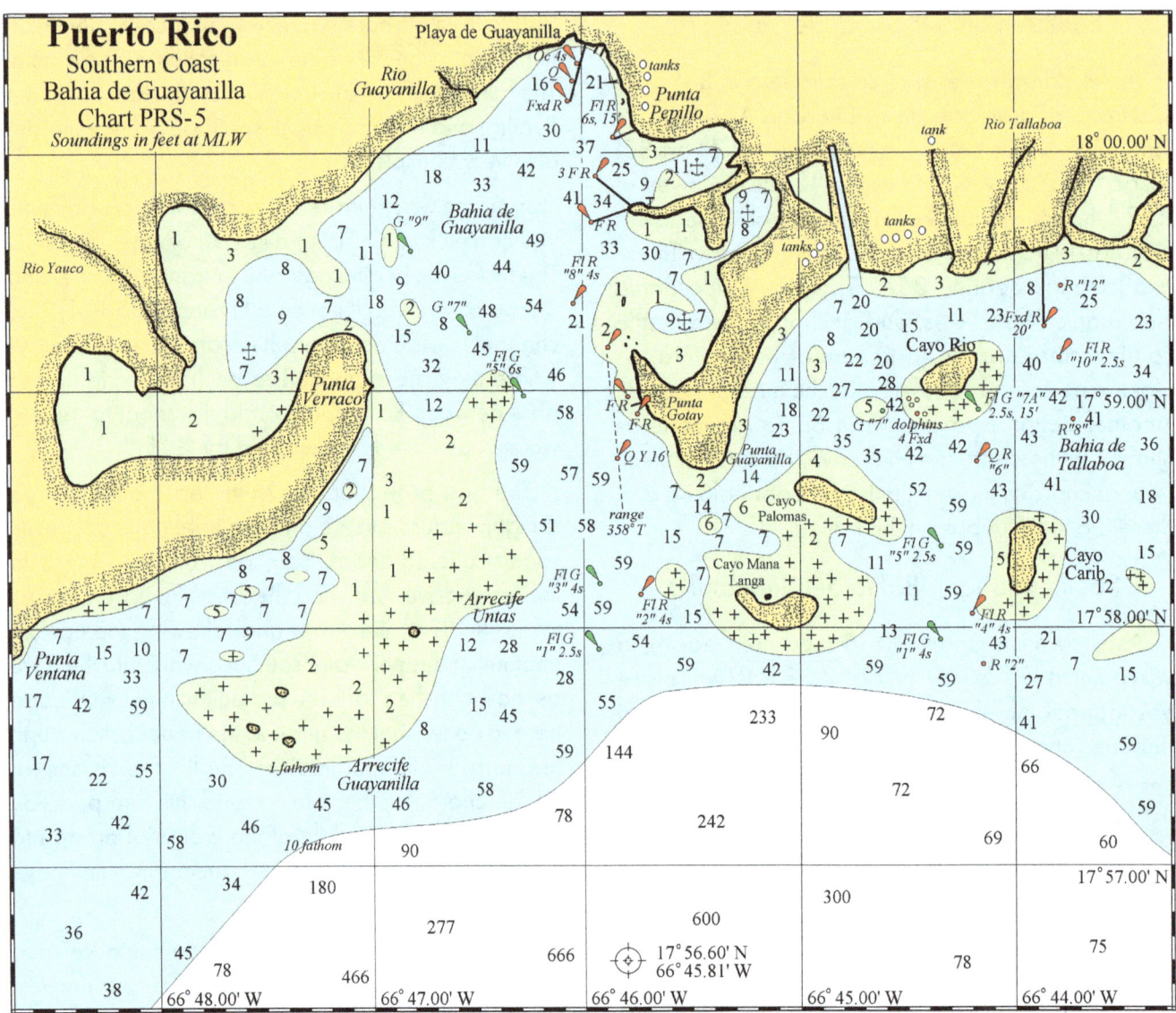

Southern Shore: *Salinas*

Salinas is one of the most protected anchorages along Puerto Rico's entire 300-mile coastline. A waypoint at 17° 56.60' N, 66° 18.10' W, will place you approximately ½ nm WSW of the entrance channel between Cayo Mata and Punta Arenas as shown on PRS-11 (see next page).

From this waypoint, the entrance is hard to distinguish at times, especially in the early morning light with a skipper's tired eyes. However, excellent landmarks are the tall high-tension power line towers on the hill to the east of the waypoint and Punta Arenas. Put these towers on your bow and head in on a course of 080° until the entrance opens up to port. Keep a bit south of Cayo Mata to avoid the shoal area off its southern tip and round up between it and the mainland to starboard keeping roughly center channel. Just to the east of Cayo Mata is a small "No Wake Zone" buoy that you can keep to port as you head northward up the channel. Once in the bay off the marina, the anchorage has excellent holding in mud but with a bit of fetch in a north/south direction and you can count on there being several unattended boats here.

If you plan to take a slip at the marina, bear in mind that the docks are low, wooden, and not floating (see photo on previous page), so tie off accordingly allowing for a storm surge.

Southern Shore: *Jobos*

This is it, the finest hurricane protection on the island of Puerto Rico for those who choose to tie off somewhere safe during the storm.

In 1998, Hurricane Georges hit Puerto Rico as a Category 3 storm and crossed the island from

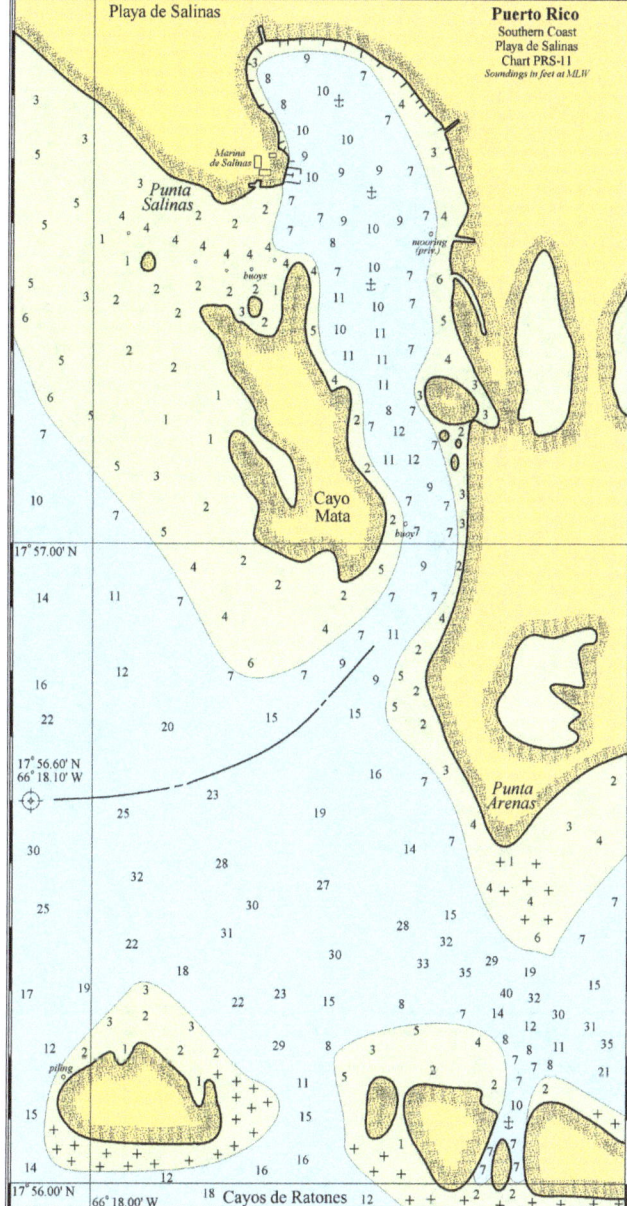

de Ratones, Cayo Morrillos, Cayos de Pajaros, and Cayos de Barca to starboard as shown on Chart PRS-12 (next page). This can only be done in good visibility as you must clear the large shoal south of Punta Arenas.

Cayo Puerca marks the opening into the anchorage and is left to starboard (east) as you enter at 17° 55.78 N, 66°14.53 W. The creeks wind west and northwest so work your way in as far as you can and as close in to the mangroves that your draft allows. Set both bow and stern anchors as well as tying to the mangroves with as many lines as you can.

Approaching from offshore you can head for a waypoint at 17° 54.75' N, 66° 13.00' W, that will place you approximately ¼ nm southeast of the entrance channel at *Boca del Infierno* as shown on Chart PRS-12. From the waypoint, head through the pass between the reef east of Cayos de Boca, and the reef southwest of Cayos Caribes. Once inside you can head towards Cayo Puerca to secure your vessel.

Eastern Shore: *Palmas del Mar*

The haul out yard at *Palmas del Mar* suffered greatly during hurricane Georges. Due to the storm surge many boats floated off their stands and the seawall was nearly destroyed. Today the marina has a new seawall that is supposed to withstand winds of 140 mph.

The marina offers slips but there always seems to be some surge present so be sure to allow for that when tying off your vessel (and don't forget to use plenty of chafe gear). Some folks like the protection that the community of *Palmas Del Mar* offers, narrow canals amid high condos. Unfortunately, one must know somebody with dock space here that will allow you to tie up for a while. The principal danger would be a strong storm surge that could wreak havoc in the canals.

As shown on Chart PRE-2 (next page), a waypoint at 18° 04.45' N, 65° 47.45' W, will place you approximately ¼ mile southeast of the entrance to the harbor. Do not stray too far north of this position as there are several shoals east of the breakwater as shown on the chart. From the waypoint head to the gap between the end of the breakwater and the shore favoring the breakwater. Strong easterly winds will

east to west, one of only two hurricanes to ever do that. Georges brought a storm surge of 10' with 20' waves reported. The storm spawned three tornadoes, two on the mainland and one off Vieques. They eye of Georges passed twenty miles south of San Juan and by the time the storm left Puerto Rico it was still a Category 2 Hurricane. Despite the severity of Hurricane Georges, boats tied off in Jobos, in the mangrove creeks west of Cayo Puerca, reported no damage, remarkable indeed! Yes, there will be other boats in these creeks but there is plenty of room.

There are two routes to the protection offered at Jobos. An inside route takes you east from Salinas paralleling the mainland shoreline keeping Cayos

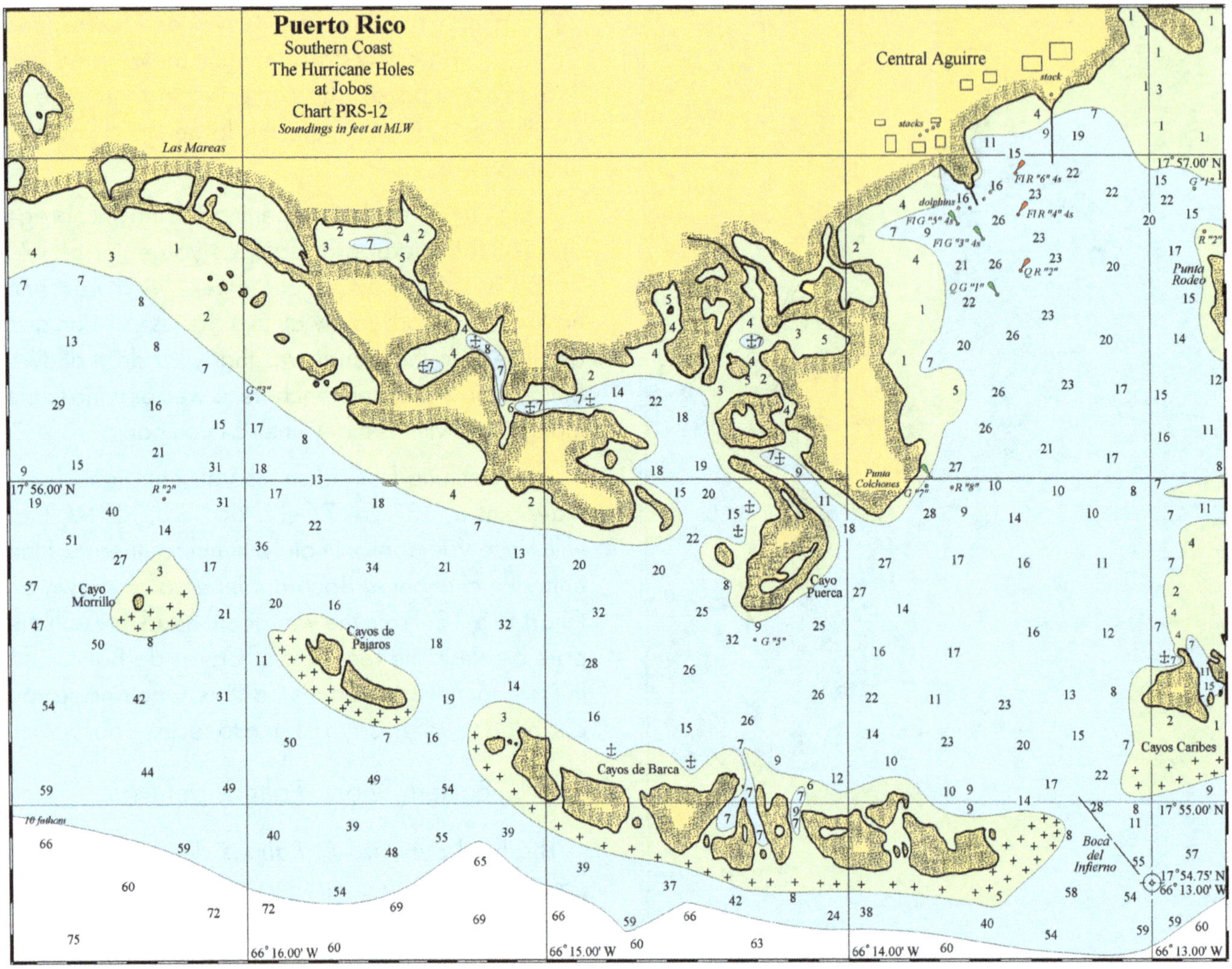

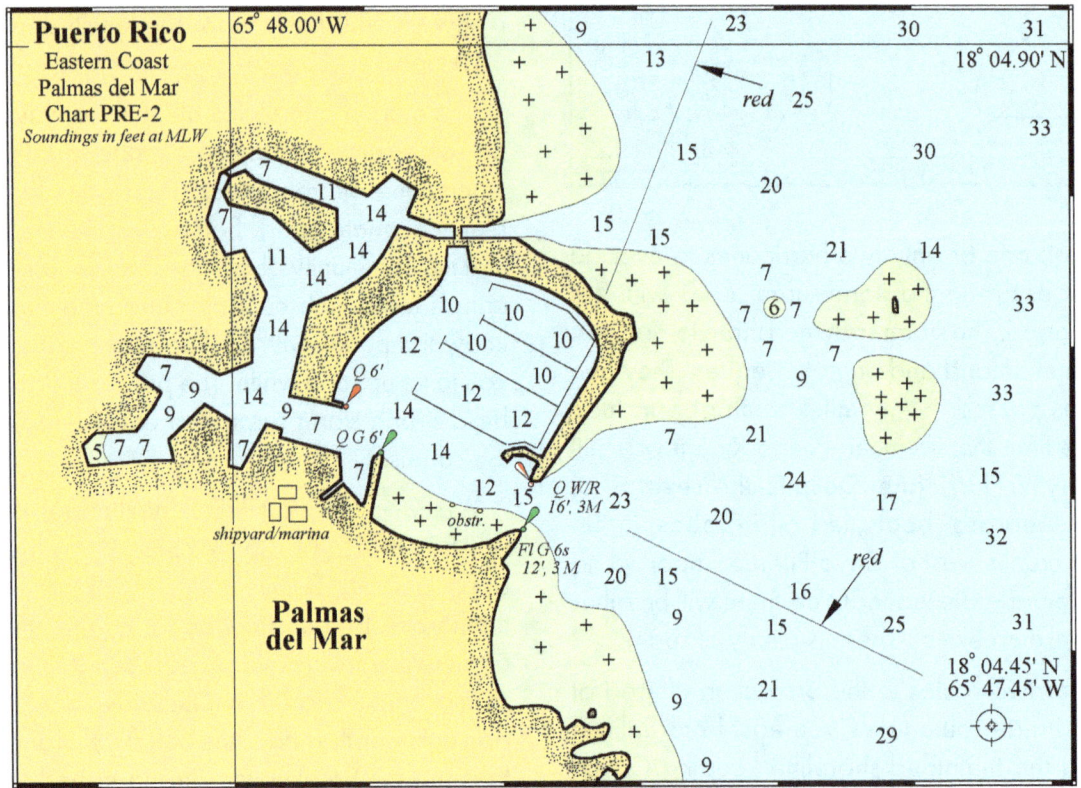

104 • THE CAPTAIN'S GUIDE TO HURRICANE HOLES

cause breaking seas in the harbor entrance and a nice surge inside the harbor. As mentioned, you can get a slip here or haul-out, it's your choice, but a haul-out makes more sense for protection.

Eastern Shore: *Puerto del Rey Marina*

Along the eastern shore of Puerto Rico lies one of the largest marinas in the Caribbean, *Puerto del Rey* (see Chart PRE-7 below). While you might not want to ride out a hurricane at a slip here, you could certainly haul out if that was your choice. The yard has an 80-ton lift with a 24′ beam capacity, a 166-ton lift with a 32′ beam capacity, a 60-ton Brownell trailer, and a multi-hull lift that can handle up to a 30′ beam. Boats hauled here sit 1,500′ inland and are approximately 30′ above sea level which eliminates any damage done from storm surge. Fees are usually 1-year in advance for a hurricane haul out.

A waypoint at 18° 17.45′ N, 65° 37.45′ W, will place you ½ mile east/northeast of the entrance to the marina in *Bahía de Demajagua*. Approach the marina entrance by passing between the breakwater and the red marker that defines the shoal to starboard, favoring the breakwater side (see Chart PRE-7 below). Enter the marina proper between the breakwater and the jetty that protrudes from shore as shown on the chart. For more information give the marina a hail on VHF ch. 16. The marina suffered minimal damage from Hurricane Irma and Hurricane Maria.

Eastern Shore: *Fajardo*

In Fajardo there are several places to haul out, and a couple of marinas that might be worth getting a slip to escape the storm. *El Conquistador Marina*, and *Villa Marina*.

If you wish to anchor out there is one small option. In the small corner of the easternmost island of Cayo Obispo (see Chart PRE-8 below) there is an 8′ channel leading in to a 90° turn with mangroves directly ahead. Put your bow as far as you can into the mangroves and set a couple of anchors on the shallow bar off your stern. Here you are protected to

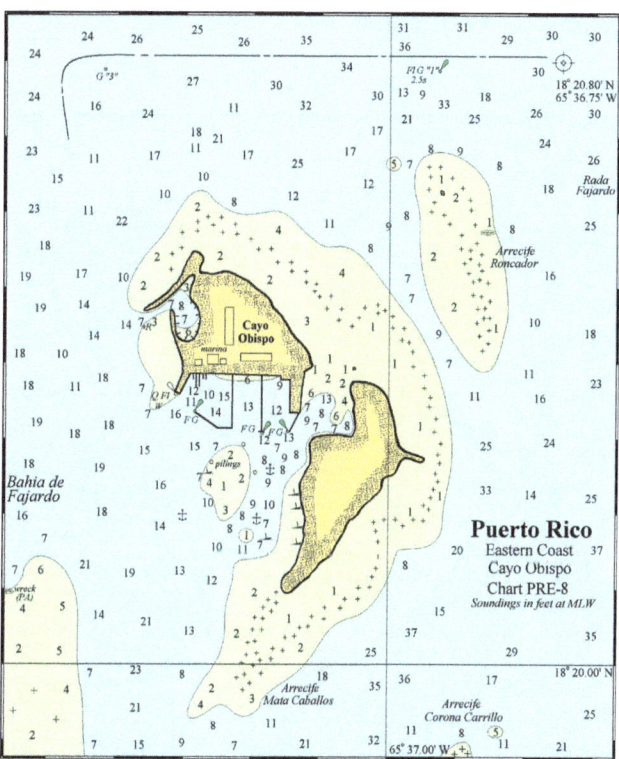

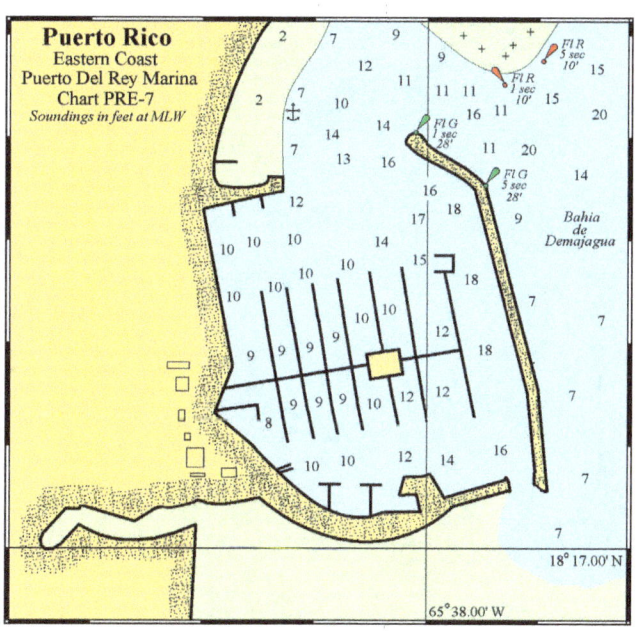

Puerto del Rey Marina

the east, south, and west by islands, and to the north by a shallow reef. Unfortunately, there is only room for one boat here and a strong storm surge could make this spot untenable. This spot is not a prime location, but is mentioned here solely as an option.

Northern Shore: *San Juan*

For small craft, San Juan offers little hurricane refuge save a slip in the marina. This is a busy commercial harbor and cruising vessels won't find a good spot to anchor here except in *San Antonio Channel* as shown on Chart PRN-5 (below). Just south of Punta Catano is the entrance to a dogleg cove that offers good protection but will have its share of unattended vessels during a storm.

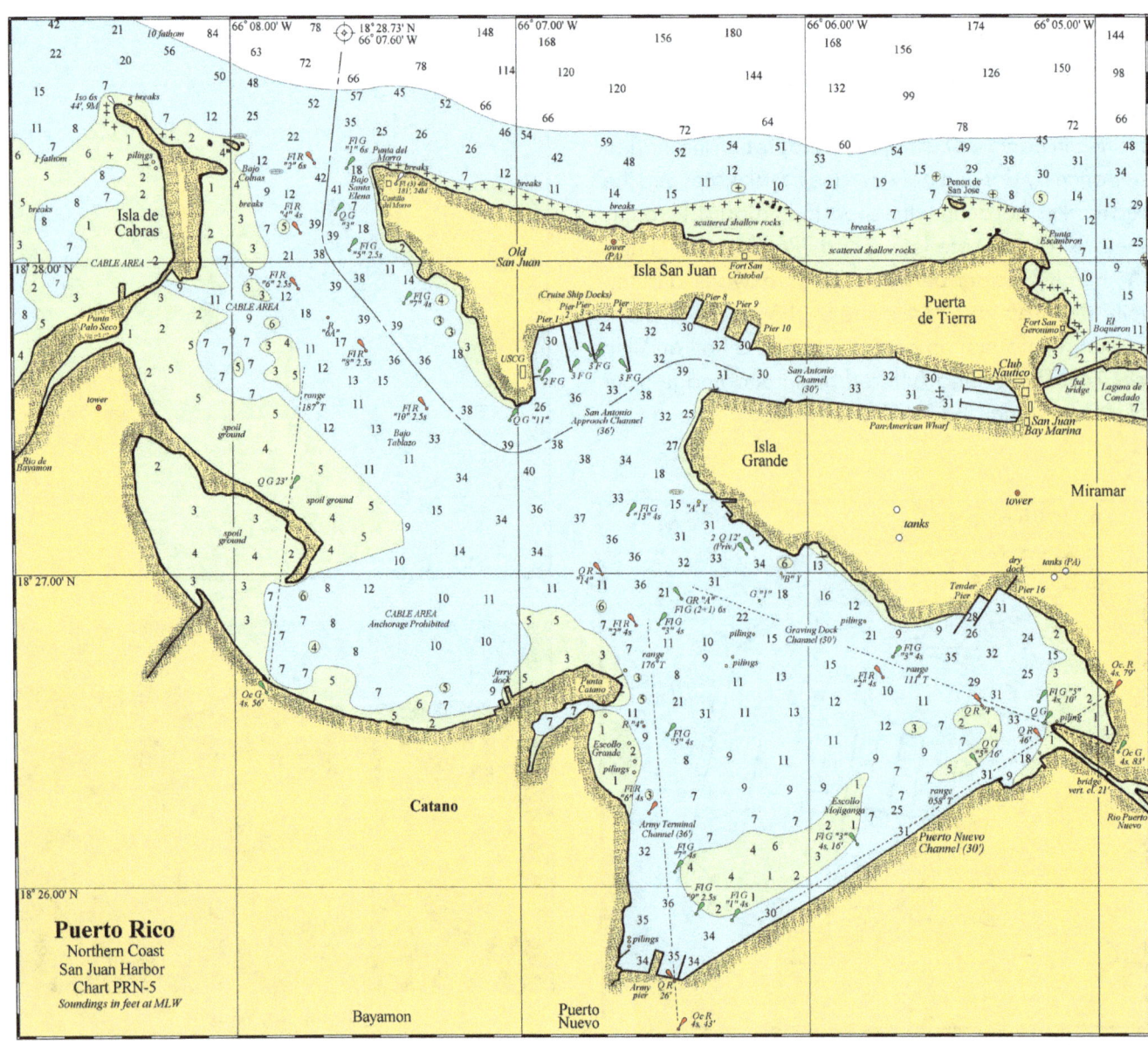

Chapter 11
The Spanish Virgin Islands

IF YOU ARE IN THE SPANISH VIRGIN ISLANDS WHERE THERE ARE SEVERAL DECENT choices for hurricane protection, two of the main ones are named *Ensenada Honda*; one at Culebra and one at Vieques. Ensenada Honda on Culebra offers good protection in its mangrove creeks where many boats survived fierce Hurricane Hugo although there was a lot of destruction to those vessels that anchored in the bay. On Vieques you can also take shelter in *Ensenada Honda*, a mangrove lagoon with three deep creeks. Remember, better get to these holes early as local boaters will have the same idea as you for seeking shelter in Culebra and Vieques! Vieques boaters also like the protection offered by Puerto Mosquito on the southern shore of Vieques. Remember, for the best protection, move EARLY!.

Culebra: *Ensenada Honda*

Culebra's *Ensenada Honda* used to be regarded as one of the finest hurricane holes in the Caribbean. Hurricane Hugo changed that notion as the Category 5 storm hit. The eye of Hugo came within 2 miles of the main town of Dewey and gusts were measured at 210-240 mph while in the harbor itself seas of 8'-12' were reported. It has been reported that some 300 boats were lost in *Ensenada Honda* during Hugo, however some DID survive.

The boats that are small and got here early survived because they were in the mangroves with as many as 20 lines out. Their hulls may have been damaged, but the mangroves cushioned them. If you can find shelter in the mangroves in the eastern side of the bay north of Punta Cabras, that is fine, if not, you might want to rethink and head for Vieques or Jobos on Puerto Rico.

To enter the bay as shown on Chart SVI-6 (next page), head for a waypoint at 18° 17.33' N, 65° 16.35' W, keeping the very visible reefs to port and R "2" (marking *Bajo Amarillo*) to starboard. Never attempt this route at night! Enter the harbor as shown on the chart and work your way to the NE to enter the protection mentioned lying north of Punta Cabras.

Vieques: *Ensenada Honda*

On Vieques you can also take shelter in *Ensenada Honda*, a mangrove lagoon on the southern shore with several deep creeks to the north. This anchorage is fair protection in a hurricane as long as you don't get any west winds (which would indicate the eye passing north of you).

As shown on Chart SVI-13 (after next page), a waypoint at 18° 06.05' N, 65° 21.90' W, will place you approximately 1 mile south of the entrance to this large bay. The entrance will have to be piloted by eye. From the waypoint, head generally north/northwest keeping the large shoal shown well to starboard. Once clear of the shoal, head towards Punta Carenero and keep to starboard the shoal north of Los Galafatos. Once clear of this shoal head southeastward and

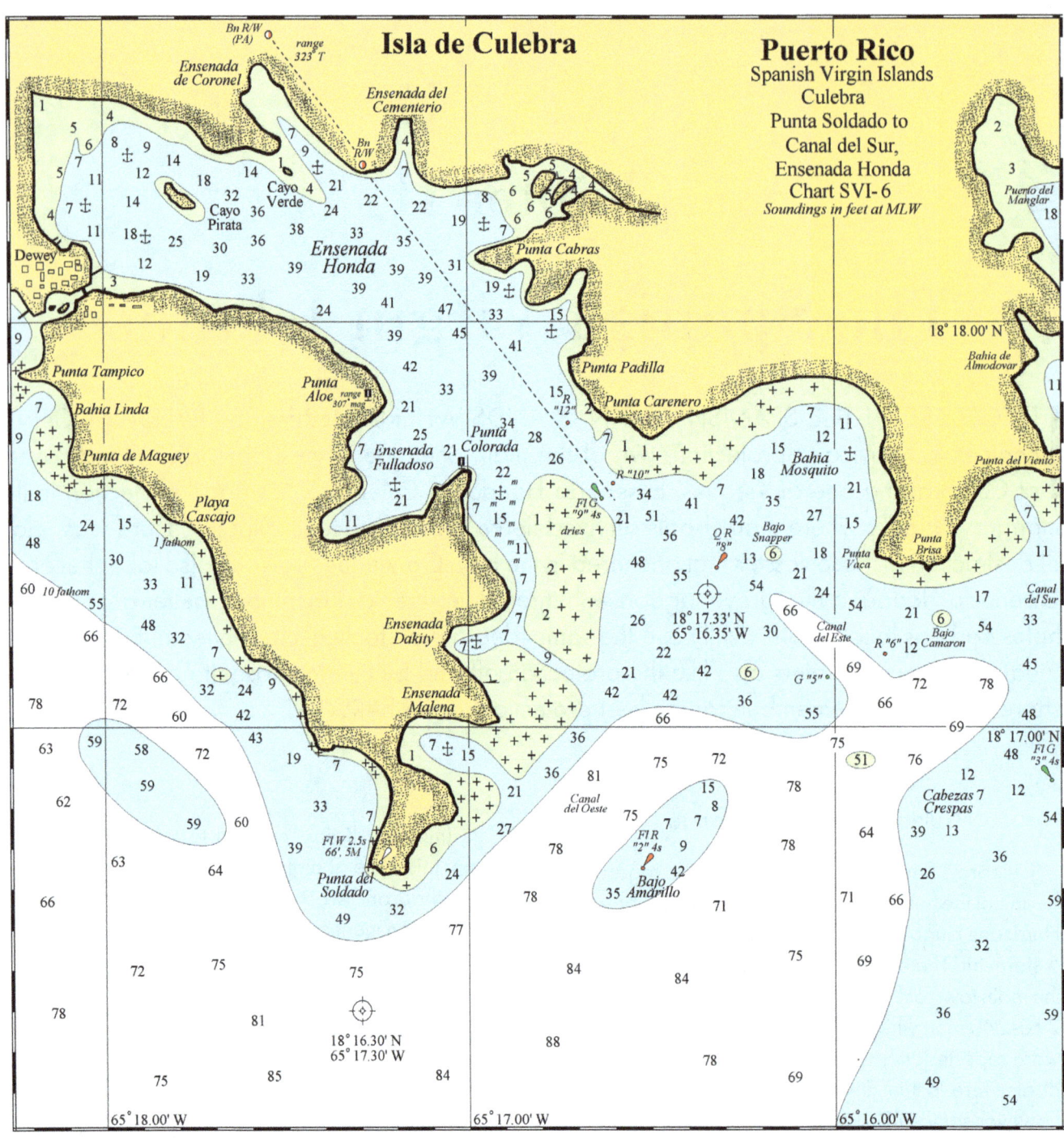

anchor wherever it pleases you, preferably in the small creeks.

Enter the large bay staying in 20' of water, leaving the breaking shoal to starboard and the inner reef to port. Pass *Los Galafatos* to the west (round these visible rocks awash leaving them to starboard). Turn east and continue 1.1 miles to what looks like the end of the bay. To the north of your vessel (port side) you'll see several openings in the densely wooded mangroves.

Enter the opening in the mangroves that is farthest to the east. You'll have 10'-15' of water that deepens to 20' once inside of the mangrove creek. Proceed in as far as your draft allows and where the canal is roughly 20'-30' wide. Find a section of the canal that lets you point your bow in the direction of the anticipated wind and set a bow anchor and stern anchor staying parallel to the lie of the creek. Try for a 10/1 rode to water depth. Using your dinghy, tie a second bow

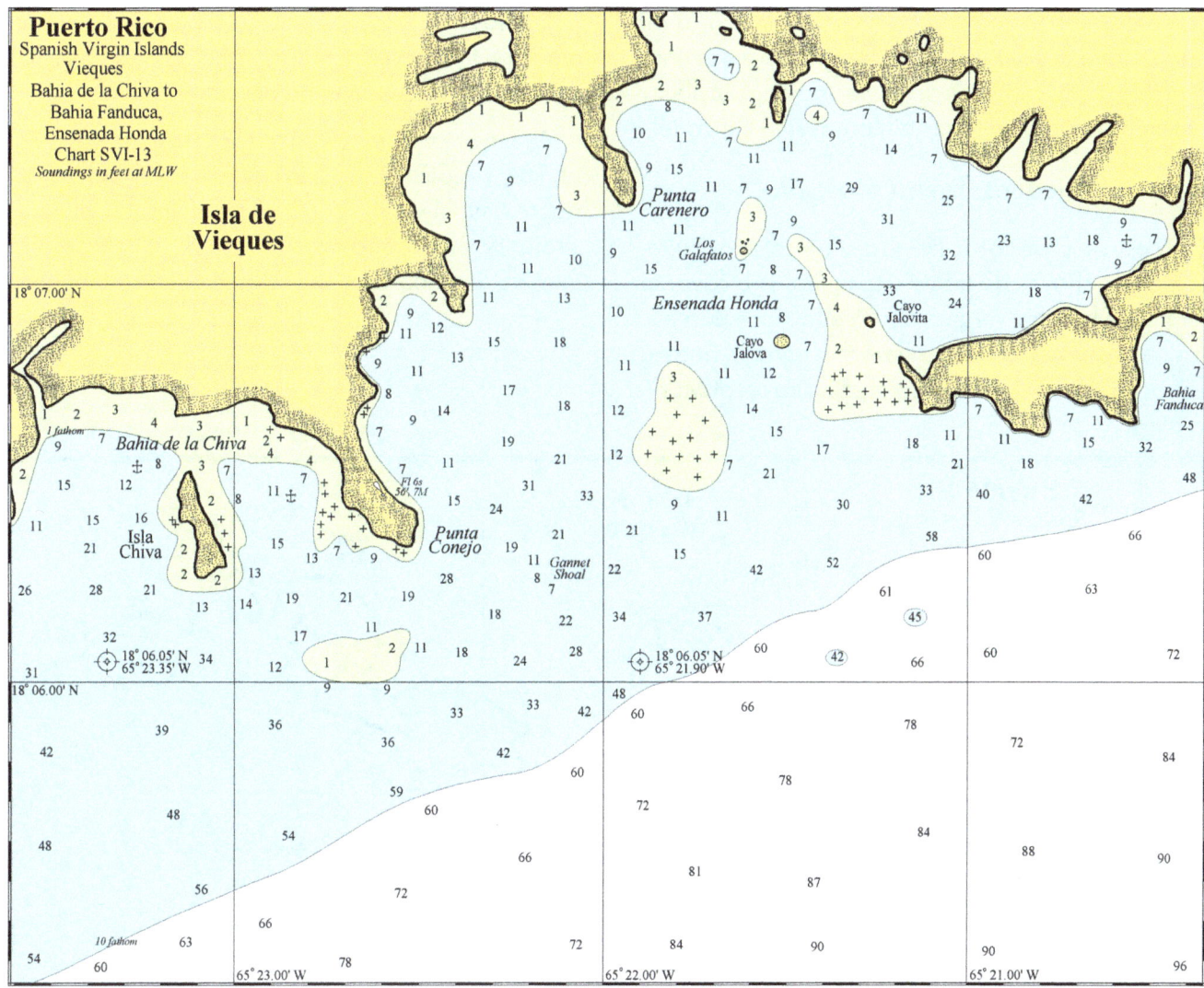

line at a slightly different angle than your anchor to the largest mangrove you can find. Tie a second stern line to another large mangrove at a slightly different angle than your stern anchor. Tie at least one midship line on each side of your vessel to large mangroves and adjust as needed during the blow.

Vieques: *Puerto Ferro*

There is a 'designated' hurricane hole located approximately 3.7 miles west of *Ensenada Honda*, Vieques. It is in one of the bioluminescent bays, called *Puerto Ferro* (and locally called *Barracuda Bay*), the entrance being marked by the remains of an old Spanish lighthouse.

As shown on Chart SVI-14 (next page), a waypoint at 18° 05.40' N, 65° 24.70' W, will place you ½ mile southeast of the entrance to this well-protected harbor. Head directly into the bay through the narrow entrance with a least depth of 7' at MLW. Keep to mid-channel as the sides shoal to 4' rapidly.

This harbor offers protection in almost all conditions. As you travel north west into the bay you will pass a small peninsula to the east (on your starboard side). Round this peninsula, staying in 20' of water to enter a second bay. This IS the designated hurricane hole, but it is a bit open to strong winds from the northwest.

There is a sign on the peninsula with a phone number to call for vehicle pickup but ironically if a storm situation exists it may not be reliable. Note that theft is a possibility here.

There is a canal up in the mangroves farther to your east from the bay that affords a tighter and thus better anchorage but there are a few derelict boats permanently moored there that might be a problem

THE SPANISH VIRGIN ISLANDS • 109

in a blow. When you find an area where your draft permits and it is 20'-30' wide, set your bow anchor and drop back to set a stern anchor, paying out plenty of line from the bow to remain parallel to the canal.

Vieques: *Puerto Mosquito*

West of *Ensenada Honda*, *Puerto Mosquito*, sometime shown as *Bahía Mosquito*, is a favorite hurricane hole for Vieques and Puerto Rican boaters. It is a very protected little cove with a narrow, dogleg entrance channel and a 5' controlling depth at MLW.

Puerto Mosquito is a beautiful bioluminescent with a ban on the usage of motors in the bay except in the case of a threatening hurricane.

As shown on Chart SVI-14 (see below), a waypoint at 18° 04.50' N, 65° 26.50' W, will place you approximately 1 nm south of the entrance which is straightforward and narrow. Work your way into the bay and find your spot to ride out the hurricane.

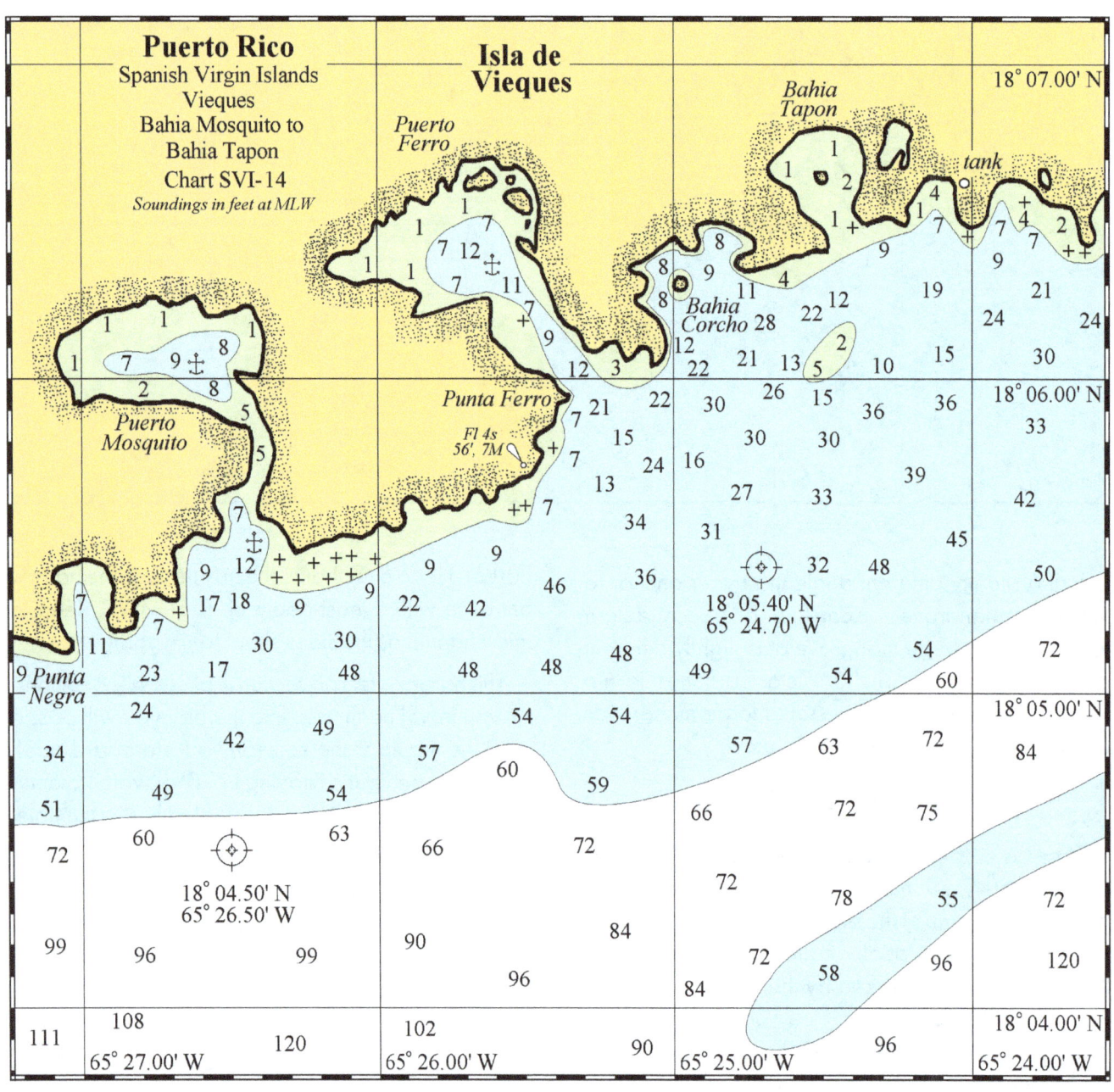

110 • THE CAPTAIN'S GUIDE TO HURRICANE HOLES

Chapter 12

The United States Virgin Islands

A VERY WELL-KNOWN SEAFARING WRITER FROM THE VIRGIN ISLANDS, CAPTAIN Fatty Goodlander, once said that the weather in the Virgins was perfect, except for a few days a year when it tries to kill you. This is quite true as the Virgin Islands seem to be in the path of a lot of hurricanes, but in a way of balancing things out, they also offer several nice holes for protection. Unfortunately for cruisers, the charter fleets usually get to the best holes first so advance warning of a storm should be acted on by the prudent mariner. St. John was heavily damaged by Hurricane Irma, St. Thomas as well but to a lesser extent. Benner Bay survived remarkably well as did Flamingo Bay near Charlotte Amalie.

St. Thomas: *Mandahl Bay*

The northern shore of St. Thomas is home to a small, virtually unknown hole called *Mandahl Bay*. I'm sure the locals will be mad at me for revealing this hole as local fishing boats usually fill this protected cove and not many cruising vessels are aware of its existence.

As shown on Chart USVI-MB (top of next page,- with photo), a waypoint at 18° 21.80' N, 64° 53.80' W, will place you approximately a mile NW of the unmarked entrance channel into *Mandahl Bay*. From the waypoint head generally SE until the entrance between the breakwaters opens up and you can work your way as far into the bay as your draft allows.

Bear in mind that this is a favorite spot for local mariners so go early and expect to be surrounded by unmanned vessels.

St. Thomas: *Flamingo Bay*

Water Island lies less than ½ mile off the southern shore of St. Thomas and most people don't realize that it is in fact the fourth largest U.S. Virgin Island.

As shown on Chart USVI-5 (next page, with photo), a waypoint at 18° 19.70' N, 64° 57.75' W, will place you approximately ¼ mile west of the entrance to *Flamingo Bay*. *Flamingo Bay* is not the hole, rather the best protection lies in the small cove east of *Flamingo Bay* (controlling depth 5').

From the waypoint head to the SE portion of *Flamingo Bay* to enter the easternmost cove that offers protection. Beware of shoals on both sides of the mouth of the entrance.

Mandahl Bay, St. Thomas

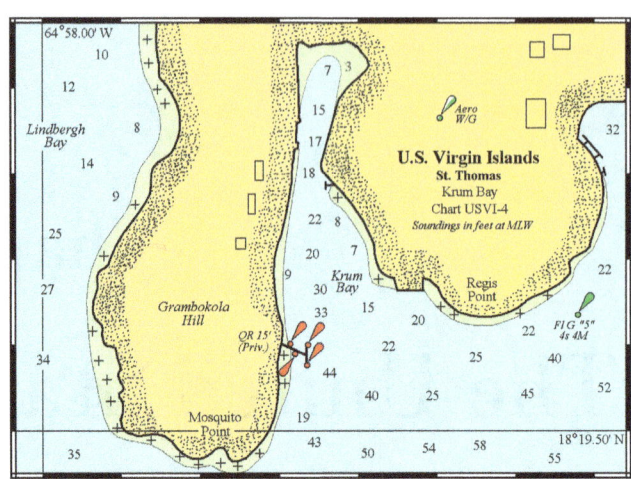

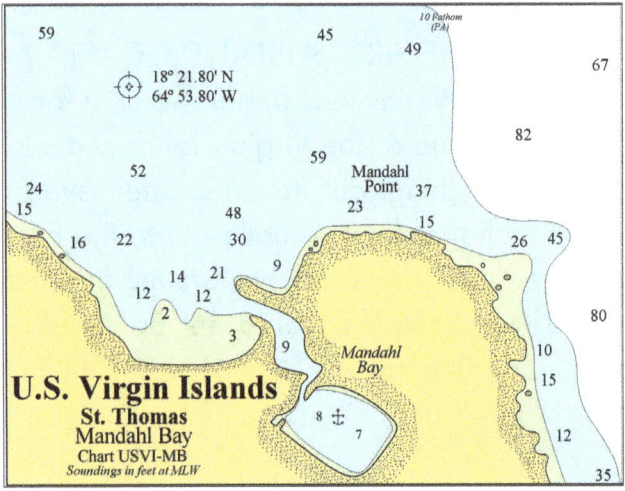

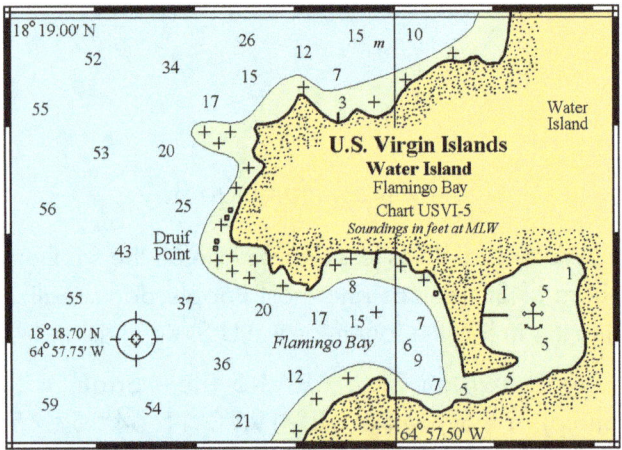

Flamingo Bay, St. Thomas

St. Thomas: *Krum Bay*

The northern tip of *Krum Bay*, as shown on Chart USVI-4 (see above), appears on paper to be a good hurricane hole, but in reality it is a small commercial area and has a lot of boat traffic on a daily basis. We cannot recommend this as a hurricane hole since there are other areas that offer better protection.

St. Thomas: *Sapphire Bay Marina*

Sapphire Beach Marina lies just "around the corner" from Red Hook and offers protected slips (67) suitable for a minimal hurricane at best (the man-made breakwater is not adequate for a Category 3 or stronger storm).

When leaving Red Hook clear Red Hook Point and head to a waypoint at 18° 20.05′ N, 64° 50.75′ W; it will place you approximately 0.1 mile ENE of the entrance to *Sapphire Beach Marina* as shown on Chart USVI-10 (see next page, with photo).

From the waypoint head southwest to pick up the markers leading into the marina, favor the point of land on the northern side of the channel as a shallow reef lies to the south of the marked entrance channel. Notify the marina (http://www.sapphirebeachmarina.com/) prior to entering the harbor on VHF ch. 16 or 12 for a slip assignment or directions to the fuel dock (to starboard near the entrance). There are no moorings available and anchoring is not permitted in the basin.

112 • THE CAPTAIN'S GUIDE TO HURRICANE HOLES

Sapphire Bay Marina, St. Thomas

Benner Bay, St. Thomas, as seen from the south

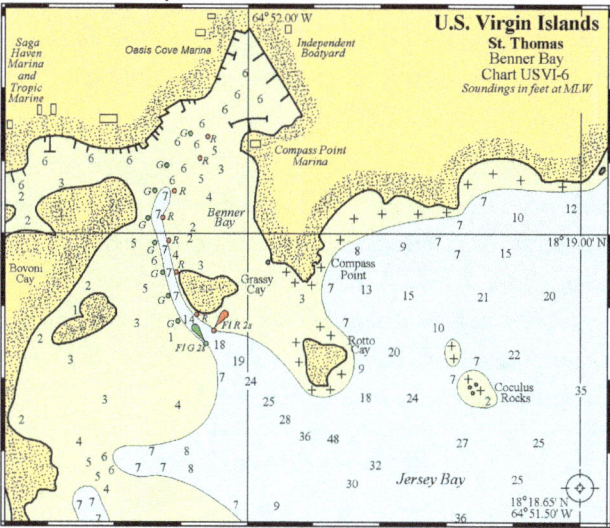

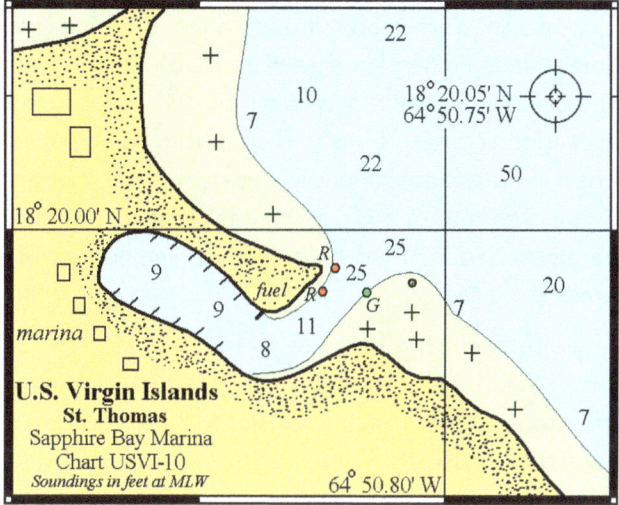

St. Thomas: *Benner Bay*

Besides being home to several marinas and a haul out yard, *Benner Bay* is the entrance to a mangrove-lined lake (sometimes referred to as *The Lagoon*) with good protection. *Benner Bay* and *The Lagoon* (controlling depth 6') have some of the best holding to be offered, but many local yachts and charter companies will be vying for berths.

The Lagoon may be deemed off limits by the DPNR (Department of Natural Resources) due to the unrestrained run-off from the nearby dump. Count on *The Lagoon and Benner Bay* to be full of day charter vessels in the event of a storm.

As shown on Chart USVI 6 (above with photo), a waypoint at 18° 18.65' N, 64° 51.50' W, will place you in *Jersey Bay*, approximately 1 mile ESE of the marked entrance channel into *Benner Bay*. From the waypoint head WNW passing south of Rotto Cay where you should be able to get a visual on the marked entrance channel beginning south of Grassy Cay.

Follow the markers into *Benner Bay* to your selected marina or yard. If you wish to enter *The Lagoon* then follow the creek leading westward from the marinas, keeping just south of the mainland of St. Thomas. Stay in the deeper water closer to shore until you are in the lake. It can be tricky getting in here, but this route can accommodate about 5 1/2' at low water. It will be crowded and tensions high. Lots of local charter cats tie off to the mangroves here with anchors out.

You can get a haul out or slip at *Independent Boatyard* (340-776-0466, http://ibyvi.com/). IBY has 80 slips, a 50-ton *Travelift*, and a 10-ton crane. Next door, *East End Boat Park* offers haul outs and dry storage for vessels under 25; LOA. Across from *IBY* is *Oasis Cove Marina* (340-244-0442, http://www.oasiscove.com/) with slips but no boatyard. Just to the west is *Saga Haven Marina* (http://www.sagahaven.com/) offering 50 slips and a select number of slips for catamarans

St. Thomas: *Charlotte Amalie*

With good protection so close by, I cannot suggest that you try to anchor in the bay at Charlotte Amalie. You can seek a haul out at *Avery's Boathouse* (340-776-0113), or at *Haulover Marine* located nearby at *Subbase* (340-776-2078, http://www.subbasedrydock.com/). Word has it that *Subbase* no longer hauls out for a storm, but you might flash some cash and see if that helps.

If available, slips might be found in Charlotte Amalie at *Avery's Boathouse* (340-776-0113), *Frenchtown Harbour*, and *Yacht Haven Grande*. *Yacht Haven Grande* (340-774-9500) is geared towards the upper end of the boating public with accommodations for vessels to 656' LOA with drafts to 18'. *Yacht Haven Grande* monitors VHF ch. 10 and 16.

St. Thomas: *Brewer's Bay*

West of Charlotte Amalie and just north of the airport runway lies Brewer's Bay and Brewer's Beach as shown on Chart USVI-BB (below). South of the beach and just on the north side of the runway is a small mangrove lined cove south and east of Range Cay. In this protected little hole, a 7' draft boat rode out both Hurricane Irma and Hurricane Maria in 2017.

As shown on the chart, a waypoint at 18° 20.30' N, 64° 59.20' W, will place you approximately ¼ nm NW of the end of the airport runway and west of the cove. To enter the small cove head west to pass south of Range Cay and the airport and tuck into the mangroves to secure your vessel.

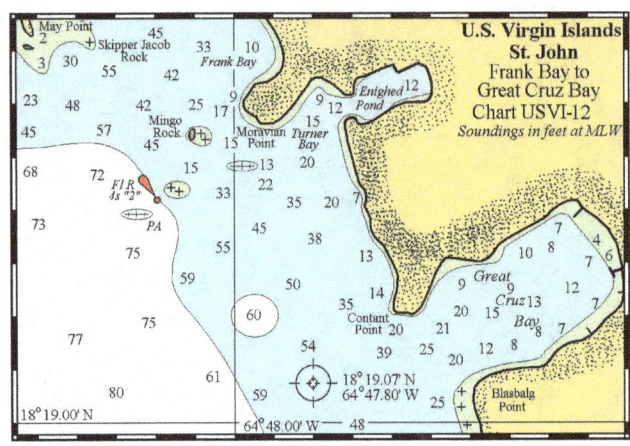

St. John: *Enighed Pond*

On the west side of St. John, just north of *Great Cruz Bay*, is *Turner Bay*, the entrance to a nice little hole called *Enighed Pond* (see Chart USVI-12 above). The pond is a busy place, it is regularly used by the inter-island car ferry from St. Thomas and has a turning basin that can accommodate a vessel with a 12' draft. If you plan to ride out a storm here that you will not be alone and may just be one of the smaller vessels seeking shelter.

As shown on Chart USVI-12, a waypoint at 18°19.07' N, 64° 47.80' W, places you approximately 1/4 nm west of the entrance to *Grand Cruz Bay*. From the waypoint head north into *Turner Bay* and then east into *Enighed Pond* when the entrance opens up to starboard.

St. John's: *Hurricane Hole*

On the eastern end of St. John is *Coral Bay* boasting one of the best hurricane holes in the Caribbean, aptly named *Hurricane Hole*. Here you'll find several deep coves with small indentations that you can head into for protection with your bow safely in the mangroves in places like *Borck Creek*, *Princess Bay*, *Otter Creek*, and *Water Bay* (probably has the best protection). But don't think this place is invulnerable, a lot of boats were damaged and lost here during Hurricane Hugo and Hurricane Marilyn, the bay is open to the south.

Hurricane Hole is closed to vessels except when a storm threatens. Moorings and anchorage spots in *Hurricane Hole* are now assigned by the *Park Service* and must be applied for in advance of the season

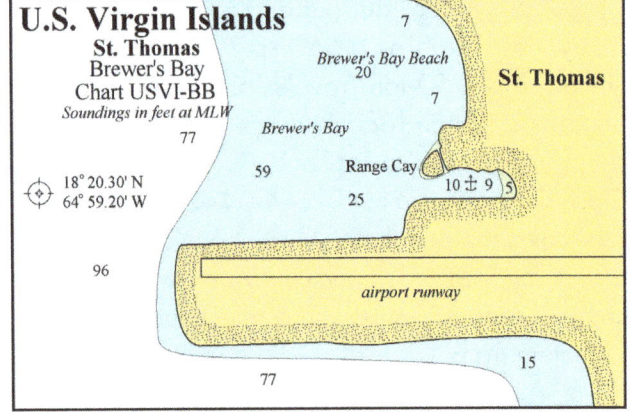

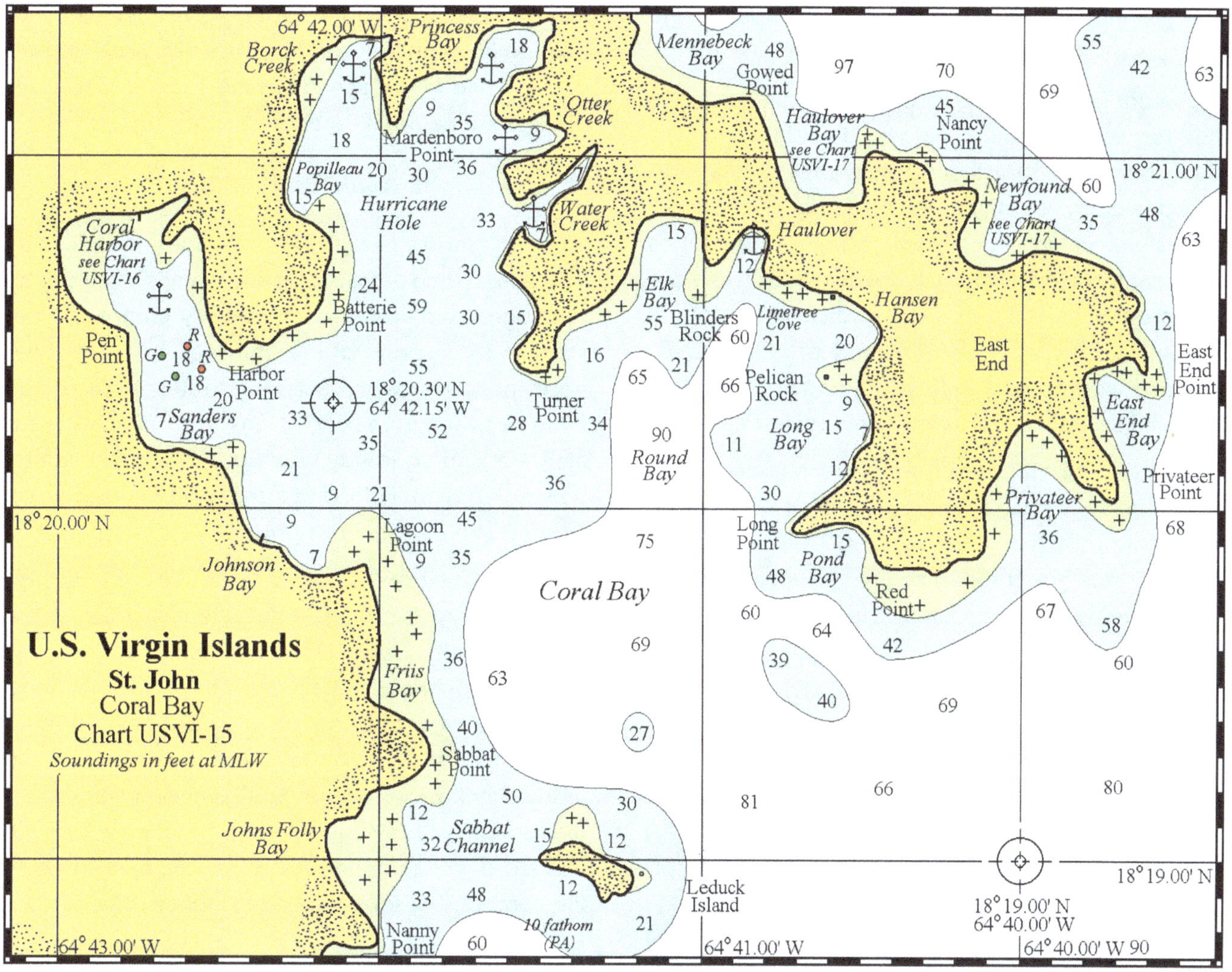

Looking northeast at *Princess Bay*, *Otter Creek*, and *Water Creek* in *Hurricane Hole*, St. John

and you must vacate the area within 48 hours of the storms passing.

The park wants NO damage to the mangroves, so their regulations require that you do not tie up to the mangroves or through them to other trees. It is illegal to tie ropes to any vegetation on park lands. The park will have staff coming through that will remove any ropes or chains fastened to the mangroves! Sand screws are also prohibited. The park suggests that vessels secure themselves fore and aft with several large anchors in an east/west orientation parallel as the winds tend to funnel though the area in those directions.

The coves of *Hurricane Hole* are easy to access. As shown on Chart USVI-15 (see top of previous page with photo), a waypoint at 18° 19.00' N, 64° 40.00' W, will place you approximately 1 mile southeast of *Coral Bay* as shown on Chart USVI-15.

From the waypoint you can head northwest passing between Leduck Island and Long Point to enter *Coral Bay* proper. Proceeding northward keep clear of Lagoon Point on the western shore of *Coral Bay* and you will find *Hurricane Hole* and its protective bays at the extreme northern end of *Coral Bay*.

St. John: *Coral Harbor*

Coral Harbor lies in the NW corner of *Coral Bay* and offers some protection in the event of a hurricane although it would not be a primary choice for me. From the waypoint at the entrance to *Coral Bay*, head northwestward passing south of *Hurricane Hole* between Turner Point and Lagoon Point, and then working your way westward to a spot midway between Harbor Point and *Sanders Bay* as shown on Chart USVI-15 (previous page).

A waypoint at 18° 20.30' N, 64° 42.15' W, will place you approximately ½ mile east of the entrance into *Coral Harbor* as shown on Chart USVI-15 (previous page). From the waypoint head generally west passing midway between Harbor Point and *Sanders Bay* where you will pick up the marked entrance channel as shown on Chart USVI-16 (bottom of previous column). Follow the channel into *Coral Harbor* and secure your vessel and enjoy the company of some good neighbors.

St. Croix: *Christiansted*

Christiansted is not recommended as a hurricane hole if you plan to anchor or get a slip here, but a haul out is possible at *St. Croix Marine* in *Gallows Bay*. St. Croix Marine (340-773-0289, http://www.stcroixmarinecenter.com/) has a boatyard with a 300-ton railway and a 60-ton *Travelift*.

The entrance into Christiansted should never be attempted at night or with strong northerly swells running. There is no perceptible current at *Christiansted Harbor*, but there is a moderate westerly flow, about ½ knot, outside the light at *Fort Louisa Augusta* at the eastern side of the harbor entrance. To the north/northeast of the harbor entrance is *Scotch Bank* leading off to the northeast, give this shoal a wide berth, it often breaks in any kind of swell. *Long Reef* lies across the harbor entrance and the entrance channel into the harbor at Christiansted is at the eastern end of the bay near the point at *Fort Louise Augusta*. As

Coral Harbor, St. John

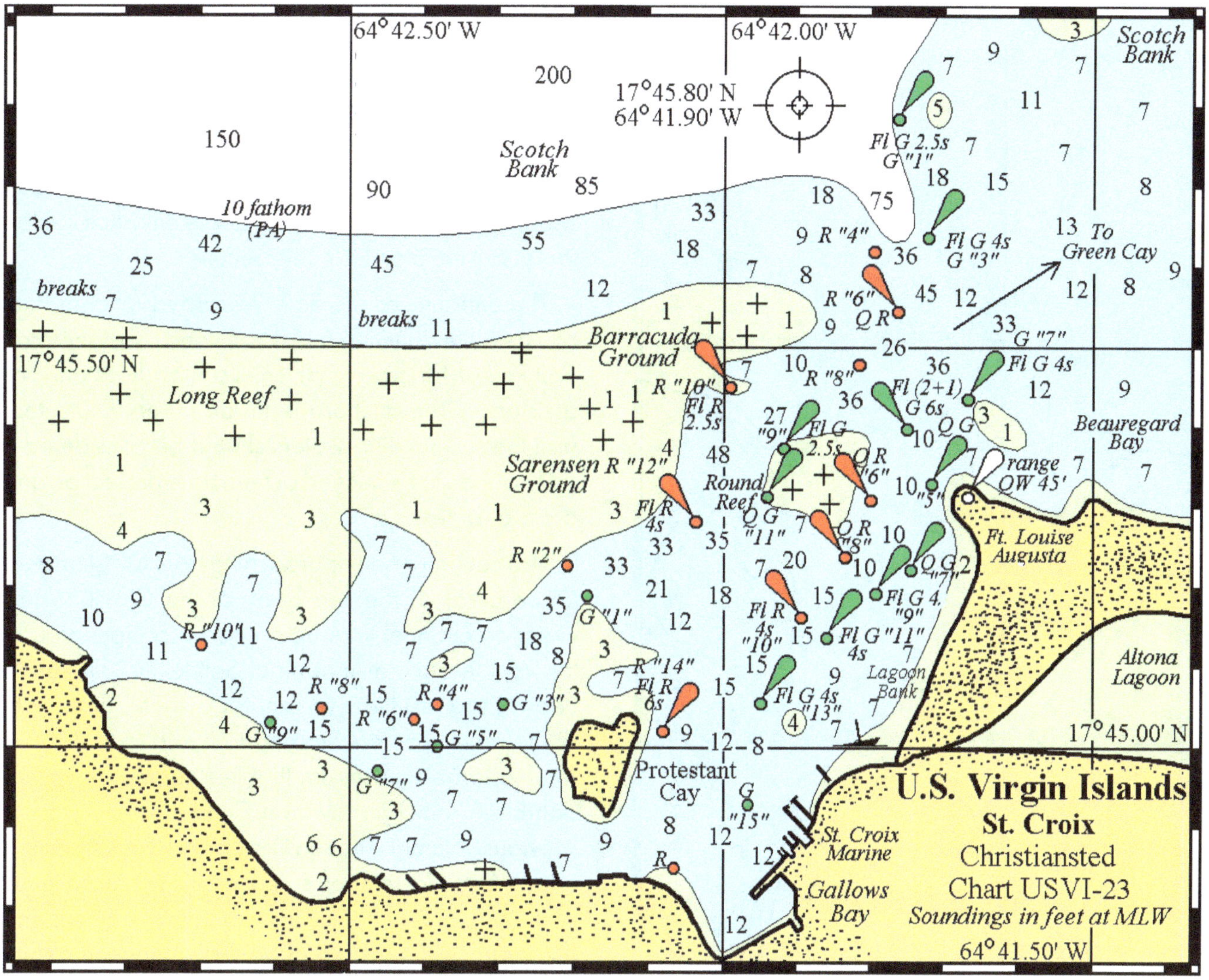

you get closer you'll pick up the entrance range on the point at *Fort Louise Augusta*.

A waypoint at 17° 45.80' N, 64° 41.90' W, will place you approximately .1 mile northwest of the marked entrance channel as shown on Chart USVI-23 (above). From the waypoint pass green buoy "1," the Christiansted sea buoy, and follow the channel southward as it approaches the point at *Fort Louise Augusta*. As you approach the point you'll notice that the marked channel splits at *Round Reef* and yachts drawing more than 10' must take round reef to port and follow the *Schooner Channel* into the harbor. Vessels drawing less than 10' can pass on either side of *Round Reef* though most pass between Round Reef and the point at Fort Louise Augusta. If you're having trouble negotiating the entrance give a hail to *St. Croix Marine* and they will be happy to talk you through it.

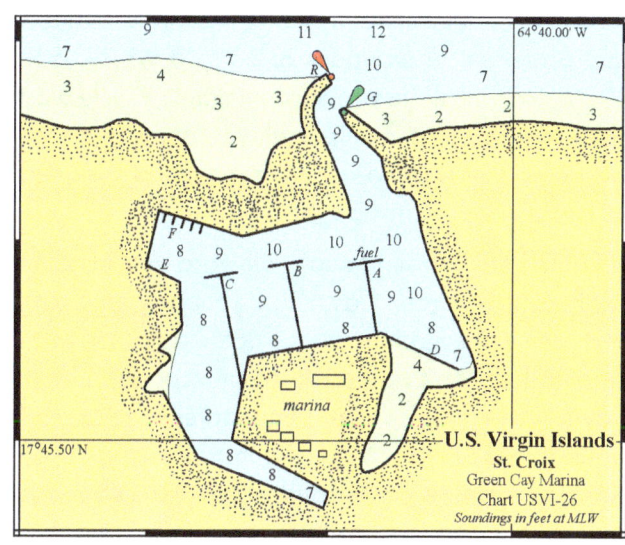

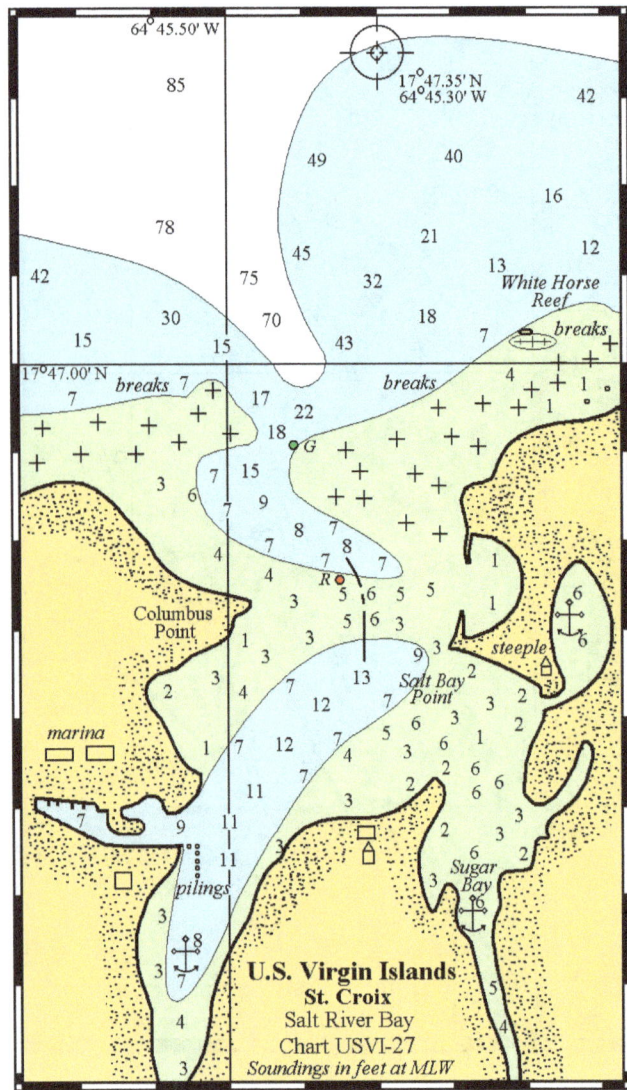

St. Croix: *Green Cay Marina*

Green Cay Marina sits on the northern coast of St. Croix, about 3 miles east of Christiansted, and is very protected. A waypoint at 17° 45.95' N, 64° 40.25' W, will place you approximately ¼ mile west of Green Cay and ¼ mile northwest of the entrance to *Green Cay Marina* as shown on Chart USVI-26 (at bottom of previous page).

From the waypoint head southward until past the southern tip of Green Cay. Work your way eastward a bit to enter the marina from the northeast as shown in greater detail on Chart USVI-26 (next page). *Green Cay Marina* (http://www.tamarindreefresort.com/marina-en.html) has 154 slips with the best protection being in the NW corner and along the SW shore. Beware of a strong surge in the marina.

St. Croix: *Salt River Bay*

Salt River Marina (340-778-9650, http://www.jonesmaritime.com/saltrivermarina.htm), is one of only two places on the island that offer any sort of true hurricane protection, the other being *Green Cay Marina* which we just discussed. The entrance can be tricky, but the marina is very secure.

The entrance to the *Salt River Bay National Park* can be quite daunting and if you need assistance call *Salt River Marina* on VHF ch. 16 for advice and/or assistance. The anchorage inside is very protected, and the surrounding waters should be considered if you happen to be in need of hurricane protection and draw 6' or less.

Use caution when approaching from Christiansted. Once clear of the sea buoy at the Christiansted entrance channel take up a course to place you at the very least, ½ mile north of Salt River Point, and I heartily suggest that you aim for a spot 1 mile north of the point. You are seeking to avoid *White Horse Reef*, a large reef system that lies north and west of Salt River Point as shown on Chart USVI-27 (top of previous column). The sea is usually always breaking over *White Horse Reef*, so the reef is not too difficult to discern, but use the utmost caution when transiting this area. Although you can pass between White Horse Rock and Salt River Point, it's safest to pass north of White Horse Rock to proceed to a waypoint at 17° 47.35' N, 64° 45.30' W, which places you approximately ½ mile north of the entrance channel.

The entrance through the reef as shown on the chart is marked by privately maintained aids that may or may not be there when you need them to be. Never attempt this passage at night or in periods of poor visibility or heavy northerly swells, and at all times keep one eye on your bearings and the water in front of you, and one eye on the depthsounder.

The end of the reef on the eastern side of the entrance is marked by a green stake. Keep the stake to port as you enter and once past the stake and visible reef turn back to port to parallel the southern side of the reef as you put your bow on the conspicuous steeple and follow that heading.

Keep an eye out for the red marker that marks the channel over the shallow sandbar, if you miss it you'll wind up aground on the same sandbar. The channel lies almost midway between Salt Bay Point and Columbus Point. A good range is the small beach house and peaked roof at the southern end of the bay. Once over the sandbar turn to starboard to work your way into the marina and anchorage at the southwestern tip of the bay.

The *Salt River Marina* offers slips, but advance notice is required if you wish to tie up here, space is limited; If you would prefer to anchor instead of getting a slip, the bay offers three great spots with lots of protection if your draft allows entry. The first is in the small bay south of the marina, the second is in *Sugar Bay* at the SE tip of *Salt River Bay*, while the third lies midway along the eastern shore of the bay with a narrow entrance SE of Salt Bay Point as shown on the chart.

Chapter 13

The British Virgin Islands

CHARTER BOATS BLANKET THE BRITISH VIRGIN ISLANDS DURING MOST OF YEAR and these boats will be your primary concern when searching for hurricane protection in the islands as the charter skippers know all the best protection and are wise enough to get there early. Some charter owners will move their fleets to Puerto Rico, on hauling out their boats at *Puerto del Rey Marina* for the hurricane season. The BVIs got hit hard by Hurricane Irma with Tortola and Virgin Gorda suffering serious damage, even Nanny Cay was destroyed, and the new hurricane haul-out facility at Virgin Gorda Yacht Harbour suffering damage. The favorite hole in the BVIs, where all the charter boats would hide, was laid to waste by Hurricane Irma, a surprise to many.

Penn's Landing at *Fat Hog's Bay*, Tortola, has about 30 'hurricane approved' mooring balls that they claim have been approved by major insurers for safe hurricane storage and they can be reserved sometimes as late as April, but they are usually full and so there never seems to be one available on a last-minute basis. Also, there is a small mangrove lined basin at the western end of *South Sound*, Virgin Gorda, with a 6' controlling depth with room for perhaps 5 cooperating vessels but it is usually full of local charter boats. If the Virgins are too crowded you might want to consider heading for Curacao or Venezuela.

Tortola: *Road Town*

As shown on Chart BVI-1 (next column), the inner harbour at Road Town offers good protection and may even have slips available, but with a few hundred charter boats secured here, and depending on the wind direction, the amount of protection could dwindle as the wind builds. You can anchor inside the breakwater between the marinas or try to find a slip. If you expected to haul at *Tortola Yacht*, they are no more, having been taken over by *The Moorings* for *TUI* usage only. Offering good protection though is the *Prospect Reef Resort* (284 494-3773) with slips (if available) for vessels up to 45' LOA with a 5.5' draft. *Road Reef Marina* (284-494-2751, VHF ch. 12) has good protection but is primarily a charter boat

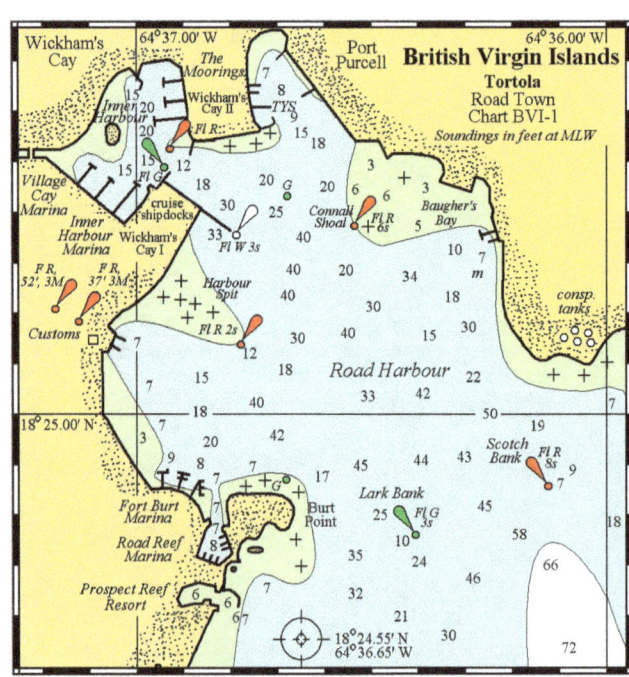

base with few transient slips so don't expect there to be room for you.

Tortola: *Paraquita Bay*

Just to the northeast of Road Town lies the finest hurricane hole in the BVIs, *Paraquita Bay* as shown on Chart BVI-3 (next column with photo). The bay has hundreds of moorings and spots must be reserved for private vessels well in advance at a cost of $850 per year (subject to change). *Paraquita Bay* will be full early so make your plans accordingly.

The mangrove-encircled bay was recently dredged and will allow slightly over 6' at high tide with 8' to 10' depths inside. After the BVI boating community witnessed the devastation of some of the other nearby islands in recent hurricanes, the charter companies got together with the government and arranged for the *Royal Navy* to blast the channel. Then some 800 moorings were placed inside the two bays. The entire bay is full of hurricane moorings and every charter boat in the BVI has a designated spot, so should a storm approach be sure to get in early as there are still some places to anchor behind (north) of the mooring field, and also to the east and southeast. You will be sharing this harbor with hundreds of other boats, most in giant rafts, tied bow and stern.

As shown on Chart BVI-3 (next column with photos), a waypoint at 18° 24.80' N, 64° 33.40' W, will place you approximately ½ mile south of Buck Island. From the waypoint head WNW until you can make out the narrow entrance channel into *Paraquita Bay*. So, if you should choose to enter *Paraquita Bay*, it would be wise to sound your way in by dinghy before entering with the big boat.

Tortola: *Maya Cove*

Maya Cove, sometimes shown as *Hodge's Creek*, lies just northeast of *Paraquita Bay*. The bay offers a bit of protection but is so full of private moorings as to be almost useless to the average cruiser.

A waypoint at 18° 24.80' N, 64° 33.40' W, will place you approximately ½ mile south of Buck Island as shown on Chart BVI-3 (next column with photo). From the waypoint head generally NNW towards the marked entrance channel into *Maya Cove*. As you proceed towards the outer markers at the mouth of the

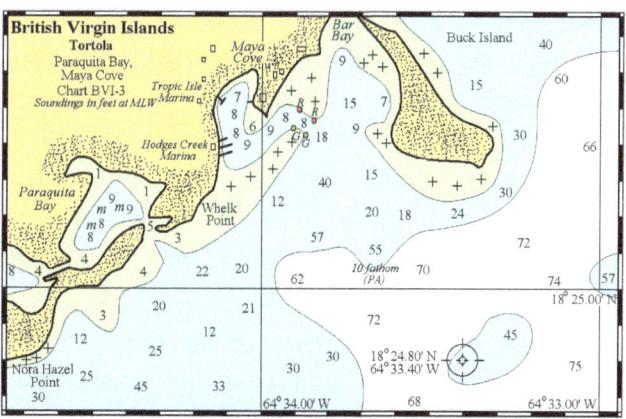

The entrance to *Paraquita Bay*, Tortola

Southwestern portion of *Paraquita Bay*, Tortola

Maya Cove, Tortola

THE BRITISH VIRGIN ISLANDS • 121

entrance channel (8' at low water) pass between them and once past the second set of markers (there were two sets at the time of this writing and of course this may change) you will need to turn to port as the channel doglegs towards the southwest and you parallel the reef to port as you enter *Maya Cove*. Never attempt to enter *Maya Cove* by passing between the green outer marker and Whelk Point.

If you wish to get a slip here, on the west side of the bay is *Hodges Creek Marina* and they monitor VHF channels 12 and 16. The marina has 200 slips and can accommodate vessels of 120' LOA with drafts to 8.5'.

In the harbor there are several moorings that are maintained by *Tropic Island Yacht Management*, and may or may not be available, check with *TIYM*, *Catamaran Company*, and *Marine Max Charters*, between the three of these companies there is usually no room for transient vessels.

Tortola: *Trellis Bay*

Trellis Bay, on the northern shore of *Beef Island*, offers a good bit of protection if the wind is forecast to come from the south, you might be able to ride out up to about a Category 3 or 4 in those conditions, but if the wind is from the north, better find protection elsewhere if the forecast is for more than a Category 1, remember, you'll be sharing this bay with a lot of local boats if you can find a spot at all (the bay is filled with private and commercial moorings).

As you approach *Trellis Bay* keep clear of the reefs off Conch Shell Point and Sprat Point as shown on Chart BVI-5 (top of next page) and in greater detail on Chart BVI-7 (top of next column with photo). Avoid the area off the end of the airstrip, it is marked by yellow buoys, no anchoring is permitted here and vessels over 8' in height are prohibited from passing through the area bounded by the yellow buoys. Vessels with masts over 30' (10 meters) must call the Beef Island airport control tower on VHF ch. 10 for permission to transit the area between Bellamy Cay and the airport runway.

Don't anchor within 200' of the western shore of the bay to keep a clear channel open for the ferry docks there. You must also watch out for the shoal that surrounds Bellamy Cay and extends a bit south of the island. The depths in *Trellis Bay* have changed

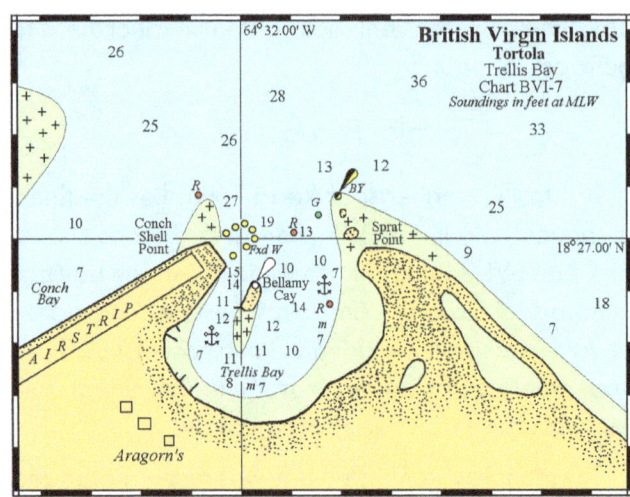

Trellis Bay as seen from the northwest, Tortola

a bit since the earthquake of 2008, it is as much as 1'-2' deeper in the SE portion of the bay.

West of *Trellis Bay*, just west of *Long Bay* is, the marked waterway separating Beef Island from the mainland of Tortola (see Chart BVI-5, top of next page) called the *Beef Island Channel*. The passage is narrow and shallow and only suitable for small, shallow draft vessels seeking shelter between the northern opening and the bridge. To the northeast of the bridge is a large mangrove encircled cove that offers good protection, but it is for shallow draft vessels only (less than a 4' draft).

Tortola: *Hannah Bay*

As shown on Chart BVI-12 (next page), a protected little cove called *Hannah Bay*, lies SW of Road Town, Tortola. *Hannah Bay* (and the off-lying Nanny Cay) is home to two marinas, *Hannah Bay Marina* and *Nanny Cay Marina*, and one haul out yard located at *Nanny Cay Marina* (http://nannycay.com/). Note that under normal conditions anchoring is not permitted in *Hannah Bay*. Both marinas monitor VHF ch. 12 and 16.

A waypoint at 18° 23.60' N, 64° 38.00' W, will place you approximately ¼ mile southeast of the

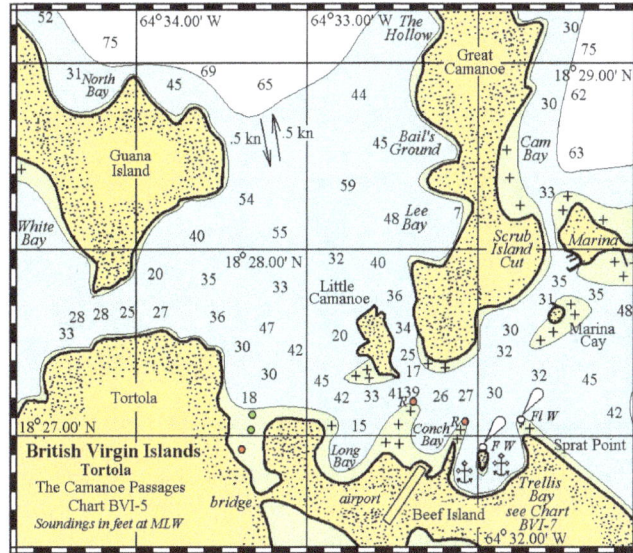

entrance to *Hannah Bay* (not shown on Chart BVI-12, next page). From the waypoint head towards the well-marked entrance channel as shown on Chart BVI-12. Follow the markers into *Hannah Bay* keeping well to the east of the green buoys, do not go between the green buoys and the shore as the water shoals there. The small *Hannah Bay Marina* is located in the northwest portion of *Hannah Bay*, and has 30 slips that can accommodate vessels to 160' LOA with drafts to 10'. Don't tie up in the outer slips at *Hannah Bay* or *Nanny Cay* if the wind is forecast to go south.

On the eastern shore of *Hannah Bay* lies Nanny Cay and the *Nanny Cay Marina* (http://nannycay.com/). The marina has 180 slips and can accommodate vessels to 160' with a maximum beam of 45'. The marina's boatyard, the *Nanny Cay Marine Centre*, has a 50-ton and 200-ton hoist with dry storage for 150 boats. For hurricane protection the boatyard builds steel cradles under each vessel and the boat is strapped down to 7' long earth screws but note that they are usually full by March unless you have a reservation.

Tortola: *Sea Cow Bay*

Just north and east of Nanny Cay is *Sea Cow Bay* as shown on Chart BVI-12 (top of next page with photo). The majority of this bay is very shallow, but a good portion has been dredged and is the home of the *Manuel Reef Marina* (http://www.manuel-reef-marina.com/, monitors VHF ch. 16). With 40 slips available (fixed concrete docks, see photo top of next page), the marina can handle vessels to 150'

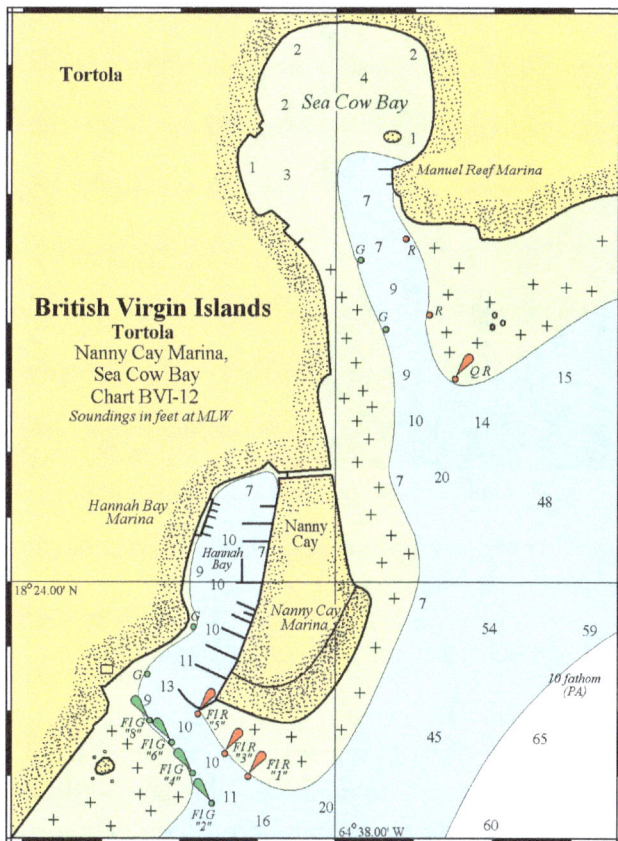

Sea Cow Bay, Tortola

LOA with drafts to 6.5' with the inner section being the most protected.

The best protection in *Sea Cow Bay* is to be found in the mangrove encircled eastern portion of the bay, especially for shoal draft vessels such as catamarans. I believe that the marina could suffer significant damage in a strong storm. A waypoint at 18° 23.60' N, 64°38.00' W, will place you approximately 1 mile south of the entrance to *Sea Cow Bay* (the waypoint is not shown on Chart BVI-12 above). From the waypoint head generally NNE until you can pick up the well-marked entrance channel as shown on the chart. Follow the markers into *Sea Cow Bay* and if

Manuel Reef Marina, Sea Cow Bay, Tortola

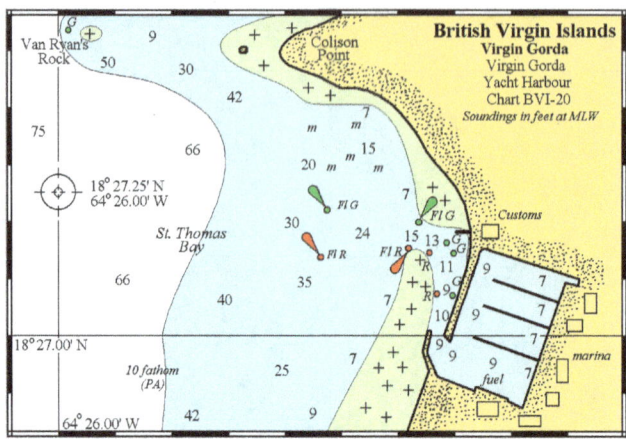

you do not get a slip, feel your way into the protection of the eastern side to anchor.

Virgin Gorda: *Virgin Gorda Yacht Harbour*

The small basin at the *Virgin Gorda Yacht Harbour*, although protected from seas (but not storm surge), is not the place I would care to ride out a hurricane at the dock, but a haul out here is certainly to be considered. The docks are fixed and the basin can get very surgy in moderate west-north/northwest winds, but the *Virgin Gorda Yacht Services* yard has a 70-ton *Travelift* and can haul and store multihulls up to 40' in length with their *Conolift* trailer. VGYS has over 40 keel pits with depths to 5' (primarily for vessels over 45' LOA). VGYS has had no damage to vessels using this system since they first employed it in 1982. In 2017, the yard introduced their new 350-ton lift and 5 new acres of boatyard!

As shown on Chart BVI-20 (top right column), a waypoint at 18° 27.25' N, 64° 26.00' W, will place you approximately ½ mile west of the marked entrance channel leading in to *Virgin Gorda Yacht Harbour* (their phone number is 284-495-5500, or you can visit their website at http://www.virgingordayachtharbour.com).

From the waypoint head east and you will pick up the marked channel that bends to starboard to access the marina entrance. On the southern side of the harbour near the fuel dock is the *Virgin Gorda Yacht Services* haul out yard and dry-storage facility.

Virgin Gorda: *North Sound*

Many tout *North Sound*, Virgin Gorda, as some of the best hurricane protection in the BVI. It is a huge, virtually landlocked bay that can accommodate a great number of boats in areas where the bottom consists of good holding mud/sand. But I find that *North Sound* is too open with too much fetch and not suitable for my list of hurricane holes, but, like so many in this book, it will do in a pinch, especially *Biras Creek* as long as the wind does not come from the W or *Gun Creek* if the wind does not go N or NE (see Chart BVI-23 below).

North Sound, Virgin Gorda, Bitter End anchorage in the background, Biras Creek in the foreground

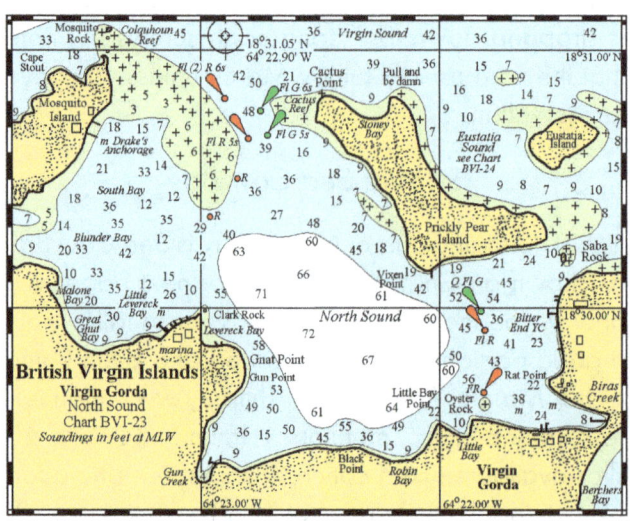

Chapter 14

The Leeward Islands

THE LEEWARD ISLANDS STRETCH FROM ANGUILLA IN THE NORTH TO Dominica in the south, and you can quickly abandon any thoughts of finding protection on Anguilla, Barbuda, Saba, St. Eustasius, Nevis, Montserrat, Barbuda, Les Saintes, Marie Galante, and Dominica. If you are cruising any of these islands you should seek shelter elsewhere. Fortunately, the Leeward Islands offer some very nice hurricane protection on St. Martin, Antigua, and Guadeloupe, and in a boatyard and new marina on St. Kitts. The Leewards were hit very hard by Hurricane Irma. Barbuda, St. Martin, St. Barth's, Dominica, and Anguilla all were leveled, and Antigua took a great amount of destruction. As of this writing Barbuda has not been repopulated and her future is uncertain. All others are rebuilding. Marinas in Simpson Bay, St. Martin met with a great amount of destruction, but they will rebuild and reopen, it will only take time.

St. Martin: *Anse Marcel, Radisson Marina*

The marina at the southeastern tip of *Anse Marcel* is very protected, but when northerly swells are running some roll works its way into the anchorage so use caution if tying up here and experiencing northerly winds and seas.

A waypoint at 18° 07.60' N, 63° 02.50' W, not shown on Chart STM-2A at the top of the next page, places you approximately ¼ mile north of the entrance into *Anse Marcel*. Enter the bay giving a wide berth to the shoal to your starboard side between the mainland and Rocher de l'Anse Marcel as you head southward.

Steer for the extreme southeastern part of *Anse Marcel* where you will pick up the marked channel leading to *Radisson Marina*. The short, narrow channel is well marked and can carry 9' at low water even though it is regularly dredged to 10'. There is only room for one boat at a time in the channel, but there is an area that is slightly wider about halfway down the channel and there is room here for most boats to pass. Before entering the channel, you should call the marina on VHF ch. 16 so they will be aware of your arrival, and then sound your horn before heading south in the channel. The *Radisson Blu Marina* offers 145 slips for vessels to 90' LOA.

St. Martin: *Oyster Pond*

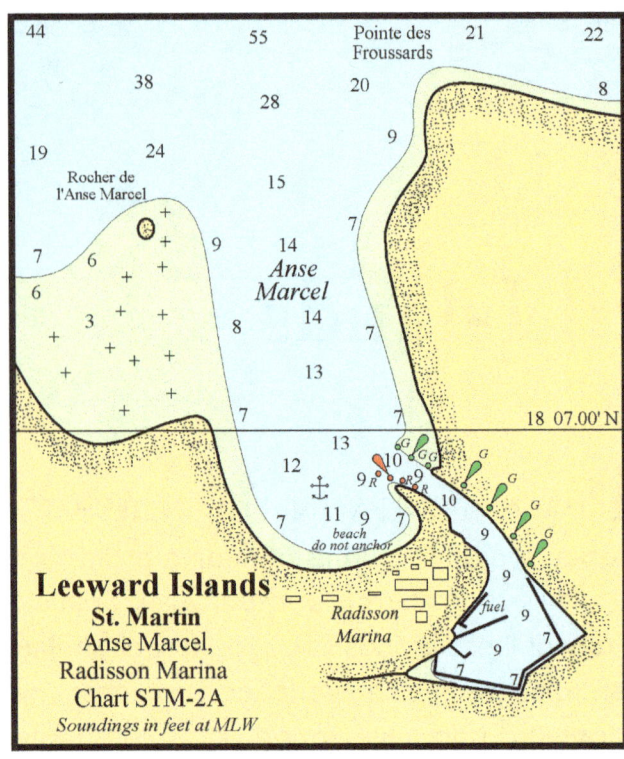

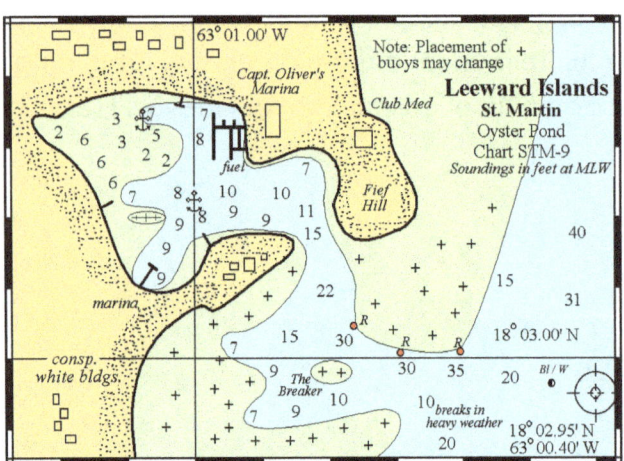

On the eastern shore of St. Martin/Sint Maarten, one of the best and most protected anchorages is located at *Oyster Pond*. Inside you'll find two marinas as well as room to anchor.

The entrance to *Oyster Pond* can be difficult and dangerous in the wrong conditions (you must wind your way through two reef systems and two smaller patch reefs) although it is well-marked as of this writing. Bear in mind that the entrance channel buoys are privately maintained and their configuration may change. Following seas add to the excitement of an entry and this passage should never be attempted at night or in heavy following seas when the entire entrance breaks.

From offshore you can work your way to a waypoint at 18° 02.95' N, 63° 00.40' W, which places you approximately ¼ mile east of the marked entrance channel as shown on Chart STM-9 (left). From the waypoint you should be able to pick up the outer blue/white buoy before sighting the first pair of red markers that mark the eastern end of the channel. Bear in mind that this outer buoy may not be there when you are, the buoy configuration here is subject to change.

On the northern side of the entrance channel are three red markers that border the southern edge of the reef off the point at Fief Hill. Keep these red marks to starboard as you enter favoring the northern side of the channel, but don't cut them too close. Keep north of the breaking area of shallow water and the reef called *The Breaker*. Once past the last red marker turn to the north/northwest to enter the harbor. There is a large shoal in the center of the harbor, but you can pass it safely by favoring the marina as you head northward. There are lots of private moorings in the harbor so finding a place to drop the hook can be a challenge.

On the eastern shore lies *Captain Oliver's Marina* (VHF ch. 16 and 67) with 160 slips (many of which are occupied by charter boats). On the southern shore is the *Great House Marina* (with 14 slips (up to 100' LOA and a 6.5' draft) available. This is a very popular hurricane hole for boaters on St. Martin.

Oyster Pond, St. Martin/Sint Maarten

St. Martin: *Simpson Bay Lagoon*

Much has been said about the protection, or the lack of it, offered by *Simpson Bay Lagoon* in St. Martin. If you're thinking of using *Simpson Bay Lagoon* for hurricane protection (see Chart STM-7 below), bear in mind that of some 1,500 boats that sought shelter here from Hurricane Luis in 1995, approximately 1,300 were lost. Still, if you can find a good spot away from other boats, you have a fair chance of survival here (especially if you haul out or get a protected slip). It seems that *Simpson Bay Lagoon* will always be a favorite for those seeking hurricane protection.

During Hurricane Luis, the *Princess Juliana Airport* recorded 87 mph sustained winds with maximum gusts of 114 mph. The Meteorological office at the *Grand Case Airport* measured a wind gust of 202 mph, while an unofficial anemometer in Marigot recorded wind gusts of 126 mph. Tropical storm-force winds brushed St. Martin for approximately 21 to 24 hours, while hurricane-force winds blew for 10 hours as the storm passed with minimal forward speed of 7-11 mph.

Two areas that offer a bit more protection for those with shallower drafts, are in *Cole Baii* (see Chart STM-6 below), home to some haul out yards as well as marinas, and Marigot (see Chart STM-4 next page), which is also home to several fine haul-out yards.

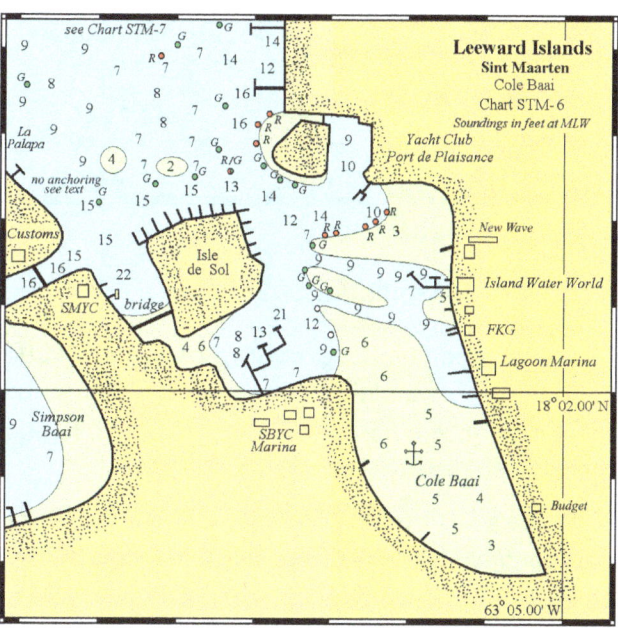

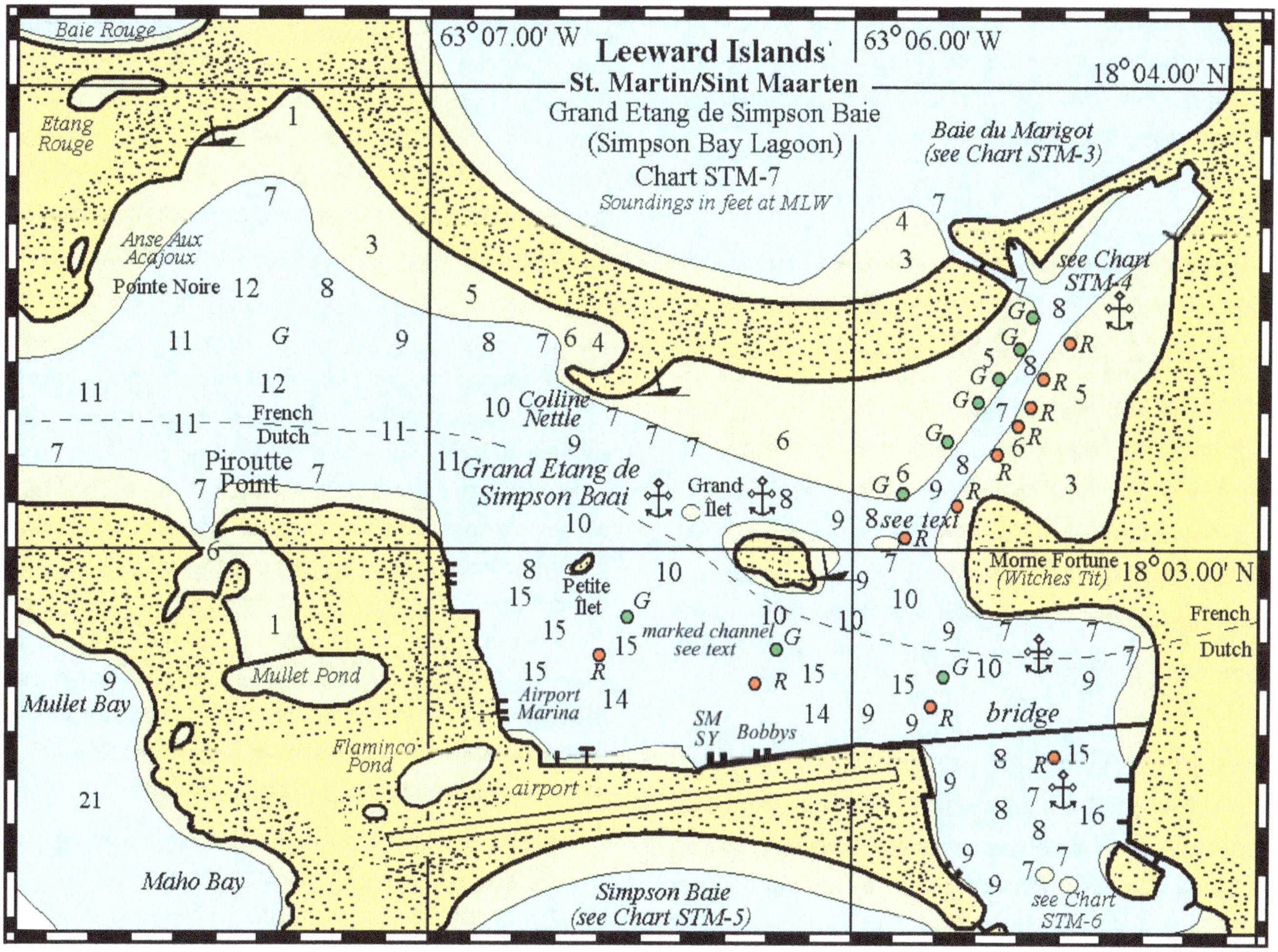

THE LEEWARD ISLANDS • 127

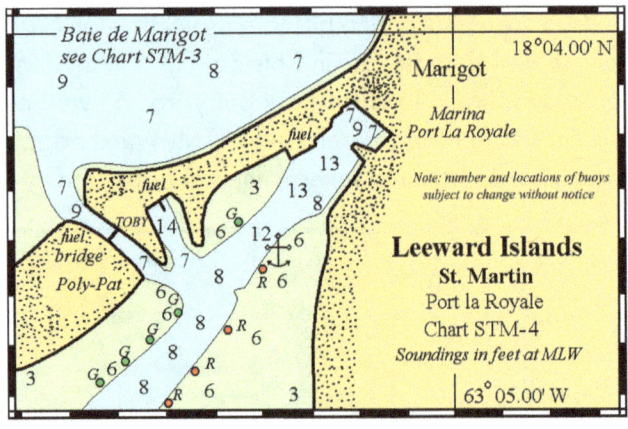
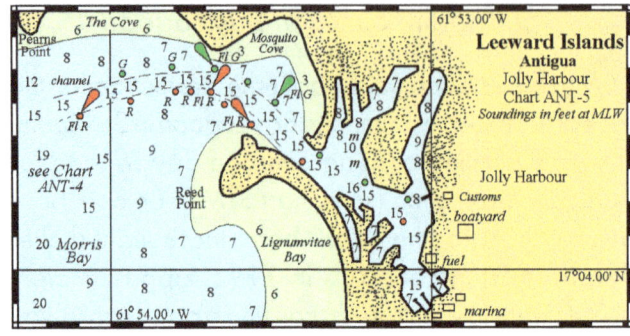

Antigua: St. John's

Antigua is a very busy boating center with LOTS of boats but, fortunately for those seeking shelter, lots of room to anchor along with plenty of places to haul out or get a slip.

St. John's Harbour on the west shore, although commercially used by cruise ships, has a small 2-3 boat anchorage located just past the cruise docks at the extreme southeastern tip of the harbour. This is not a recommended hole as any northeasterly seas will turn the anchorage into a washing machine making holding untenable. It would be best to go south a few miles to Jolly Harbour.

Antigua: Jolly Harbour

Jolly Harbour is a well-protected man-made canal system with a marina and boatyard that is very experienced in hauling boats for hurricanes (with keel holes and tie downs). So good is the protection here that over the last few years the area has only suffered minor damage in hurricanes. The area is well-protected from seas but not from surge.

A waypoint at 17° 04.60' N, 61° 55.40' W, will place you at the western edge of Morris Bay, approximately 1 ¼ miles west of the marked entrance channel to Jolly Harbour as shown on Chart ANT-5 (top of next column).

From the waypoint head generally east keeping south of the Five Islands until you pick up the outer marker just south of Pearns Point. The channel markers may change a bit between the time of this writing and the time of your arrival so keep that in mind, either way, it will be red, right, returning as the dredged channel (15' depth, 75' wide) enters the Jolly Harbour complex as shown.

If you're entering Jolly Harbour to tie up at the marina or haul out, follow the channel as it doglegs past Mosquito Cove into the harbour complex. There are several white mooring balls as you pass the first canal to port, but I would not trust a mooring in a hurricane situation. Anchoring is not permitted in the Jolly Harbour complex except in an emergency such as a hurricane. The yard suffered minimal damage from Hurricane Irma in 2017.

If you plan to take a slip in the marina, go past the Customs dock and the boatyard where you'll find the marina waiting for you. The Jolly Harbour Marina and Boatyard (http://www.jolly-harbour-marina.com/) is a full-service complex offering 152 slips that can accommodate vessels up to 200' LOA. The yard has a 70-ton lift and hurricane storage for 200 vessels with a few keel holes for deep draft vessels.

Antigua: Falmouth Harbour

Falmouth Harbour is a good anchorage in nearly all conditions, but the harbour is large and has a bit of fetch so it should not be considered as a hurricane hole although there are slips to be had there. The focus of activity in Falmouth Harbour is centered around three marinas, the Antigua Yacht Club Marina and the Falmouth Harbour Marina which are located on the eastern shore of the harbour, and the Catamaran Marina situated on the northern shore. For those reasons we are including Falmouth Harbour in this guide but will not suggest anchoring here, better to head to English Harbour or Indian Creek.

As shown on Chart ANT-7 (above), a waypoint at 17° 00.15' N, 61° 46.95' W, will place you approximately ¼ mile south of the entrance to

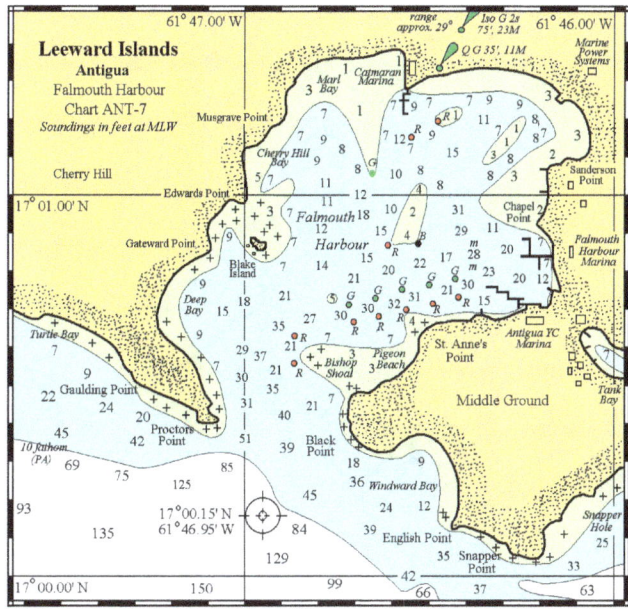

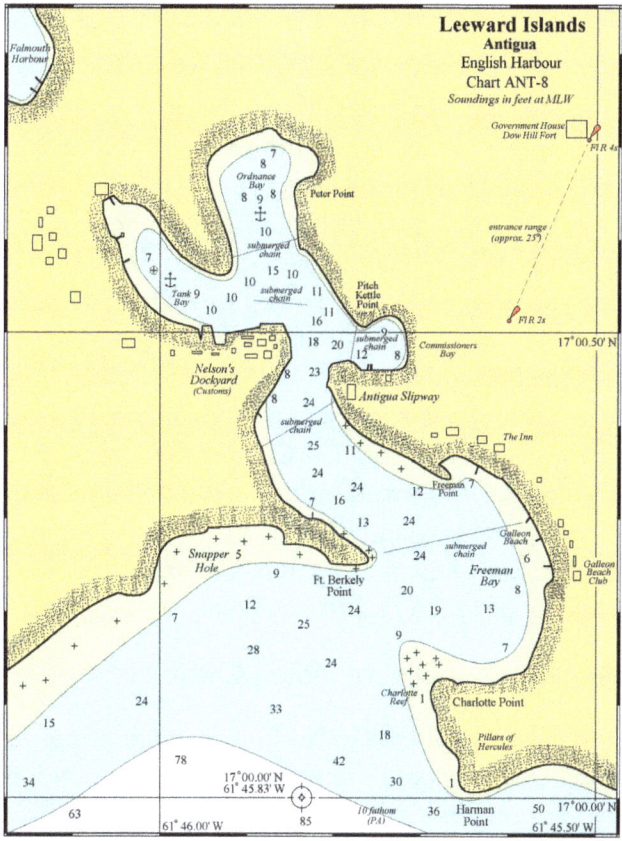

Falmouth Harbour. From the waypoint you can head north keeping an eye out to starboard to clear *Bishop Shoal*, which is usually marked with a large buoy. However, as is often the case, the buoy is often missing just when you need it the most. But don't panic if the buoy is not there, the shoal is usually easy to see except on the calmest of days when it does not break.

If you're headed to *Catamaran Marina*, keep the red buoy to starboard and then pick up the range on the northern shore of the harbour. If you're headed instead to the *Antigua Yacht Club and Marina*, pass *Bishop Shoal* and turn to starboard where you'll pick up a line of red and green buoys to guide you to the marina.

Antigua: *English Harbour*

English Harbour is probably everyone's first choice for hurricane shelter on the southern shore of Antigua. *English Harbour*, quite a bit smaller than *Falmouth Harbour*, narrow and not as open, is surrounded by high hills offering excellent protection with several places to haul out or get a slip if that is your choice.

As shown on Chart ANT-8 (top of next column), a waypoint at 17° 00.00' N, 61° 45.83' W, will place you approximately ½ mile southwest of the entrance into *English Harbour*. From the waypoint head northward toward Ft. Berkeley Point until you clear the reef lying north of Point Charlotte at which point you can head east into *Freeman Bay*. There is a makeshift range if you need one, and a good landmark is the inn on the hill behind Freeman Point as shown on the chart. Line up this inn with the largest and westernmost beach house on *Galleon Beach* and you can enter on an approximate heading of 039° magnetic.

A turn to port and you will enter *English Harbour* proper where you may anchor off the western edge of the entrance channel south of *Nelson's Dockyard*, in *Tank Bay*, or in *Ordnance Bay*. If you anchor near *Antigua Slipway* anchor as close as possible to the

Antigua Slipway, *English Harbour*

THE LEEWARD ISLANDS • 129

slipway or the mangroves as there is a trough here and holding is poor there in rough weather.

There are four hurricane chains in the harbour that were placed by the British Navy during the development of *Nelson's Dockyard* to give incoming vessels a mooring to catch and stop their boat. There is one between *Antigua Slipway* and the shore to the west about 160 yards away where it leads to an anchor on the beach. Another chain leads from the *Antigua Slipway's* mangroves northward for 80 yards to a clearing in the mangroves. A third chain located in *Tank Bay* runs from the *Clarence House* jetty to the *Powder Magazine's* dock and has large boat moorings identifying its location. The fourth chain is in *Freeman's Bay*, running from the large anchor on *Galleon Beach* to *Ft. Berkeley Point*.

Antigua: *Indian Creek*

A mile east of *English Harbour* is *Indian Creek*, a small and very well-protected harbour surrounded by land on three sides. From *English Harbour* head east keeping about ½ mile offshore until the entrance to *Indian Creek* opens up as shown on Chart ANT-9 (top of next column).

The main hazard to navigation is Sunken Rock just south of the entrance to *Indian Creek*. Sunken Rock has less than 6' of water over it, lies about 100 yards off Indian Point, and can be very difficult to see if the seas are calm (they usually break over the rock), never attempt to enter *Indian Creek* at night. You can pass

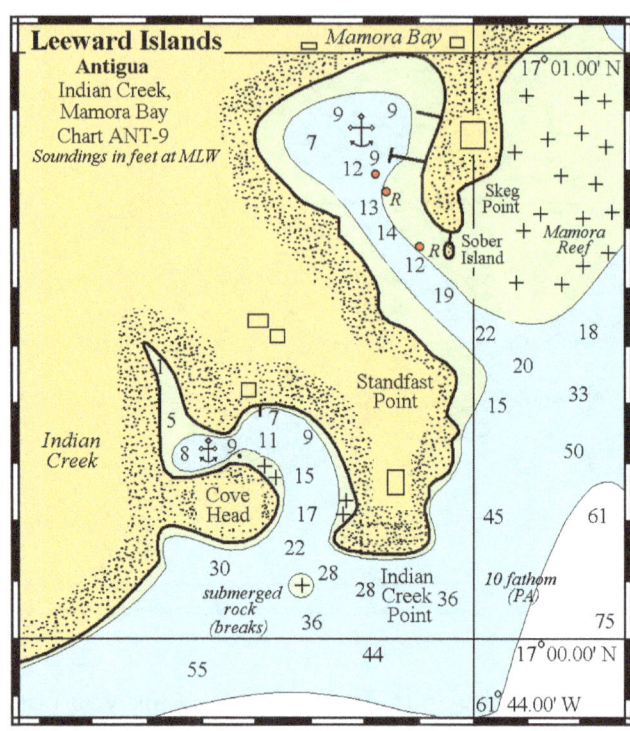

on either side of Sunken Rock, but there is more room on its leeward or western side.

Once past Sunken Rock stay in mid-channel as you enter the creek. Very quickly you will turn to port to anchor in the small cove in 7'-9' of water in the lee of Cove Head, and if you draw less than 5' you will be able to work your way further up into the creek. Make sure your anchors are set in the rocky bottom and tie some lines to the mangroves.

Antigua: *Mamora Bay*

Mamora Bay lies about ½ mile east of *Indian Creek* and is home to the upscale *St. James's Club Marina*.

Indian Creek, Mamora Bay in background

While this bay and marina may offer some small bit of protection, were the wind to come from the south it could be a death trap (it would mean that the eye is moving south to east of the island so check your forecasts).

From *Indian Creek*, pass south of Indian Creek Point before heading northeast where you will pick up the well-marked entrance channel leading into *Mamora Bay* as shown on Chart ANT-9 (previous page). The buoys marking the channel are maintained by the *St. James's Club* and may not be there when you are, so proceed cautiously (and don't' confuse *Mamora Bay* with *Willoughby Bay* which lies a mile eastward and has a reef almost all the way across the entrance to the bay). Proceed down the channel as it winds northwest past Sober Island into *Mamora Bay* where you can tie up at the marina or anchor off the docks in a sand/mud bottom that offers good holding. The exclusive *St. James's Club Marina* monitors VHF ch. 68 and ch. 11 and has room for 15 or more mega-yachts.

Antigua: *Ayres Creek*

Ayres Creek is as close to returning-to-nature as you'll find on Antigua. Here you'll feel as if you are up some tropical river with lovely trees growing right down to the water's edge and a colony of West Indian whistling ducks that live in the area around the creek. The creek also offers some of the best hurricane protection on the eastern tip of Antigua at *Nonsuch Bay* although it is slightly exposed to the northeast.

As shown on Chart ANT-11 (see below), the entrance to mangrove fringed *Ayres Creek* lies at the extreme western edge of *Nonsuch Bay*. A waypoint at 17° 03.40' N, 61° 39.45' W, places you approximately ¾ mile southeast of the entrance channel to *Nonsuch Bay* between Green Island and the mainland of Antigua. Never attempt this entrance at night, you will need daylight to avoid the patch reefs.

As shown on the chart work your way between Green Island and the mainland (do not pass too close to Neck O' Land and the conspicuous wreck) until you reach the area of *West Bay* where you can pass *Middle Reef* on either side. At this point you can point your bow west until the opening for *Ayres Creek* appears before you. Enter the creek and anchor as far up the creek as your draft allows availing yourself of the mangroves if you'd like.

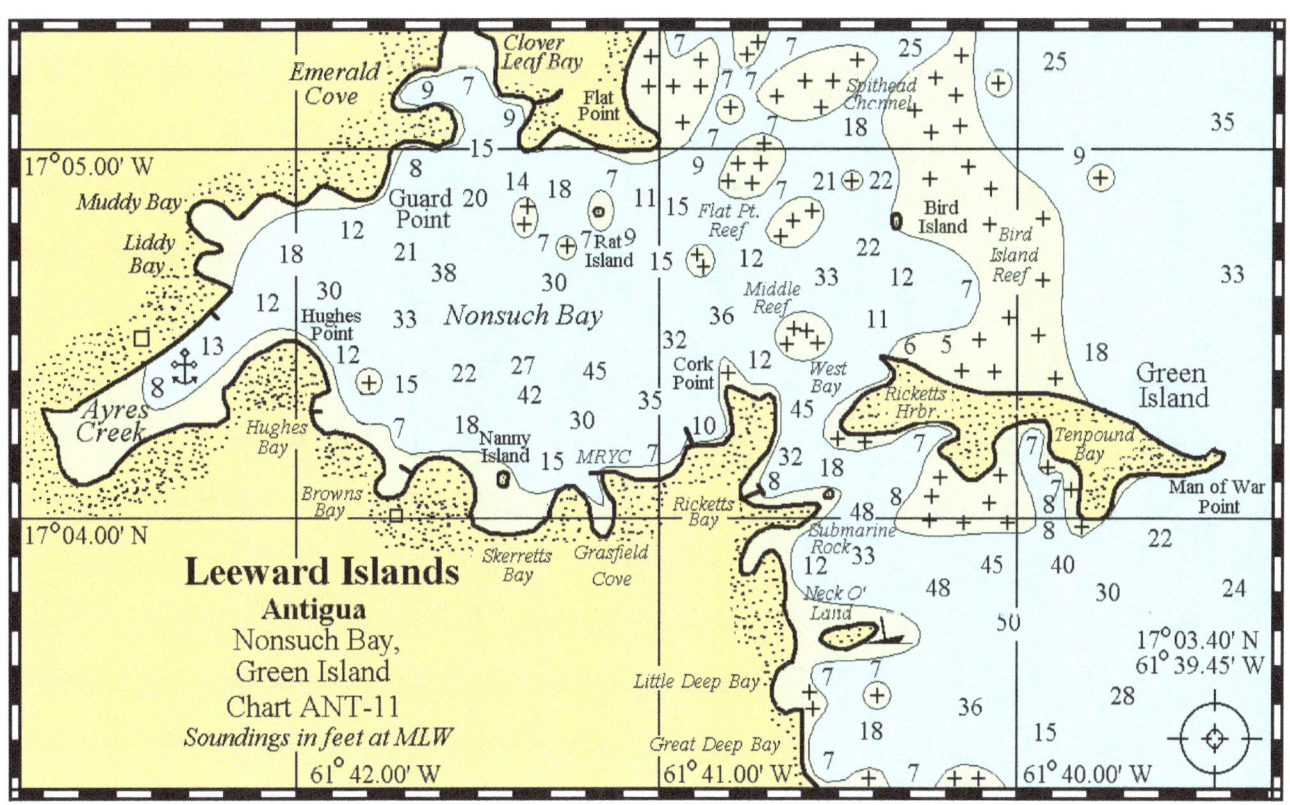

Antigua: *Parham Harbour*

At the northern end of Antigua, many folks like *Parham Harbour*, but it is a bit open. Deep-water seas could not work their way in, but the area has a long fetch that would permit seas to build up inside the protected harbor. However, there is one small spot to find some shelter, as long as the wind does not come out of the west (meaning the eye of the hurricane is north of you), and two boatyards if you require a haul out.

On the southwestern shore of Crabbs Peninsula (see Chart ANT-13 below), is a small anchorage under the lee of Old Fort Point. Unfortunately, this spot is open to the west so take that into account before securing your vessel here. There are several small well protected creeks in this area, one allowing 5.5' draft in the southeast corner of the mangroves. Set bow and stern anchors and tie at least two extra lines into the mangroves. There will be other boats using these creeks so arrive EARLY.

On the northern shore of Crabbs Peninsula is the *North Sound Marina and Boatyard*. The yard has secure stands with fixed tie-downs, a 150-ton lift with a 32' width and a 40' lift height, as well as an 18-ton mobile crane and a sea wall with several hundred feet of stern-to or alongside docking which is not recommended for a hurricane here although the boats in the yard suffered no damage when the eye of Hurricane Irma crossed Barbuda 25 miles to the north in 2017.

Northwest of Barnacle Point (see Chart ANT-13), is *Shell Beach Marina* that can accommodate vessels with drafts to 10' and a 35-ton lift.

St. Kitts: *St. Kitts Marine Works*

If you are near St. Kitts you have three choices for hurricane protection. You can get hauled out at *St. Kitts Marine Works*, get a slip at *Christopher Harbour Marina*, or go elsewhere for protection, to Antigua or Guadeloupe perhaps. We cannot recommend *Port Zante Marina* in Basseterre for hurricane protection. Under normal conditions the marina can be home to a lot of surge, you would not want to be there during a hurricane.

Just south of the NW tip of St. Kitts, south of Sandy Point and *Brimstone Hill Fortress*, is an area called New Guinea and a small village called Half Way Tree. You probably won't recognize the village from offshore, but you will see the jetty that marks Regiwell "Reg" Francis' *St. Kitts Marine Works*, a yard that can accommodate vessels fleeing hurricanes.

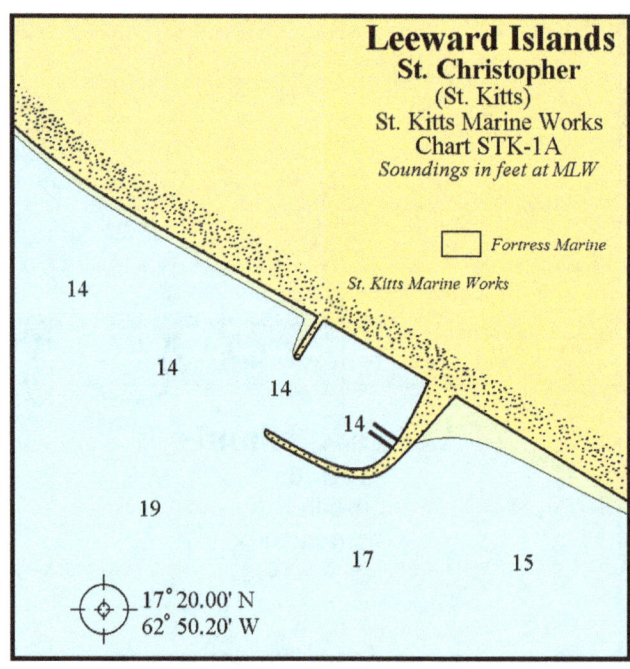

As shown on Chart STK-1A (below), a waypoint at 17° 20.00' N, 62°50.20' W, will place you approximately ¼ mile south/southwest of the jetty that protects the boatyard complex. From the waypoint pass around the end of the jetty to enter the basin of *Telca Marina*.

St. Kitts Marine Works (869-662-8930; http://www.skmw.net/) has a 164-ton lift, and with a depth of 14' at the haul out slip, they are able to haul large vessels to 120' in length. The yard can also accommodate catamarans with beams to 35'. The staff has built keel holes and can tie down vessels for hurricane season as well as removing your mast with an 85-ton crane.

St. Kitts: *Christophe Harbour Marina*

In *Ballast Bay*, as shown on Chart STK-3 (below), you will find the entrance to the huge superyacht marina, *Christophe Harbour Marina* (http://www.christopheharbour.com/). This huge project is well-protected and has the ability to handle yachts to 250' LOA. The marina requests that visiting yachts contact the office on VHF ch. 71 when about 5 miles out. The entry channel is 105' wide and reported to be 18.5' deep. There is no place to anchor here, no way to get hauled out, but you can avail yourself of one of their slips.

Guadeloupe: *Marina de Rivière Sens*

Guadeloupe offers all manner of hurricane protection for cruisers. Everything from slips to haul outs, to mangrove lined creeks well inland, the choice is yours.

On the southeastern coast of Guadeloupe sits the capital, Basse Terre, and the local marina, *Marina de Rivière Sens*. Damage from storms have caused large portions of the seawall at the entrance to the marina to fall into the channel, and although these are easily seen and avoided, we cannot recommend this as hurricane protection, but it is included here for those that might find themselves in the unfortunate position of needing to use the marina for shelter.

As shown on Chart GUA-5, a waypoint at 15° 58.85' N, 61° 43.40' W, will place you just south

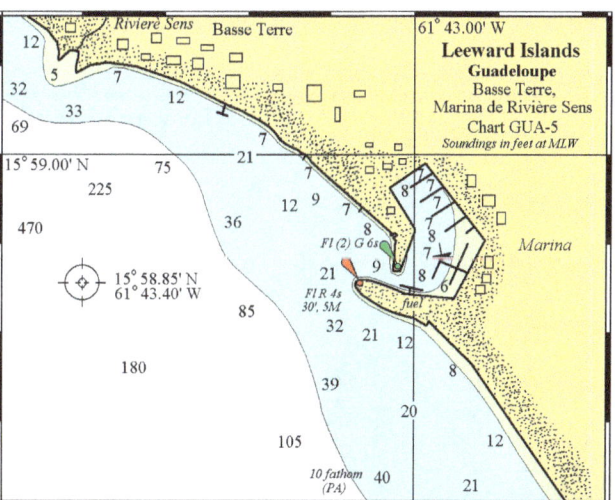

Marina de Rivière Sens, Basse Terre, Guadeloupe

THE LEEWARD ISLANDS • 133

of Basse Terre, approximately ¼ mile west of the entrance into the *Marina de Rivière Sens* (http://www.marina-rivieresens.com/en/). There are no shoals or hazards to navigation save for the shallows at the mouth of the *Rivière Sens*.

The marina entrance is well-marked and can accept a 7' draft although part of the inner basin shallows to less than 6' as shown on Chart GUA-5 (bottom of previous page). From the waypoint head eastward until you can turn to starboard to enter the marked entrance channel.

Guadeloupe: *Marina Bas du Fort*

The marina, *Port du Plaisance de Bas du Fort*, usually just called *Port du Plaisance* or *Marina Bas du Fort* (and sometimes shown as *Marina Gosier*), has a well-marked 100-meter-wide channel with a depth of 10 meters (until you make the turn into the marina proper where the depth rises to 14') that lead you to 1,200 slips (including those in *Lagoon Bleu*). The marina can accommodate vessels to 240' in length with drafts to 14.5'. The staff at the marina

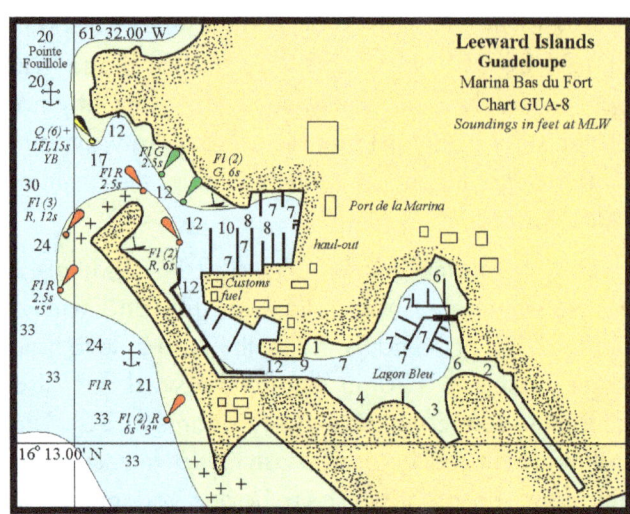

is bi-lingual and take reservations by email as well as by phone (0590-93 66 20). *Marina Bas du Fort* monitors VHF ch. 9 and you are advised to call while you are still 15 minutes out so their dockmaster can come out to assist you.

As shown on Chart GUA-7 (bottom of previous column), a waypoint at 16° 12.50' N, 61° 31.75' W, will place you approximately ¼ mile south of the entrance channel to *Marina Bas du Fort* as well as the southern entrance of the *Rivière Salée*. From the waypoint head northward passing between the red and green buoys as you follow the buoys into the channel as shown on Chart GUA-8 (above).

The slips at the *Marina Bas du Fort* are primarily stern or bow-to Med-style. For the best protection head for the inner lagoon at *Lagon Bleu*.

Guadeloupe: *Rivière Salée*

The *Rivière Salée* is basically a large saltwater mangrove swamp that separates the two "halves" of Guadeloupe, Basse Terre and Grande Terre. Traversing of this waterway is limited to vessels with mast heights of less than 80' and drafts of less than 7'. This body of water offers very good hurricane protection in a narrow, mangrove lined waterway, well inland yet still susceptible to a strong storm surge.

The *Rivière Salée* passage is well-marked and the bridges timed to make your trip on the river painless. There are two bridges on the river, *Gabarre* in the south, and *Alliance* in the north and they open regularly Monday through Saturday, but they do not open on Sunday. The *Marina Bas Du Fort* (0590-93

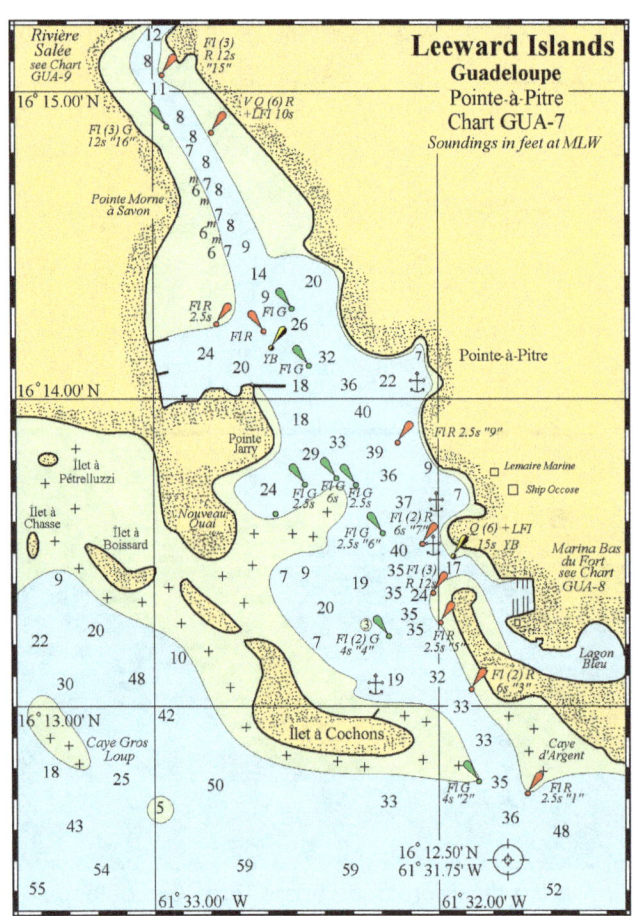

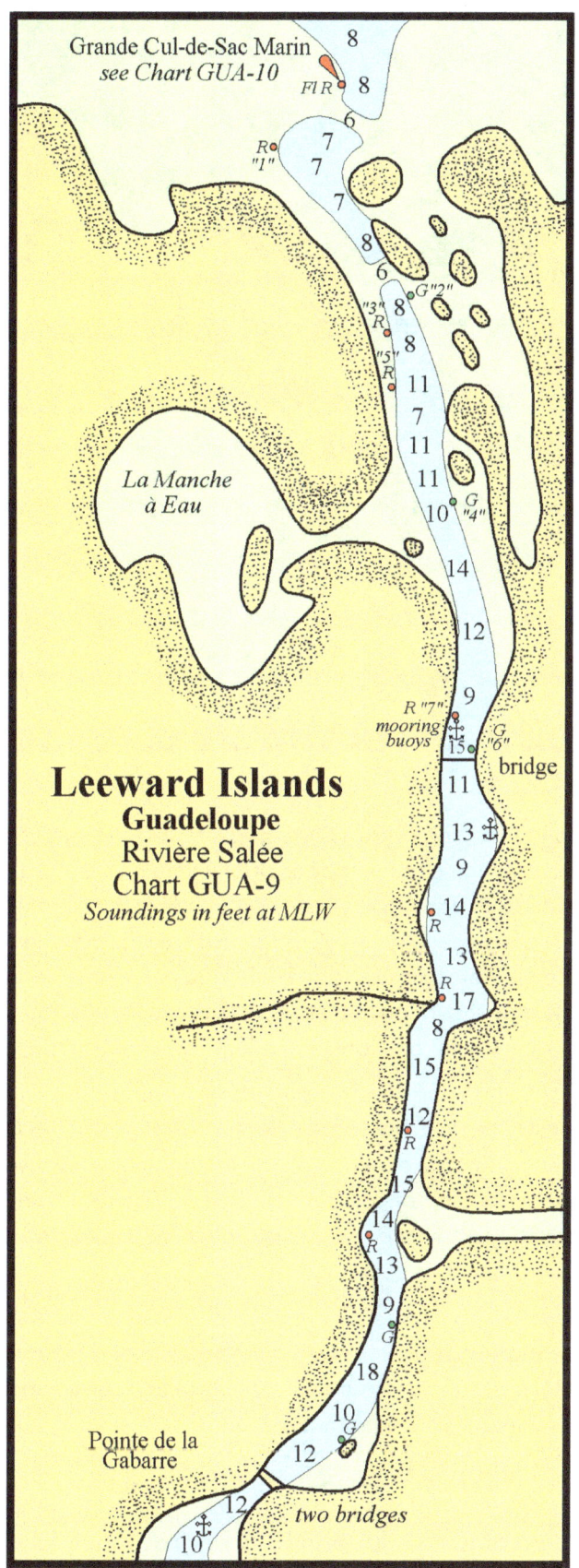

66 20 or VHF ch. 19) has the latest information on this passage, or you can call the DDE for bridge opening information at 0590-21 26 50.

If you are northbound, you'll need to be at *Gabarre* (see Chart GUA-9 left) at least 15 minutes before the scheduled opening at 0500, running lights on and engine running. A good plan is to anchor south of the bridge. There are no longer any moorings here or on the north side of the bridge. If you choose to stay in the marina or one of the anchorages and attempt to head north in the channel in the early morning hours to make the opening, you might find yourself confused by the many lights between *Marina Bas du Fort* and the *Gabarre bridge*.

The lights here are red-right-returning until you reach the *Gabarre Bridge* as shown on Chart GUA-9 (previous column). Remember, when traversing the waterway, northbound boats have right of way, but don't expect the southbound boats to know that! Use caution when transiting the *Rivière Salée*, especially at the bridges.

Gabarre is actually two bridges and they both open simultaneously (if somebody says that there are actually three bridges on the *Rivière Salée*, this is why). As you pass through *Gabarre* keep to port a bit as you pass the first green buoy and head north on the *Rivière Salée* (see Chart GUA-9) until you find a spot to anchor and ride out the oncoming hurricane.

You'll find the waterway well-marked with red and green buoys, make sure you keep the green buoys to starboard as you head north (remember: red-right-returning). Keep mid-channel if you are unsure of the depths on the edges of the channel.

The northern bridge, *Alliance*, usually opens about 20-30 minutes after *Gabarre*, when all the northbound boats arrive (usually about 0520).

There are some moorings north of the bridge where you can tie or anchor while you have breakfast and wait for the sun to rise. When the sun has risen to your satisfaction, you can drop the mooring (or raise your anchor) and find a place to ride out the hurricane.

If you are heading south from *Grand Cul-de-Sac Marin* (see Chart GUA-10 below) and Chart GUA-9, you're advised to tie to one of the three moorings north of *Alliance* the night before your planned transit

THE LEEWARD ISLANDS • 135

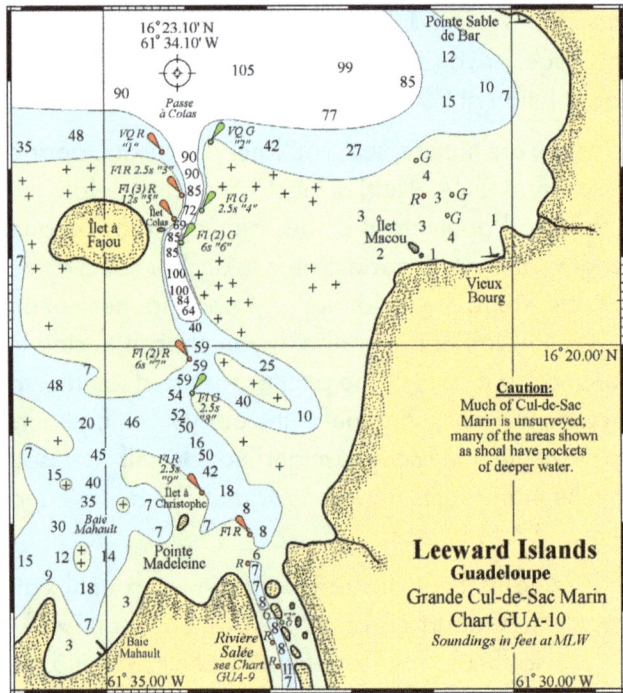

Marina de St.-François, Guadeloupe

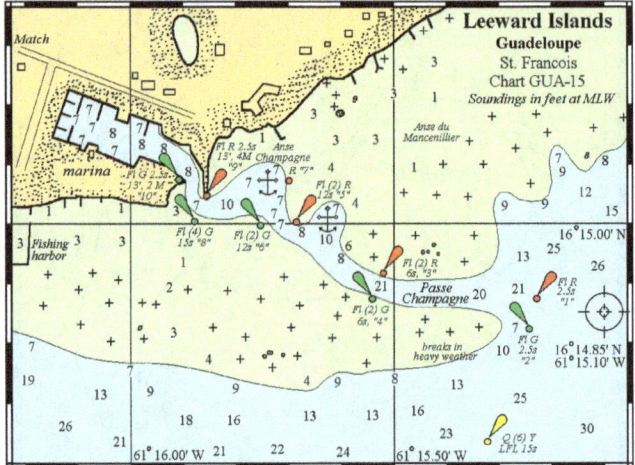

and be ready to go, engine running and running lights on at 0415 unless you find a spot that looks suitable for protection and do not require passing the bridge. *Alliance* opens at 0430, which allows southbound boats to reach *Gabarre* by 0500. When heading south, remember to keep the red buoys to starboard (red-right-returning). When *Gabarre* opens at 0500 allow the northbound boats to pass before heading south.

Guadeloupe: *Marina de St.-François*

St.-François lies about 8 miles east of Sainte-Anne and offers a very nice marina, all against the backdrop of cosmopolitan St.-François. As shown on Chart GUA-15 (top of next column), a waypoint at 16° 14.85' N, 61° 15.10' W, will place you approximately .1 of a mile east of the buoyed entrance channel (*Passe Champagne*) leading into the marina at St.-François. There is no anchoring here, you will have to get a slip, or, if you are fortunate and your vessel small enough, you might be able to haul out.

From the waypoint head westward between the red and green channel markers keeping the red to starboard; you will be entering with following seas and winds so use caution and remember that you'll be heading out into them when you leave. Never, I repeat NEVER, attempt this passage with strong following seas and winds and don't try to cut too close to the channel edges at any time as the shoals encroach right up to where you perceive the channel to lie between the buoys. To access the marina follow the markers to pass inside the jetty to enter the marina area as shown on the chart. Yachts over 6 ½' should proceed with caution if the tide is low or ebbing.

The *Marina de St.-François* lies inside the well-protected harbor and offers 200 slips that can accommodate a draft to 8'. The marina (http://www.marina-saint-francois.com/) monitors VHF ch. 16 and can be reached by phone at 0590-88 47 28.

On the southeast side of the marina area is a ramp used for haul outs. The trailer used will accommodate a vessel up to 14 tons with a draft of 5.5' and a beam of 21'. For more information about hauling out, contact Dominique at the marina.

Marie Galante and Desirade

Both Marie Galante and Desirade have small docks behind the protection of a seawall, but they cannot be recommended as a hurricane shelter as they are

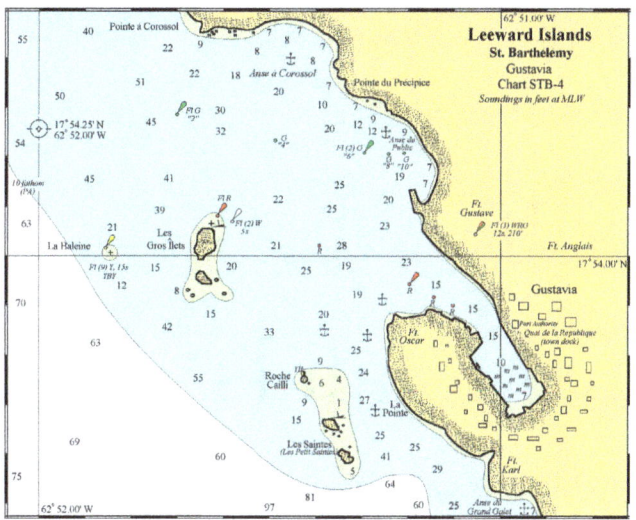

Gustavia, St. Barths

far too open and do not provide enough shelter, however, if you wish to tie up your boat and get a hotel room, and you are not picky about how secure your vessel is, you will not mind the accommodations on these two islands.

St. Barths: *Gustavia*

Just south of St. Martin lies St. Barths and the harbor of Gustavia. More than a few people boast that Gustavia is a good hurricane hole and it should definitely be considered if no other options are available. Certainly, there are better choices for hurricane protection than Gustavia with its large and very crowded mooring field, but as the old adage advises..."Any port in a storm."

As shown on Chart STB-4 (above), a waypoint at 17° 54.25' N, 62° 52.00' W, will place you approximately ¾ nm WNW of the harbor entrance channel. From the waypoint head generally south of east passing between green #2 and Les Gros Îlets, which is marked by a flashing red light off its northern tip. Take up a course towards the harbor mouth where you will pick up a line of red buoys that will guide you into the harbor at Gustavia.

If you intend to enter the inner harbor at Gustavia you are required to call the *Gustavia Port Authority* (*Port of Gustavia*) on VHF ch. 16 and then switch to their working channel, ch. 12, after you have made contact with them. It is permissible to call two hours prior to your arrival to alert them of your impending arrival and to secure a berth.

As you head down the marked entrance channel keep an eye out for a danger buoy on the southern edge of the channel near Gros Ilet. The buoy marks the wreck of a freighter in 10'-15' of water.

In the harbor of Gustavia you'll find four rows of moorings. The innermost will accommodate vessels of 30' in length, the next outside row will accommodate vessels of 40', the next row will accommodate vessels of 50', and the outermost row will accommodate vessels with a length of up to 60' LOA.

If you need to haul out, visit the *St. Barth Boatyard* where the owners can haul vessels to 50' LOA and 18 tons (the yard has an 18-ton fork lift and a 45-ton crane). The boats are lifted from the water and trailered the short distance to the yard where they are secured with tie downs. Unfortunately, the yard cannot store catamarans due to the width of the road on which they must operate their trailer, however they can haul a catamaran at the dock.

Chapter 15
The Windward Islands

LIKE THE LEEWARD ISLANDS, THE WINDWARD ISLANDS HAVE A FEW ISLANDS lacking protection, and a few with excellent protection for those wanting to anchor, get a slip, or haul out. St. Lucia has tiny *Rodney Bay* and *Marigot Bay*, which will be packed. Castries has a tiny keyhole harbor, but commercial vessels will present a danger to cruising vessels seeking shelter. Martinique has several nice choices for protection, and St. Vincent even has a marina/haul-out yard available. You will find that isolated Barbados has a spot or two if no other alternative presents itself. Further south, Carriacou and Grenada offer several more alternatives to the shelter seeking skipper.

Martinique: *Cohe du Lamemtin*

Heading south from the Leeward Islands, the first protection that you will find along the western shore of Martinique is in *Baie de Fort du France*. And the best protection in *Baie de Fort du France* is without a doubt in the small cove lying just north of the runway at the airport at *Cohé du Lamentin*. For those with shallow draft vessels, there are several small rivers and streams leading into *Baie de Fort du France* where one could find shelter.

At the northeastern tip of *Baie de Fort-de-France* lies the anchorage at *Cohé du Lamentin* and the *Marina de Port Cohé* as shown on Chart MAR-5 (top of next page), with the entrance to the marina shown in detail on Chart MAR-7 (next page). The bay is sometimes used for hurricane protection, I would prefer the creeks off the bay instead. Entrance to the anchorage in *Cohé du Lamentin* is fairly easy as shown on Chart MAR-5. From the waypoint at the western end of *Baie de Fort-de-France* head a bit south of east until you can pick up the entrance channel buoys northeast of Pointe du Bout. Follow the channel into the bay keeping the red buoys to starboard and the green ones to port.

If you are approaching from *Baie des Flamands* or *Baie des Tourelles*, give *Banc du Ft. St. Louis* a wide berth and keep well off the shoreline between Pointe des Carrières (Pte. des Grives) and Pointe des Sables. There is talk of a new, large marina being planned for the area between Pointe des Carrières (Pte. des Grives) and Pointe des Sables so keep an eye out for it. Once in *Cohé du Lamentin*, anchor wherever your draft allows, the water is not very clear here (with garbage often floating by) and the bottom muddy with holding that varies from poor to almost good. Keep a sharp eye out for the shoal shown as *Sèche San*

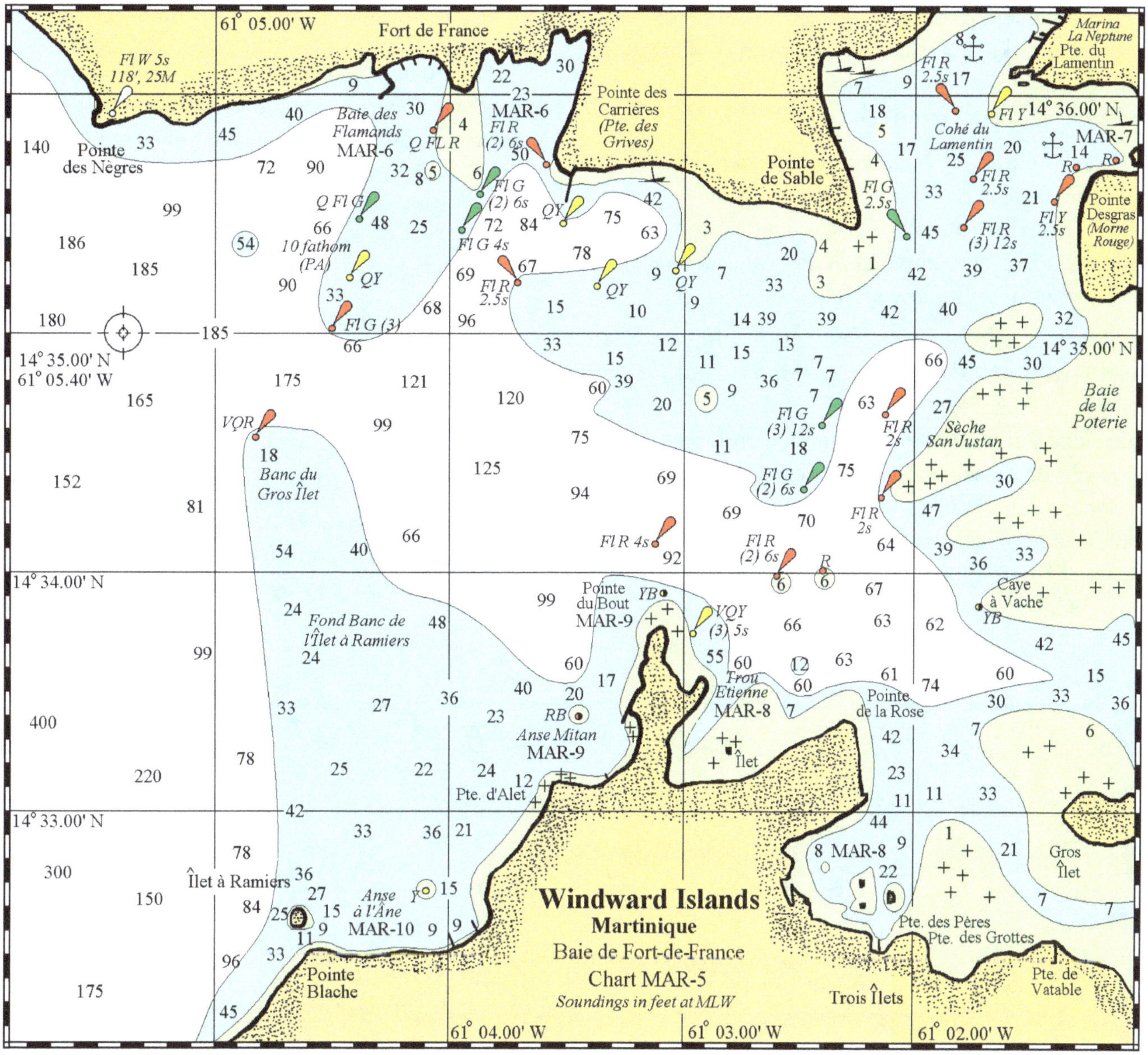

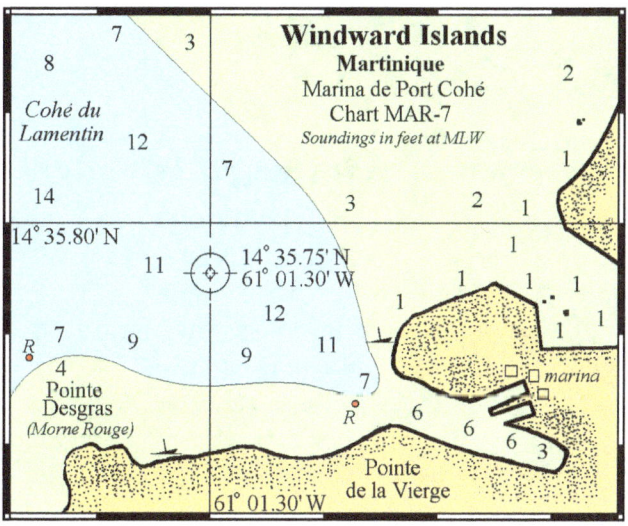

Justan and the unmarked shoal just north of it as well as any unmarked wrecks in the surrounding waters.

The best hurricane protection here is at *Marina de Port Cohé* and the channel next to it. To enter the channel leading to *Marina de Port Cohé* make for a waypoint at 14° 35.75' N, 61° 01.30' W, which will place you ¼ mile west/northwest of the entrance channel as shown on Chart MAR-7. As you approach the narrow and difficult to discern entrance, make sure that you keep the red buoy off Pointe Desgras (sometimes shown as Morne Rouge) to starboard and the conspicuous wrecked barge north of the entrance channel to port.

THE WINDWARD ISLANDS • 139

When entering favor the northern shore (keeping the red marker to starboard) and then, keeping mid-channel, you should not have a problem. The controlling depth here is a bit over 5½' at MLW.

Fresh water is available at the marina (and sometimes slips) and the airport is less than 1½ miles away. The *Marina de Port Cohé* usually charges boats a fee to enter their channel during a hurricane and the waters there are always full.

There is another small marina in the extreme northern part of *Cohé du Lamentin, Marina La Neptune*. Primarily a sailing school, the marina offers absentee yacht owners a safe haven and somebody to tend their vessel while they are away; however, the marina is usually filled with local vessels. As shown on Chart MAR-5 (previous page), the marina lies just north of Pointe du Lamentin with a controlling depth of a bit less than 6'.

Martinique: *Cul-de-Sac du Marin*

At the southern end of Martinique there are several small coves that are open to the south, but Le Marin (Cul-de-Sac du Marin) offers the only true hurricane protection including slips, haul out facilities, and a wonderful protected little cove for anchoring. This area is VERY popular with boaters so make your plans and move EARLY.

Although the primary anchorage and mooring field is usually crowded, there are two small mangrove-surrounded coves that are deep and well-protected. The innermost cove is called *Baie des Cyclones* and lies SSE of the marina docks just under Pte. Malé as shown on Chart MAR-17 below. The second cove, which is a bit more exposed, lies just inside Pte. Marin and south of Îlet Baude, southwest and more to seaward of Baie des Cyclones.

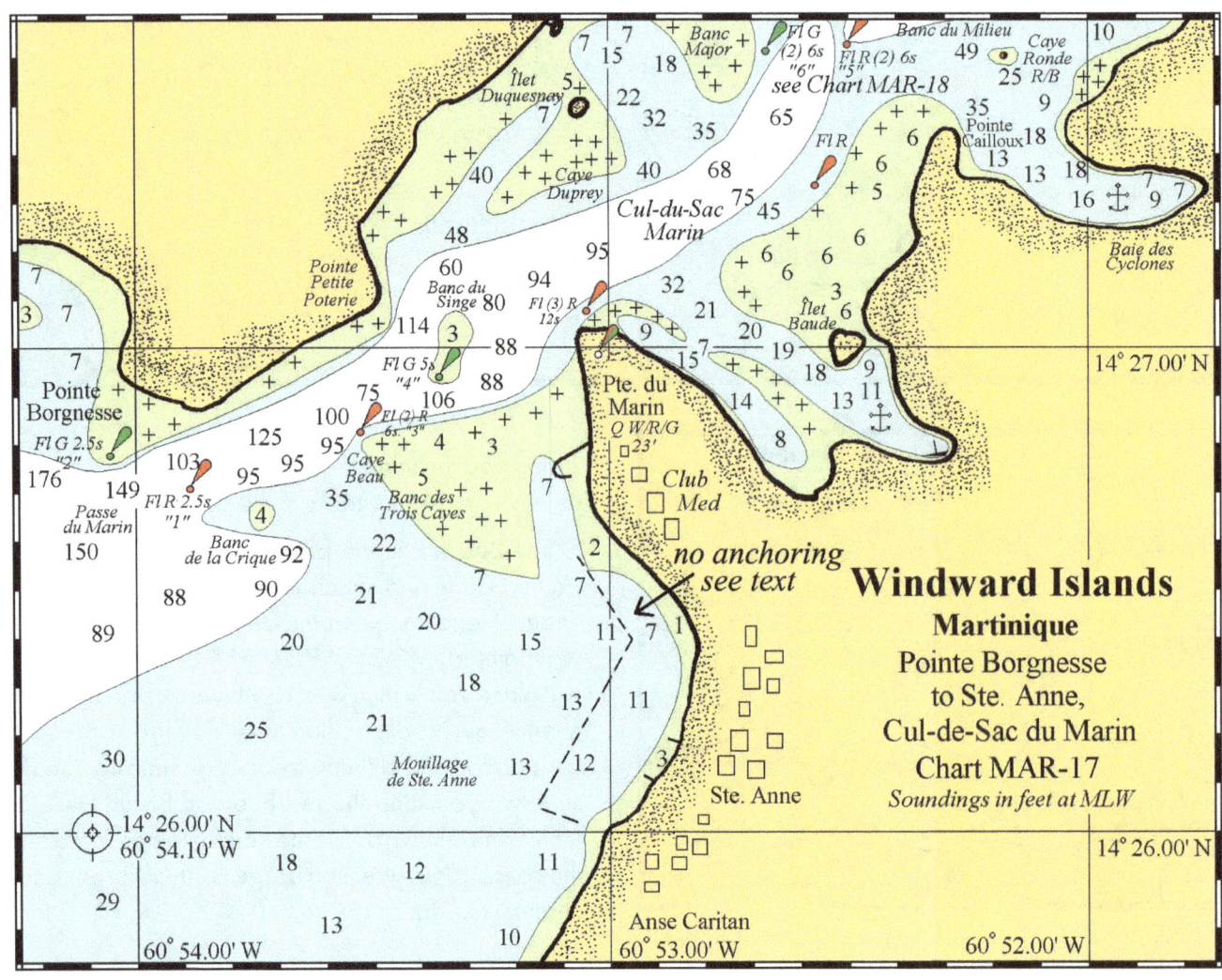

As shown on the Chart MAR-17, a waypoint at 14° 26.00' N, 60° 54.10' W, will place you approximately ¾ mile south/southwest of the entrance channel, *Passe du Marin*, to the facilities and anchorages of *Cul-de-Sac du Marin*. From the waypoint head east of north to enter the channel between markers "1," which marks the tip of *Banc de la Crique*, and "2." From the anchorages at Ste. Anne you may enter the channel east of *Banc de la Crique*, between *Banc de la Crique* and *Banc de Trois Cayes* just off Club Med and Pointe du Marin by keeping the red light "3" at Caye Beau to starboard.

The channel leading into *Cul-de-Sac du Marin* is deep for most of the way before you reach the crowded anchorage off *Port de Plaisance du Marin Marina*. From the waypoint pass between "1" and "2" keeping "3" to starboard and passing on either side of *Banc du Singe* (marked by a green light) though the proper channel keeps the shoal to your port side and the red light off Pointe du Marin to starboard. If you take *Banc du Singe* to port and then clear Pointe du Marin, you can turn to starboard to work your way into the small bay south and southeast of *Îlet Baude*. This is a good spot to ride out just about any weather, even a minimal hurricane.

Continuing into the harbor toward the marina at Le Marin, with the passage to Îlet Baude on your starboard beam, you will take the flashing red light that sits southwest of Pointe Cailloux to starboard. Once clear of Pointe Cailloux you can turn to starboard to anchor north of Pointe Cailloux with all the other boats anchored off the marina just south of *Banc du Milieu* (see Chart MAR-18 at top of next column), or proceed into the *Baie des Cyclones* for excellent protection.

East of *Banc Grande Basse* (Chart MAR-18) is a narrow, mangrove lined creek that would be a good spot to hide from a hurricane, but you'll have to sound your way in by dinghy first. At the northwestern tip of Cul-de-Sac du Marin is a small anchorage area west of the yellow-buoyed no-anchor zone. Here is the marked channel leading to the *Carenantilles* yard that can accommodate vessels with beams to 23' on their 65-ton *Travelift* and catamarans up to 60' LOA+ with any beam on their hydraulic trailer. The yard also has another *Travelift* rated at 440 tons.

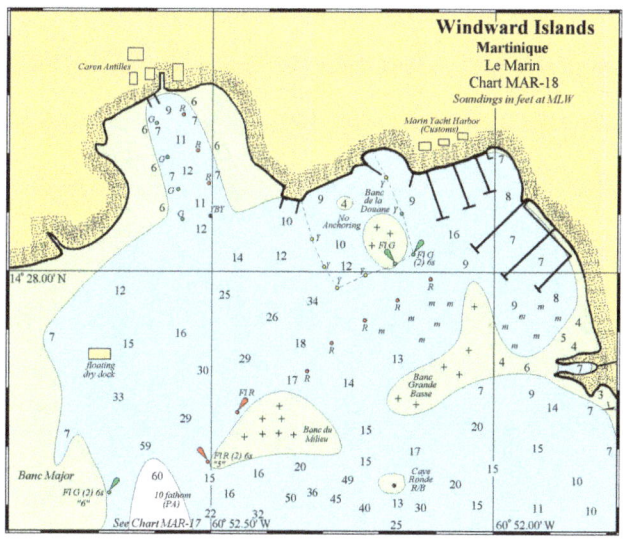

Dock Cleaner Ecologique (*Switch Charters*) runs the dry dock and they can handle vessels up to 70 tons, 40' in width, with a draft of up to 6'.

Port de Plaisance du Marin Marina, sometimes just called *Marin Yacht Harbor* or *SAEPP*, monitors VHF ch. 16/09 and offers 600 slips w and 70 moorings, and can handle up to 4 mega yachts with drafts to 13'.

Martinique: *La Marina Hâvre du Robert*

The only protection to be found here is *La Marina du Robert* which lies up a small creek at the southwestern end of *Hâvre du Robert* and is set in the old sugar mill of Le Robert as shown on Chart MAR-22 (top of next page). Small and basic, the marina offers a mobile crane for a haul out (the controlling depth in the creek is 4' at MLW) and co-author Stephen J. Pavlidis has seen boats as large as a Catana 44 hauled-out there).

As shown on Chart MAR-22, a waypoint at 14° 40.45' N, 60° 50.10' W, will place you approximately ½ mile east of the *Passe de Loup Garou*. From the waypoint head west in the pass into *Hâvre du Robert*. You will pass between the two outer markers, "1" and "2" and once abeam of Îlet Madame you will need to turn a bit more to the southwest to split the next two markers "3" and "4." Work your way to the southwest where the marina opens up south of *Gros Loup* on the chart.

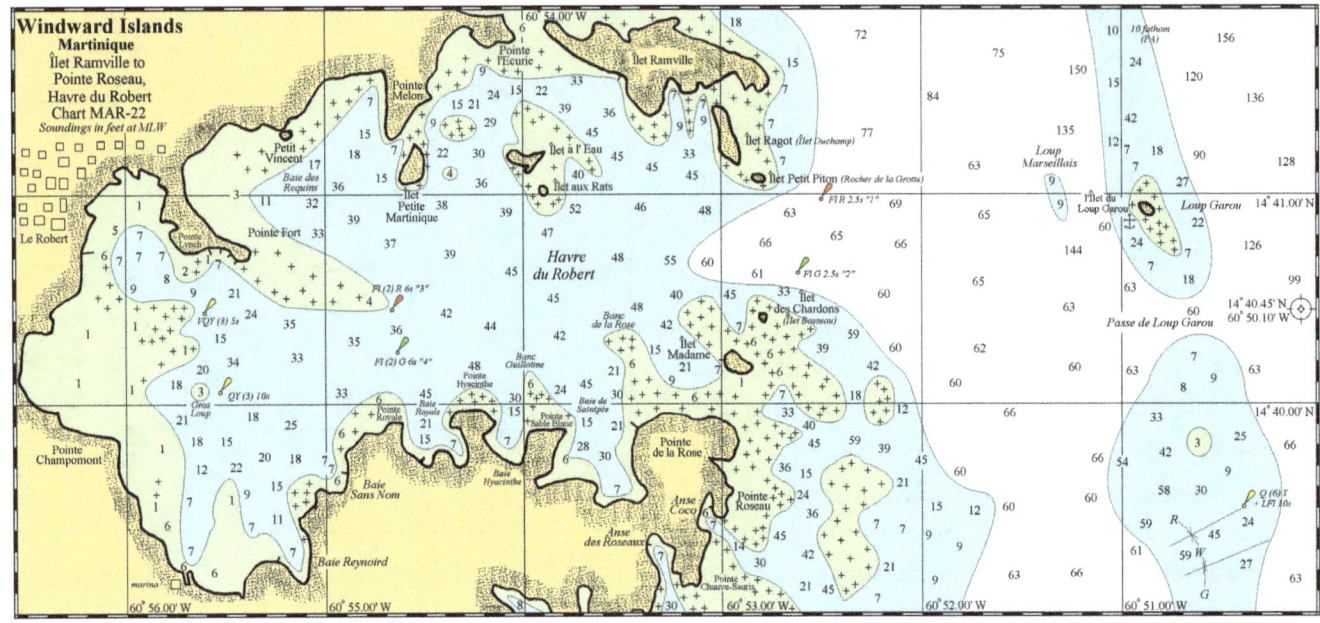

Barbados: *Port St. Charles*

Barbados...there is not enough hurricane protection to convince the authors to stay in Barbados with a hurricane coming. However, inside the lagoon at the *Port St. Charles* complex and the *Carenage* in Bridgetown offer the closest thing to real protection on Barbados (there is a haul out yard in Bridgetown) and should be used only if you do not have time to get to better protection at Martinique, St. Lucia St. Vincent, Carriacou, or Grenada.

The Port St. Charles complex is designed for the wealthy, have no doubt about that. It has been described in its brochures as the "ultimate reward." The complex has been designed to be eco-friendly with a garbage incineration system and an inner lagoon that is designated as a marine sanctuary.

A waypoint at 13° 15.75' N, 59° 39.25' W, will place you approximately ½ mile west of the entrance to the *Port St. Charles* complex and marina. Before attempting to enter the marina contact the dockmaster on VHF ch. 16 or 77 for instructions on entry.

From the waypoint head a bit south of east keeping the southern tip of the marina breakwater (marked by a light, Fl G) to port. As shown on Chart BAR-2 (above), pass between the breakwater to port and the line of red buoys to starboard that mark the western edge of *Tom Snooch Reef* and you'll see the marina's docks to port. The *Port St. Charles Marina* can accommodate

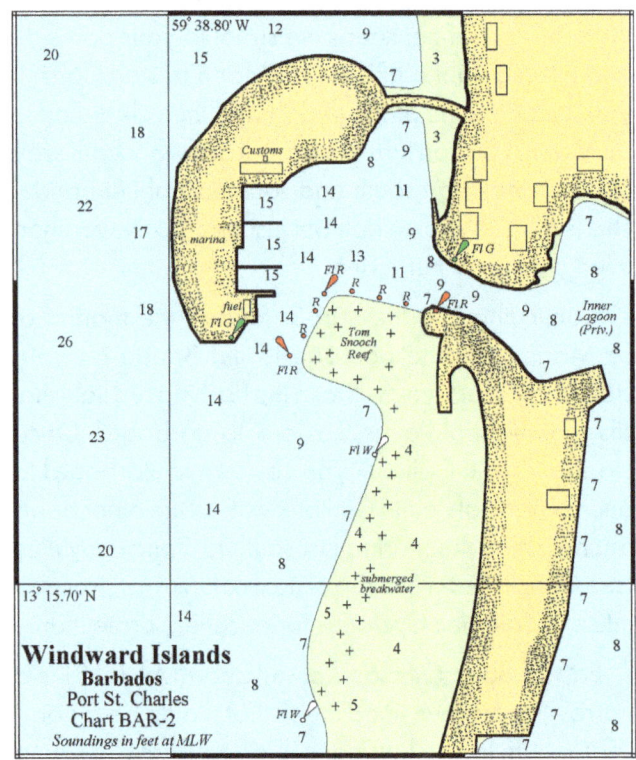

six transient vessels with lengths to 250' LOA and drafts to 13.5'.

The inner lagoon offers the best protection, however it is private, and anchoring is not permitted as the bottom is very hard and scoured. However, boats needing shelter can enter and tie up to a vacant seawall as per the dockmaster of the *Port St. Charles* complex.

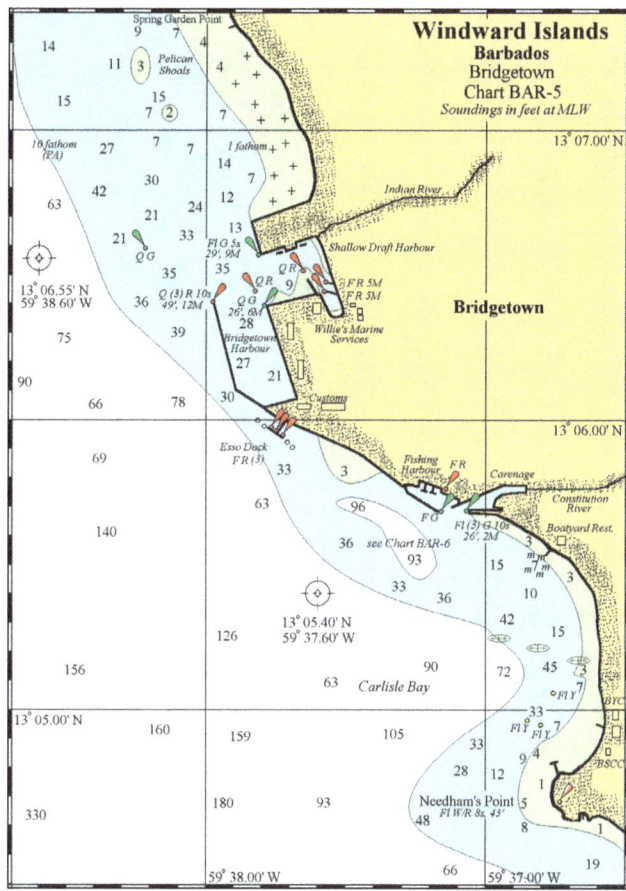

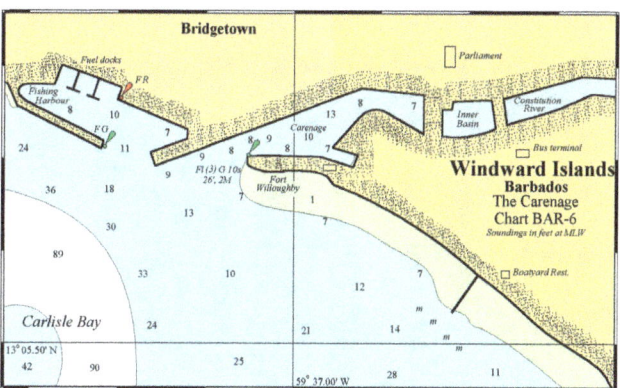

Barbados: *Bridgetown*

As shown on Chart BAR-5 (above), a waypoint at 13° 06.55′ N, 59° 38.60′ W, will place you approximately ½ mile west of the entrance to *Bridgetown Harbour* and the *Shallow Draft Harbour*. From the waypoint, head generally eastward passing between the green-lighted buoy (QG) and the end of the jetty, which is also lit (red). Do not enter the harbor before contacting Customs on VHF ch. 16.

Most of Bridgetown's marine services are centered in and around *Bridgetown Harbour*. If you need to haul out, *Willie's Marine Services* is located at the southern end of *Shallow Draft Harbour*, well inside *Bridgetown Harbour*. Owner Willie Hassell, an experienced cruiser himself, has a 45-ton *Travelift* that can handle yachts with drafts to 7′.

If a haul out just won't do, you might be able to tie to a bit of seawall in the *Carenage* (see Chart BAR-5 and in greater detail on Chart BAR-6, next column).

As shown on Chart BAR-5, a waypoint at 13° 05.40′ N, 59° 37.60′ W, will place you approximately one mile west of the *Carenage*. From the waypoint head generally eastward and you will come to the entrance to the *Carenage* and *Constitution River* as shown on Chart BAR-6.

Here you'll notice quite a few boats tied up along the seawall, you can check with the *Barbados Port Authority* for slip availability.

St. Lucia: *Rodney Bay*

Rodney Bay on the western shore is a completely protected harbor. It will be crowded so arrive early. *Rodney Bay* offers a marina that has haul out services as well as an anchorage with moorings.

A waypoint at 14° 05.20′ N, 60° 58.50′ W, will place you approximately ¼ nautical mile northwest of horseshoe-shaped *Rodney Bay*. From the waypoint you can also approach the entrance channel to the inner lagoon and *Rodney Bay Marina* as shown on Chart STL-3 (see top of next page).

Enter the bay and head past the marina to anchor in the small lagoon to the southwest just past *Waterside Landings* (an informal and friendly marina offering 15 slips). Protected from seas here, your primary concerns would be the holding in the dredged lagoon, and flying debris from nearby condos. If you prefer the protection of mangroves, you should move south to *Marigot Bay*.

IGY Rodney Bay Marina boasts 253 slips (on floating docks) and can accommodate vessels up to 285′ LOA with drafts up to 14′. Please note that Dock 1 is for mega-yachts. As you turn to the south, on the northern shore of the inner lagoon lies the 4.5-acre *Rodney Bay Marine Boatyard*. The boatyard monitors VHF ch. 68 and has a 75-ton *Travelift*, a 40-ton boat

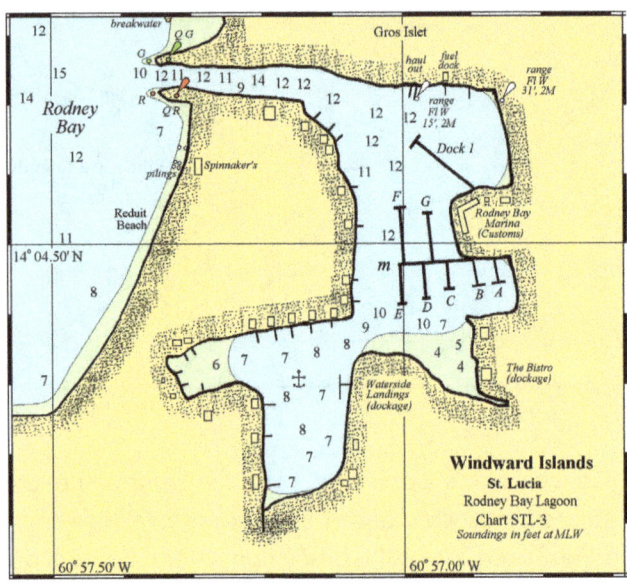

Marigot Bay, St. Lucia

trailer, and can accommodate vessels to 85', LOA with a 27' beam, and 13' draft.

St. Lucia: *Marigot Bay*

Marigot Bay, located on the western shore of St. Lucia, is well-sheltered and famous for being the classic hurricane hole. The inner lagoon, sometimes called *Hurricane Hole*, has moorings as well as a marina with a LOT of charter vessels that call the place home, so be sure to arrive EARLY. Charter fleets tend to wait on confirmation of a hurricane's path so there usually is a small window of opportunity to grab a mooring or anchor before they arrive, tucked up nice and snug into the mangroves far from the entrance channel. Marigot Bay offers high hills all around, except to the west, to seaward, and good protection from seas. The biggest danger here, as usual, comes from unattended boats, mostly charter boats, that would be left to survive or not.

As shown on Chart STL-5 (below), a waypoint at 13° 58.10' N, 61° 02.10' W, will place you approximately ¼ mile west of the entrance to this most impressive harbor. The entrance to *Marigot Bay* is marked by a conspicuous red-roof high on a hill on the south side of the entrance to the bay. The entrance lies approximately 1 ¼ miles south of the conspicuous green tanks of the *Hess Oil* plant at *Cul de Sac Bay* (which itself lies approximately two miles south of Castries). The entrance is easy to miss, so easy that a French fleet passed right by an anchored British fleet that camouflaged their rigging with palm fronds.

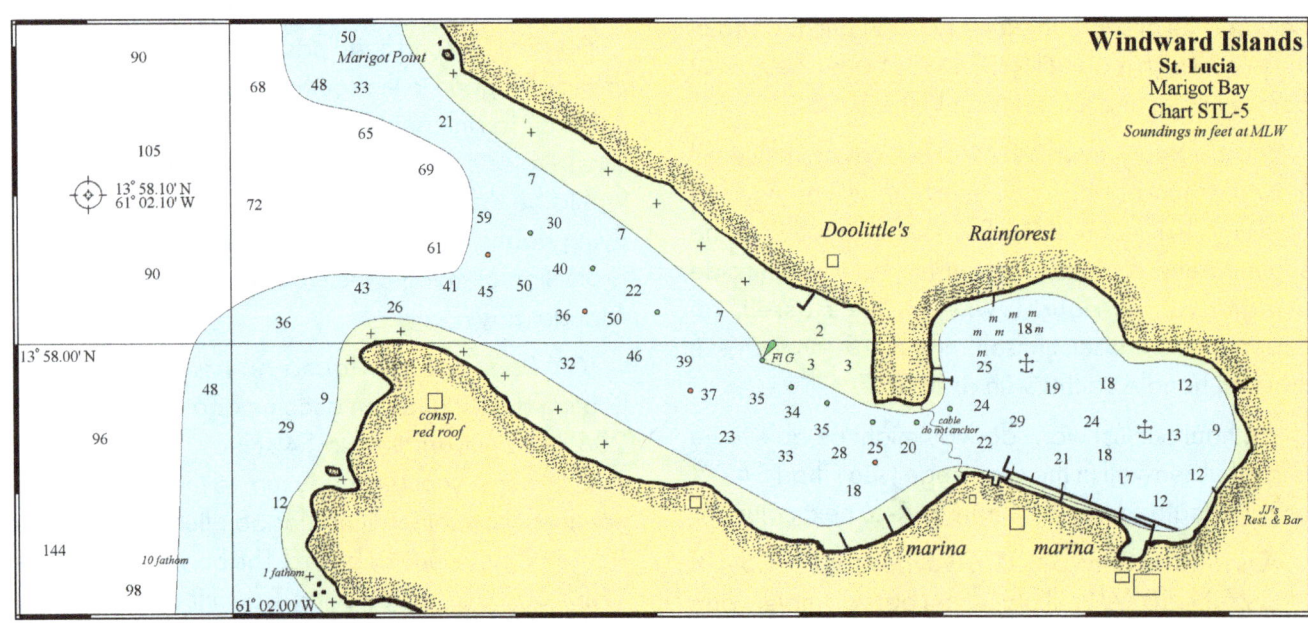

144 • THE CAPTAIN'S GUIDE TO HURRICANE HOLES

From the waypoint head in on a southeasterly course, the exact course is not important here, you simply wish to avoid the shoal off the northern shore of the outer cove. As you head in, keep center channel until the headlands are abeam, then favor the southern shoreline following the dark blue water towards the entrance. There is a lighted buoy (Fl G) south of the resort and just before you reach the spit of land separating the inner and outer coves; keep this buoy well to port. Pass between the spit of land and the southern shore and find a spot in the mangroves, grab a mooring, or get a slip.

On the south side of the inner lagoon sits the *Marigot Bay Marina*. The marina offers moorings in the lagoon, stern-to dockage, and a few alongside slips and can accommodate vessels to 250' LOA.

Located on the southern shore of the outer bay is *Chateau Mygo*, a marina offering six stern-to berths but the marina lies outside the protection offered by the inner lagoon.

St. Vincent: *Ottley Hall*

There is only one place in St. Vincent and the Grenadines that should be considered for hurricane protection, *Ottley Hall Marina and Shipyard*. The yard is located west-northwest of Kingstown at Ottley Hall Bay on the southwestern shore of St. Vincent at the foot of 636' high *Fort Charlotte* just below the conspicuous large white satellite dishes on the hillside that are easily seen from offshore.

As shown on Chart STV-8 (next column), a waypoint at 13° 09.45' N, 61° 15.10' W, will place you approximately ¼ mile southwest of the entrance to the marina. From the waypoint, pass between the jetty/seawall/dock on your port side and the point of land on your starboard side and enter the marina. Be sure to call on VHF ch. 16 before entering the marina. The marina is deep and there is a bit of a surge at times so keep that in mind if you're thinking about a stay here, for hurricane protection you do best to haul out here.

Ottley Hall Marina and Shipyard is a large, full-service marina with 22 slips for vessels up to 200' in length. The yard has a covered dry-dock and can haul vessels up to 200' LOA with a 36' beam, and 1,000-ton displacement. The yard also has a 500-ton lift, a 40-ton *Travelift*, a 20-ton crane, and offers

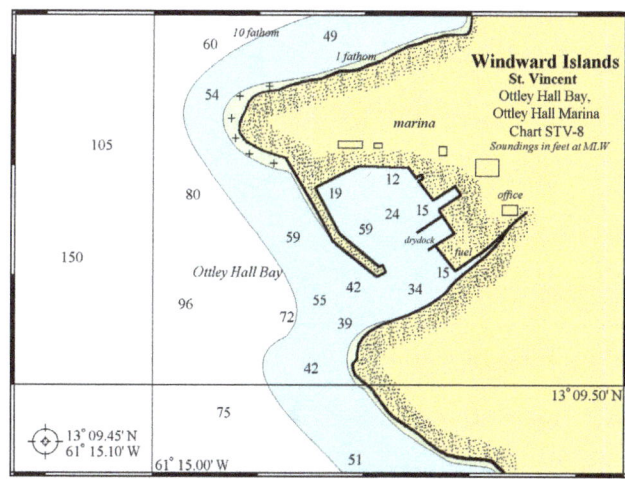

Ottley Hall Marina and Shipyard, St. Vincent

storage on the hard. You can reach the marina offices by phone at 784 457-2178 or by email at ottleyhall@vincysurf.com.

Canouan: *Glossy Bay Marina*

South of Charlestown Bay is Glossy Bay Marina located at the southwestern tip of Canouan just south of the airstrip as shown on Chart GND-7 (top of next page). Glossy Bay Marina is a luxury superyacht marina with 120 berths including 36 super-yacht berths (up to 100+ meters) with drafts to 17'. The marina is also a Port of Entry for St. Vincent and the Grenadines. The marina opened early to accommodate vessel seeking shelter from Hurricane Irma in 2017.

As shown on Chart GND-7, a waypoint at 12° 42.00' N, 61° 21.60' W, will place you approximately ¼ nm W of the jettied and well-marked entrance. From the waypoint head slightly south of west to enter between the jetties. You will have a minimum of 17.5' of water in the entrance and in the basin at MLW.

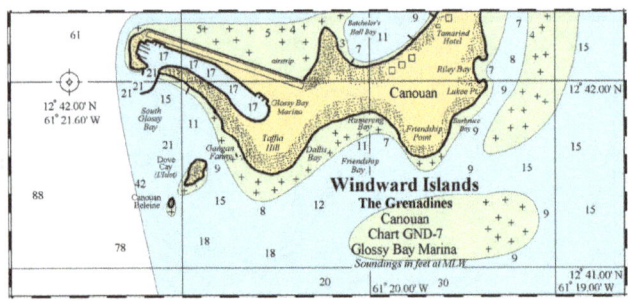

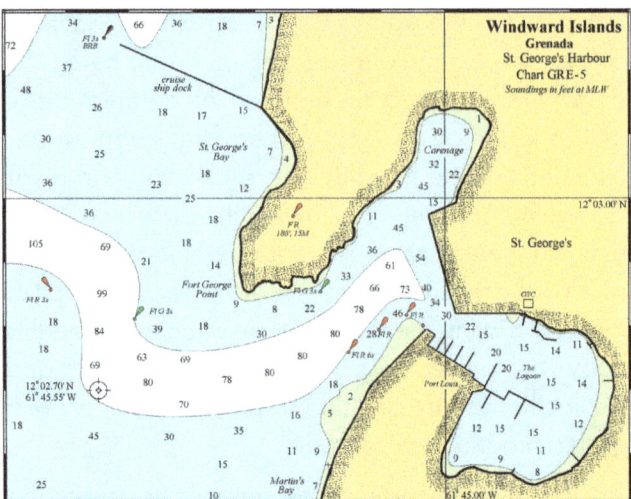

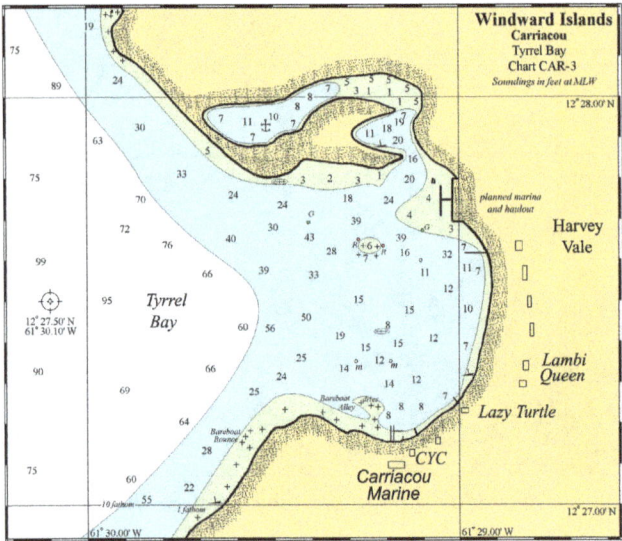

St. George's, Grenada, *Lagoon* in left background

This unique marina has dockhands that are internationally trained in marina operations, first aid, and even firefighting. Each superyacht has a fueling hydrant at each slip as well as a pumpout (for smaller vessels there is a fuel dock selling duty-free fuel). Metered water and electric is first class (European and US from 16 amps to 630 amps) as is high-speed internet access, all of this guaranteed by a backup power plant in case the island's primary power grid goes down.

Carriacou: *Tyrrel Bay*

On the north shore of *Tyrrel Bay* is a very protected mangrove lagoon, a great place for shallow draft boats with lots of mosquito repellent.

As shown on Chart CAR-3 (see above), a waypoint at 12° 27.50' N, 61° 30.10' W, will place you approximately ¾ mile west of the anchorage area in *Tyrrel Bay*. When entering *Tyrrel Bay*, keep an eye out for the reef on the northern side of *Tyrrel Bay* that has almost 6' over it at MLW and is marked by a red buoy.

If you prefer to haul out, check with *Carriacou Marine* on the southern shore of *Tyrrel Bay*. The yard can be contacted on VHF Channel 16/68 if needed. They have a 50-ton lift which can accommodate boats with a beam of up to 18' and a length of up to 60' LOA.

At the extreme northern end of *Tyrrel Bay* is the entrance to a mangrove lined lake that is a good spot to secure your vessel in the event of a hurricane. The entrance is straightforward as shown on Chart CAR-3, but once past the second point of land to port, you must favor the northern shore of the lake, that is where the deeper water lies (about 5' at MLW). Vessels are not allowed to anchor here unless it is an emergency; there are oyster beds here that must not be damaged. The holding is fair, and you will be in the company of a lot of other boats, most unattended. In the 1960s, a hurricane wiped out the fleet anchored in this lagoon.

Grenada: *St. George's*

Grenada is a major destination for cruisers in the Caribbean and the island is home to several nice marinas, boatyards, and mangrove lined hurricane holes for protection.

Some cruisers tout St. George's on the western shore of Grenada as a viable hurricane hole. I've never thought about riding out a storm in St. George's, it's a bit too crowded for my tastes, as well as not having any mangroves around. I much prefer the southern shore of Grenada if I'm in search of protection. Your best bet for protection in St. George's is to get a slip at the *Grenada Yacht Club* (*GYC*; VHF Ch. 16 & 06; http://www.grenadayachtclub.com/), or at Port Louis Marina (VHF ch. 16 & 14)). Anchoring in the *Lagoon* can be tricky as there is some old construction debris littering the bottom as well as the remains of a century of old pilings. The *Lagoon* is also subject to some odd currents that are most noticed on nights with no wind when you'll find yourself swinging round and round your anchor just as your neighbor is doing to his. One last note, in 1955, the eye of Hurricane Janet passed between Grenada and Carriacou and the storm destroyed most of *St. George's Pier* (820' long) and the *GYC* clubhouse.

As shown on Chart GRE-5 (top of previous page), a waypoint at 12° 02.70' N, 61° 45.55' W, will place you approximately ¼ mile west/southwest to the entrance to *St. George's Harbour* and south of the center of the two outer buoys.

From the waypoint the approach is fairly easy, simply head for the harbour mouth keeping between the outer red and green lighted buoys as you work your way inside the entrance to the harbor. To port lies the *Carenage* which is off limits for anchoring, to starboard lies the *Lagoon* where you'll want to drop the hook or get a slip.

To enter the *Lagoon*, follow the red buoys keeping them to starboard. You will run parallel to the ship dock until past its southern end when the channel doglegs to the east, to port. When you arrive at the last red buoy you'll see a green marker to port, just off the *Grenada Yacht Club*. Pass between the green marker and the last red marker to starboard and you are in the *Lagoon*. Keep clear of the two red buoys.

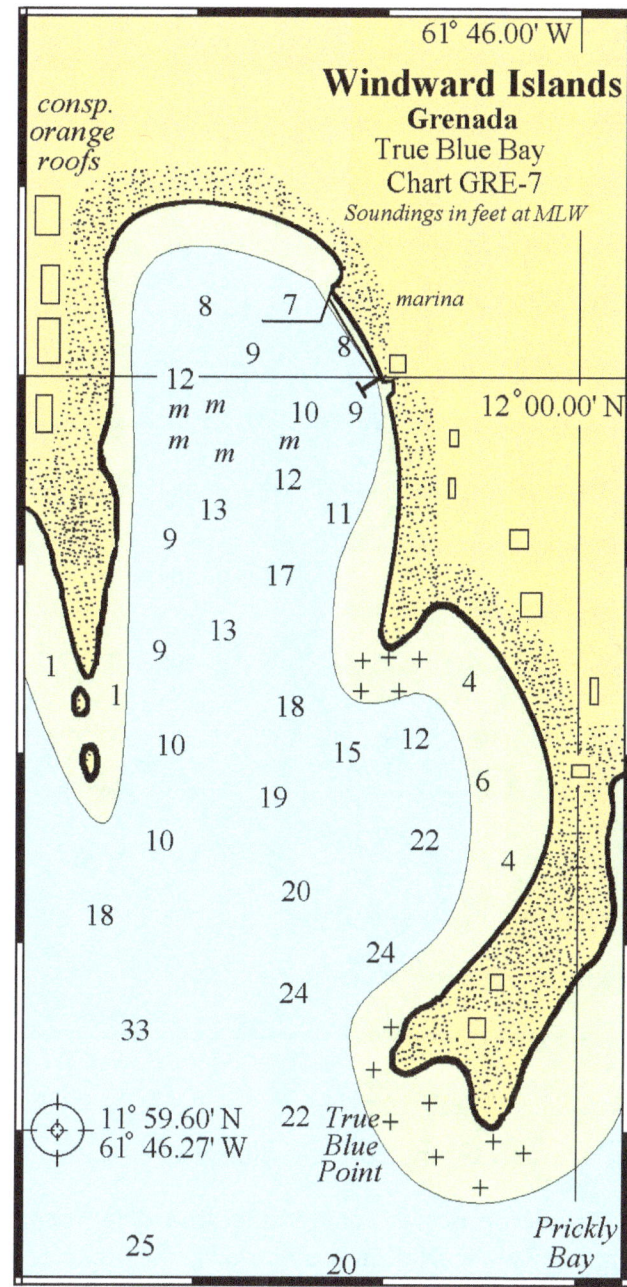

THE WINDWARD ISLANDS • 147

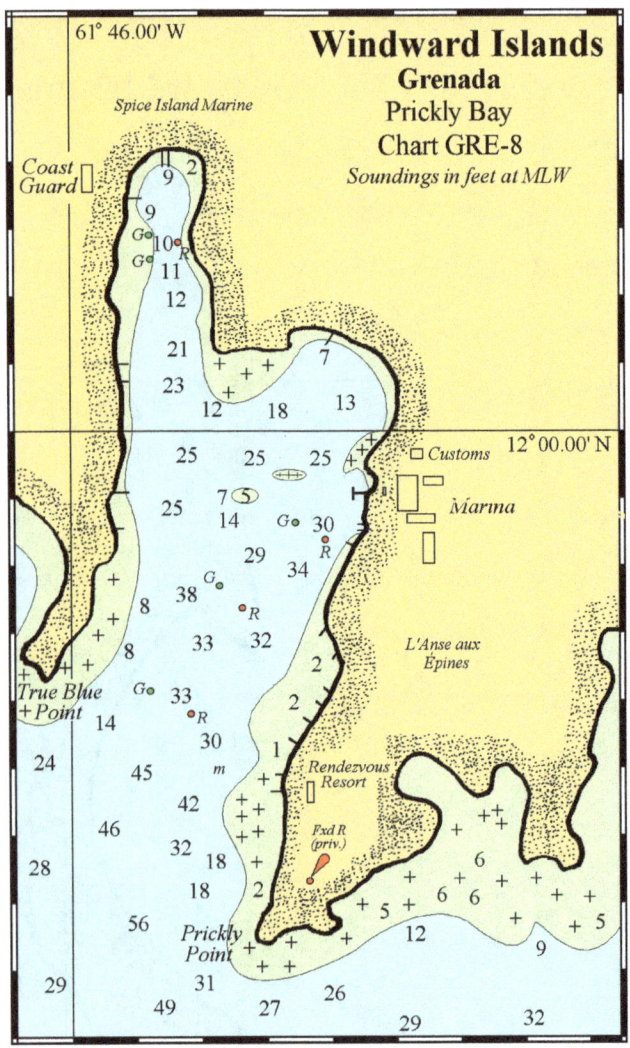

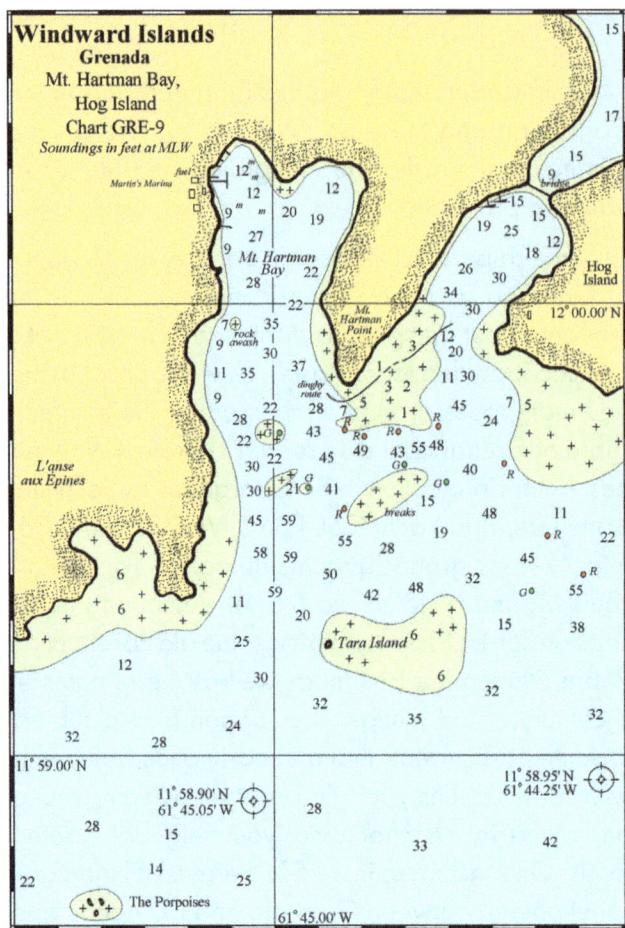

Grenada: *Prickly Bay*

Grenada: *True Blue Bay*

True Blue Bay is home to the *True Blue Resort* (http://www.truebluebay.com/), the base for *Horizon Yacht Charters* and they offer dockage as well as a few moorings. Strong southerly winds can make this bay a death trap so keep an eye on your forecasts if you choose to get a slip here.

As shown on Chart GRE-7 (see top of previous page), a waypoint at 11° 59.60' N, 61° 46.27' W, will place you approximately ¼ mile west of True Blue Point and southeast of the entrance to *True Blue Bay*. From the waypoint, steer towards the center of the bay before turning to the north to pick up a mooring or anchor off the marina.

Grenada: *Prickly Bay*

Prickly Bay is not a good spot to anchor for a hurricane, but there are slips available at *Prickly Bay Marina*, and a great yard, *Spice Island Marine Services*, perfect for a hurricane haul out. The problem with anchoring here is that the bay is open to winds from the south, which would occur should the eye of the hurricane move west of *Prickly Bay*; always know where the eye is forecast to move and the direction that the wind is expected to blow.

A waypoint at 11° 59.10' N, 61° 46.15' W, which will place you approximately ¼ mile south/southwest of the entrance to *Prickly Bay*. From the waypoint, head into the harbor keeping approximately halfway between True Blue Point and Prickly Point as shown on Chart GRE-8 (top of this page). The entrance is wide and deep and offers no hazards at this point, but you must avoid the shoals lying just off those two points, and especially the reefs along the eastern shoreline.

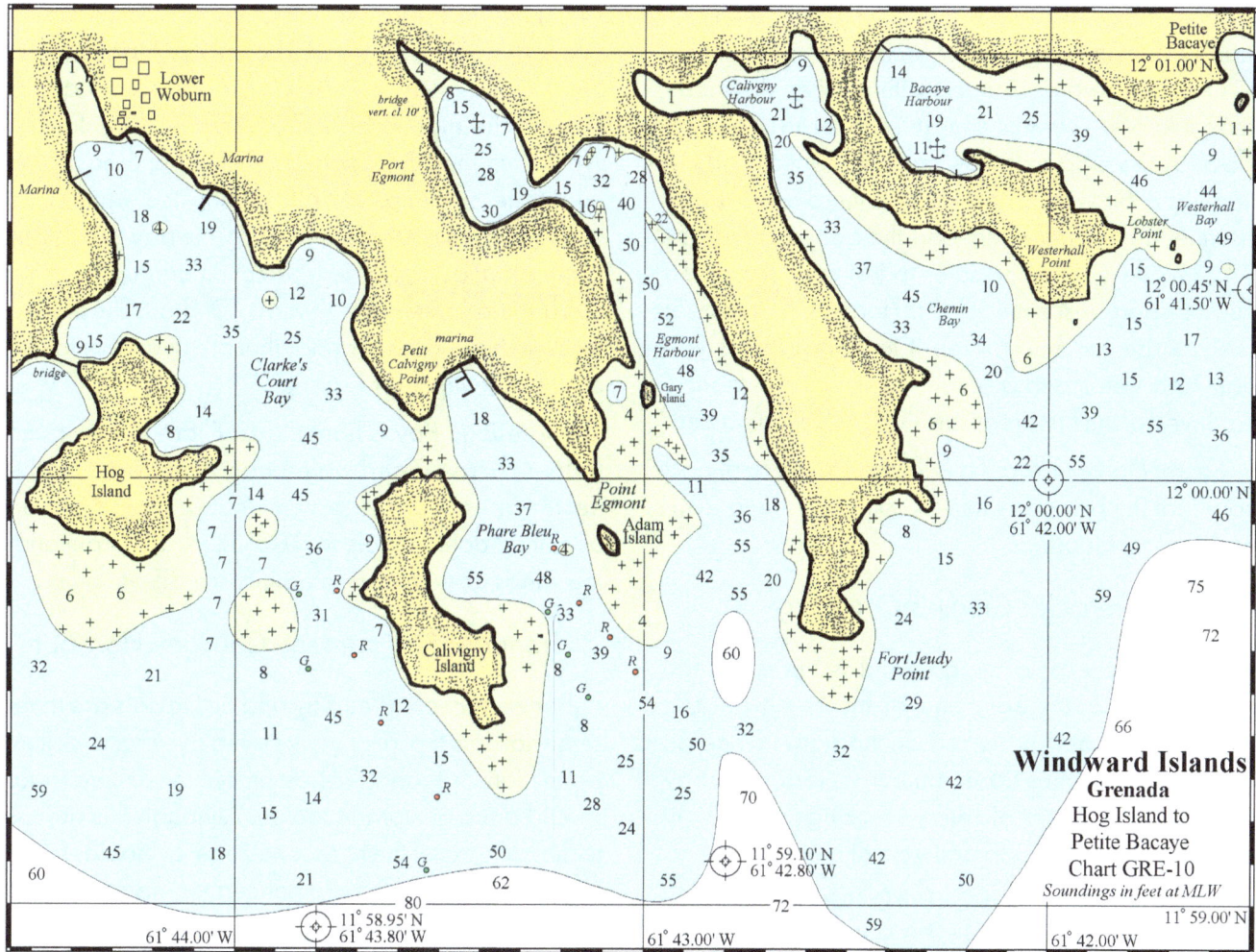

Prickly Bay Marina (http://pricklybaymarina.com/) offers dockage for vessels to 200' LOA with drafts to 17'. *Spice Island Marine Services* (http://spiceislandmarine.com/) is located in the extreme northwestern tip of *Prickly Bay*, just past the *Coast Guard* station. The yard has room for 200 vessels and their *Travelift* can handle boats to 70 tons with a 25' beam. The current depth at the dock is 10'. The yard offers hurricane preparation with mast removal and tie downs.

Grenada: *Mt. Hartman Bay*

Cruisers can also find a bit of protection tucked into the mangroves in Mt. Hartman Bay and behind Hog Island although these two anchorages are open to southerly winds. *Mt. Hartman Bay*, sometimes called *Secret Harbour*, lies just a bit east of *Prickly Bay*. Home to *Secret Harbour Marina* (http://www.secretharbourgrenada.com/), you can get a slip or mooring here, but there are no haul out facilities and you must once again beware of southerly winds.

Entrance to Mt. Hartman Bay should never be attempted at night, and caution must be exercised if visibility is poor and the reefs are not visible. As shown on Chart GRE-9 (top of previous page), a waypoint at 11° 58.90' N, 61° 45.05' W, will place you approximately ¼ - ½ mile south of the entrance that begins west of Tara Island. If approaching this waypoint from offshore, extreme care must be used as you must avoid the Porpoises, a small series of low-lying rocks that are difficult to see in good weather, and almost impossible to see in heavy weather.

From the waypoint head generally northward keeping the red buoy that marks the western edge of the Tara Island shoal to starboard as shown on Chart GRE-9. Bear in mind that any and or all of these markers may not be there when you arrive so don't panic if you don't see them, the reefs are visible in good light and with no glare on the water such as is sometimes encountered on cloudy days. The entry gets deep as you pass Tara Island, 60' in places, and once past Tara Island keep an eye out for the large

THE WINDWARD ISLANDS • 149

shoal to starboard that is marked by two small red markers that may or may not be there. Ahead of you will be a large buoy topped with a green daymark, keep it to port. It marks a large shoal area. Actually, the buoy sits atop a shallow spot of about 6' and the reef lies a bit west of it. Although there is a narrow channel between the reef and the buoy that carries 27'-32', it's easier and safer to just keep the green buoy to port. Just south of Mt. Hartman Point you'll see a large red buoy, keep it to starboard as you pass between it and a second green marker north/northwest of the first green one that you've just passed.

Secret Harbour Marina is located in the northeast corner of the bay and offers 53 slips (depths to 20'), and several moorings.

Grenada: *Clarke's Court Bay*

Just to the east of Hog Island is the entrance to Clarke's Court Bay, one of the south coast's most popular and protected anchorages. Once an anchorage for ships taking on cargo, today the bay is primarily the haunt of cruisers seeking to get away and get a bit of solitude and yet not be isolated.

As shown on Chart GRE-10 (top of previous page), a waypoint at 11° 58.95' N, 61° 43.80' W, will place you approximately ½ mile south of the entrance to *Clarke's Court Bay*. From the waypoint head generally northward passing to the west of Calivigny Island keeping a sharp eye out for the reefs that you'll want to take to port, passing between the reefs and Calivigny Island. Once past the reefs and the shoal off the eastern tip of Hog Island, you can find a good spot to anchor your vessel in the mangroves around the bay.

Tying off in the mangroves, *Port Egmont*, Grenada; note bridge in left background

On the northwestern side of the bay, across from Woburn, is *Clarke's Court Boatyard & Marina* (VHF, ch. 16 & 74; https://www.clarkescourtmarina.com/). The marina boasts 56 slips and the largest *Travelift* in the SE Caribbean. The marina has some hurricane moorings for rent during the winter and spring. On the northeastern side of the bay is *Whisper Cove Marina* (http://whispercovemarina.com/) that can handle vessels to 60' with a 10' draft. The marina also has hurricane moorings that are available to rent during the winter and spring.

Phare Bleu Bay is home to *Le Phare Bleu Marina* (http://www.lepharebleu.com/) located at the northern end of the bay with 50 slips that can accommodate vessels to 100' LOA. Use caution if the winds are forecast to go into the SE-S.

Grenada: *Port Egmont, Calvigny Harbour*

Between *Phare Bleu Bay* and St. David's are three small harbors that are rarely visited but offer excellent protection, arguably the best protection on Grenada, should a major storm threaten. Although this area is technically out of the hurricane zone, at least as far as the insurance carriers are concerned, named storms are not unknown and protection should be sought as soon as possible in the event of the approach of a storm.

The first of these anchorages, when approaching from the west, is Port Egmont, whose entrance is through *Egmont Harbour* as shown on Chart GRE-10 (top of previous page). A waypoint at 11° 59.10' N, 61° 42.80' W, will place you approximately ½ mile south of the entrance to *Egmont Harbour*. From the waypoint head generally north avoiding the reefs east of Adam Island and the reefs lying off the land to the east, on your starboard side; favor the Fort Jeudy Point side of the channel. You'll continue up *Egmont Harbour* past Gary Island until you get to the head of the bay, a pleasant anchorage in itself, and turn to port to enter *Port Egmont* (the prime choice of co-author Stephen J. Pavlidis who has ridden out a storm here).

Once inside *Port Egmont* you can anchor almost anywhere or tie off into the mangroves if you choose. Don't get too near the small bridge at the northern end, you would not want to blow into that should the worst case scenario occur. The bottom is a bit grassy in places here so make sure your anchor is set well. Both *Port Egmont* and *Calivgny Harbour* are mangrove-lined and offer deep water right up to the mangroves in many places.

A bit further east lies *Calivigny Harbour* at the north end of *Chemin Bay*. This is another enclosed anchorage, very similar to *Port Egmont*, only smaller. From the waypoint head south of *Egmont Harbour* as shown on Chart GRE-10. Work your way south of Fort Jeudy Point keeping an eye out for the dangerous reefs south of the point, to a waypoint at 12° 00.00' N, 61° 42.00' W, which will place you approximately ¼ mile south/southeast of the entrance to *Chemin Bay* and south of Westerhall Point.

From this waypoint you must thread the needle between two shoals that can be difficult to see at times. Never attempt this passage in heavy following seas! Once inside *Chemin Bay* head up the bay until you can round the point to starboard and anchor in Calivigny Harbour.

Still eastward is another small bay, *Bacaye Harbour* (at the northwestern end of *Westerhall Bay*). Although not as sheltered as the two already mentioned, this bay is certainly worth consideration if the other harbours are full. From the waypoint south of Westerhall Point as shown on Chart GRE-10, make for a waypoint at 12° 00.45' N, 61° 41.50' W, which will place you approximately ¼ mile south/southeast of the entrance to *Westerhall Bay*.

From this waypoint keep the rocks off Lobster Point well to port as you enter *Westerhall Bay* and then turn to the northwest to again thread the needle between two reefs as you work your way into *Bacaye Harbour*.

Grenada: *St. David's Harbour*

St. David's Harbour is a very popular stop for cruisers on the southern coast of Grenada, due in no small part to the good reputation enjoyed by *Grenada Marine* (http://grenadamarine.com/), the local yard.

Approximately 1¼ miles east of Westerhall Point lies the entrance to the harbour at St. David's.as shown on Chart GRE-11 (bottom of next column). Anchoring here during a hurricane is not the smartest move, even a slip can get rough, but this is a great spot to haul out.

A waypoint at 12° 00.40' N, 61° 40.60' W, will place you approximately ½ mile south of the entrance channel into St. David's Harbour. From the waypoint head northward towards the mouth of the entrance channel lying between two reef systems and you will pick up lighted markers to guide you in as shown on the chart.

Grenada Marine (monitors VHF ch. 16/14) offers 60 slips that can handle drafts to 26'. The yard has a 30-ton *Travelift* that will haul boats up to 32' wide with a 12' draft. Hurricane tie-downs are available and mast removal is no problem. *Grenada Marine* buried 750 steel and concrete anchors in its yard and custom made many boat cradles that can accommodate vessels to 80' LOA.

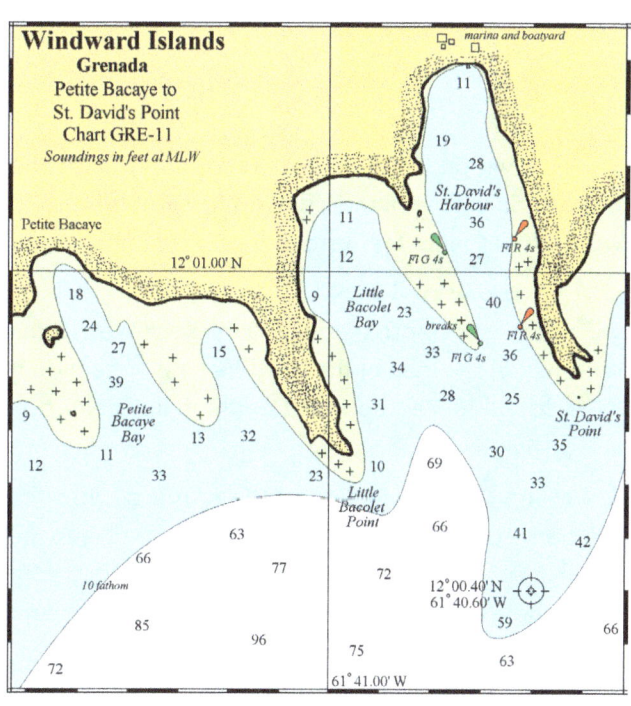

Chapter 16

Trinidad and Tobago

TRINIDAD AND TOBAGO ARE GENERALLY CONSIDERED TO BE OUTSIDE of the normal hurricane zone, but if the truth be known, hurricanes, though extremely rare, are not unknown here. Between 1850 and 2000, only two hurricanes (the last one was Hurricane Flora in 1963) and six tropical storms (five since 1990 with Bret crossing the southwestern peninsula of Trinidad as recently as June of 2017) hit Trinidad/Tobago. In 1995, Hurricane Iris passed 180 miles north of Trinidad, but brought heavy southerly winds that did a lot of damage along the *Gulf of Paria*. Hurricane Lenny, although far to the north of Trinidad, brought huge seas to the northern and western coasts of Trinidad and Tobago in November of 1999. Although your chances of getting hit by a hurricane in Trinidad or Tobago are small, don't get complacent. There is not a lot of hurricane protection on these islands, only one tiny spot on the island of Tobago, *Bon Accord Lagoon*, and several small coves on Trinidad located at Port of Spain, *Scotland Bay*, *Winns Bay*, and the *Carenage*. You can also haul out at any of several yards in Chaguaramas. There aren't any haul-out facilities or marinas in Tobago.

Trinidad: *Scotland Bay*

As shown on Chart TRI-4 (next column), *Scotland Bay* lies on the mainland side of *Boca de Monos* and works its way into the surrounding mountains.

A waypoint at 10° 41.75' W, 61° 40.20' W, will place you approximately ¼ nm southwest of the entrance. As you enter the bay, the best protection is to be found tucked up in the northern tip of the bay in 20'-40' of water.

For the best protection work your way into the northern tip of the bay. Drop your hook between the small beach on your left and the concrete slab to your right where people sometimes camp. The bottom here is mud and the holding is good.

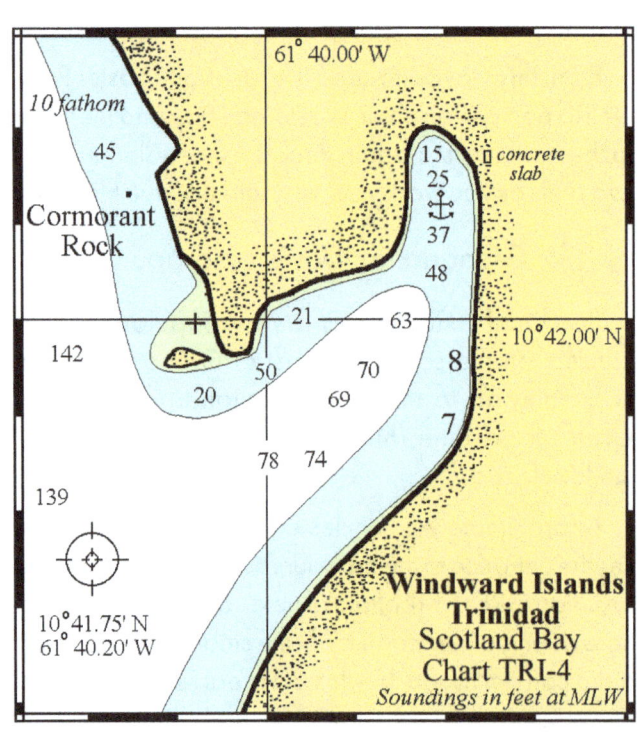

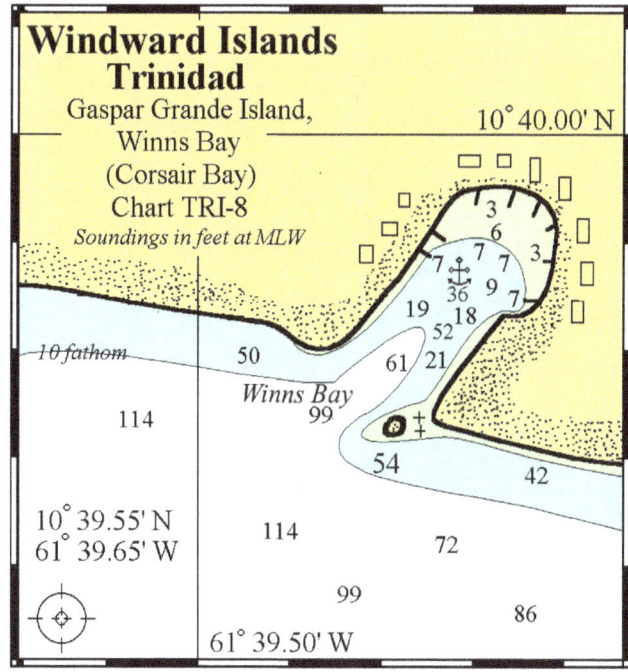

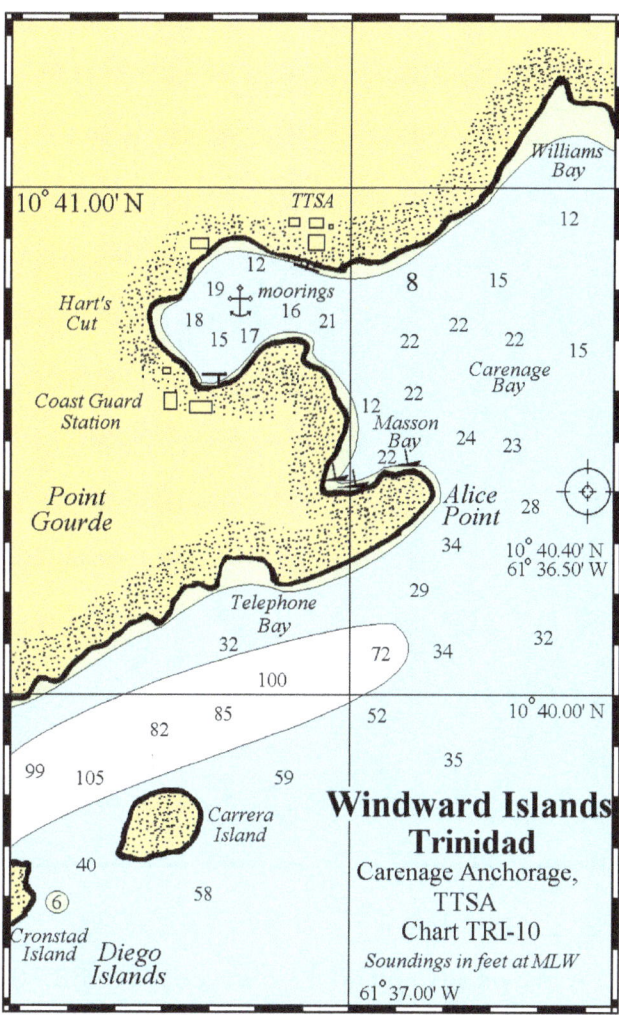

Trinidad: *Winns Bay*

On the south shore of Gaspar Grande Island, lies *Winns Bay*, a tiny anchorage marked by a large fig tree on the small rock lying off the point.

As shown on Chart TRI-8 (see above), a waypoint at 10° 39.55' N, 61° 39.65' W, will place you approximately ¼ mile southwest of the entrance. Head in towards the bay between the off-lying rock and the point of land to the west of it. The rock is very conspicuous, there is a small stand of fig trees on the rock.

The water will shallow towards the northern end of the bay so don't head too far in, keep an eye on the depthsounder. The bay is surrounded by several vacation homes and there are no facilities ashore here. In moderate winds from any southerly direction it can get rolly in here so if the wind is forecast to go into any southerly direction you need to find better shelter.

Trinidad: *The Carenage*

A waypoint at 10° 41.40' N, 61° 36.50' W, places you approximately ½ mile southeast of the mooring field off *TTSA* as shown on Chart TRI-10 (top of next column). Some shelter can be found in *Masson Bay* and in the cove by *Hart's Cut*, just off the *Coast Guard* station docks.

These anchorages are a bit open to the east, but it is southeast winds that would do the most damage in here so know your wind forecast before you seek shelter here. *TTSA* has a haul out facility here with a 20-ton lift. Check with them if you require a haul.

Trinidad: *Port of Spain*

Just south of *Grier Channel*, the entrance to the commercial docks in Port of Spain, Trinidad, is the *Sea Lots Channel*, which leads to a small cove that offers fair protection in the event of a major windstorm. The cove is used primarily by commercial fishing vessels, and there are several large wrecks lining the shore, but a cruising boat could find some shelter in the southern end of the cove between the wrecks and the mangroves. You'll have to get here early though as the commercial boats will certainly be heading here the minute they realize a storm is on its way. Please note that the primary danger here would be from other boats.

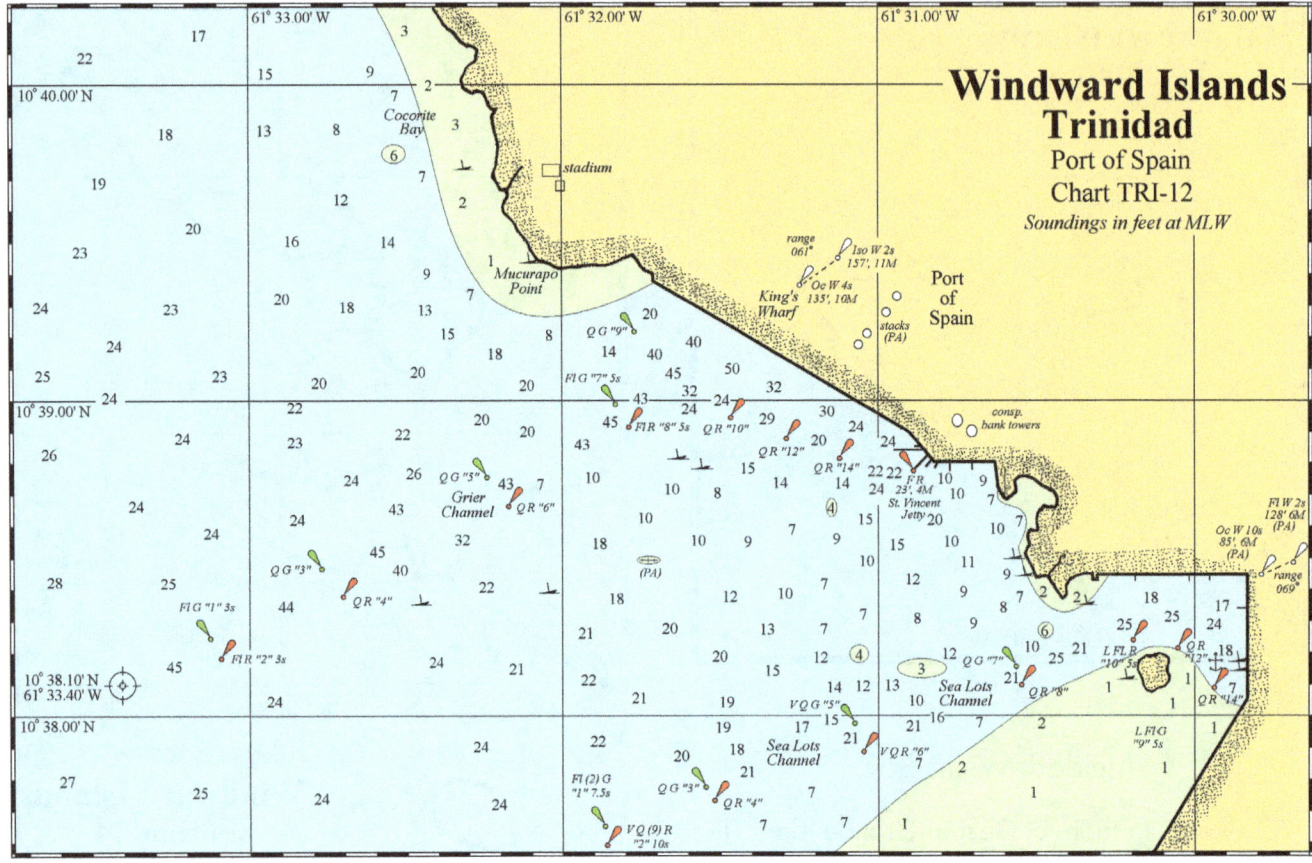

As shown on Chart TRI-12, a waypoint 10° 38.10' N, 61° 33.40 W, will place you approximately ¼ mile west/southwest of the entrance to *Grier Channel*, just outside the first buoys.

From the waypoint at the entrance to *Grier Channel*, head southeast and you will pick up the well-marked *Sea Lots Channel*. Follow the channel until you pass the small cay and can turn to starboard to tuck up in the cove.

Tobago: *Bon Accord Lagoon*

Bon Accord Lagoon is off-limits to boaters during the year, but it is available in case of a hurricane. The entrance is tricky, a winding path over a shallow reef before reaching the safety of the small, protected, mangrove-line cove. Here you will find good holding in 10' and great protection from all directions. Entrance directions will be given here with the understanding that the anchorage is only to be used for shelter from a hurricane. As of 2016, the markers defining the reef area were reported as missing.

As shown on Chart TOB-4 (see next page), a waypoint at 11° 11.30' N, 60° 50.70' W, will place you approximately ½ mile northwest of *Gibson Channel*, the entrance channel to *Bon Accord Lagoon*. You'll have to use your eyes to pilot through here, no waypoints can be given. If approaching from Pigeon Point, keep an eye out for *False Channel*, easily recognized, that dead ends quickly in the reef system.

Gibson Channel can accept a 4' draft a bit above low water, and 6'-7' at high water (remember that the tidal range in Tobago is roughly 3'). From the waypoint given, pilot your way through the deep-water pocket and then proceed on the dogleg leading into *Bon Accord Lagoon*. Go slow, and use your eyes.

If you're not happy with the protection in Trinidad and Tobago, and if there is time, you can head southwest to Venezuela for better protection. Hurricanes are usually not a problem for boaters in the waters of Venezuela which is why so many cruisers flee the Leeward and Windward Islands, as well as Trinidad and Tobago, if a storm threatens. Those heading south from the Leewards or Windwards usually make for

Isla Margarita as most storms stay north of there. But for those wanting a bit more protection, the waters of the *Rio Macareo* or the *Rio Orinoco* will give shelter to Trinidadian cruisers who don't wish to travel a long distance for shelter. Those heading south from the more northerly islands can also find shelter at Puerto La Cruz and in one of my favorite places, *Laguna Grande*.

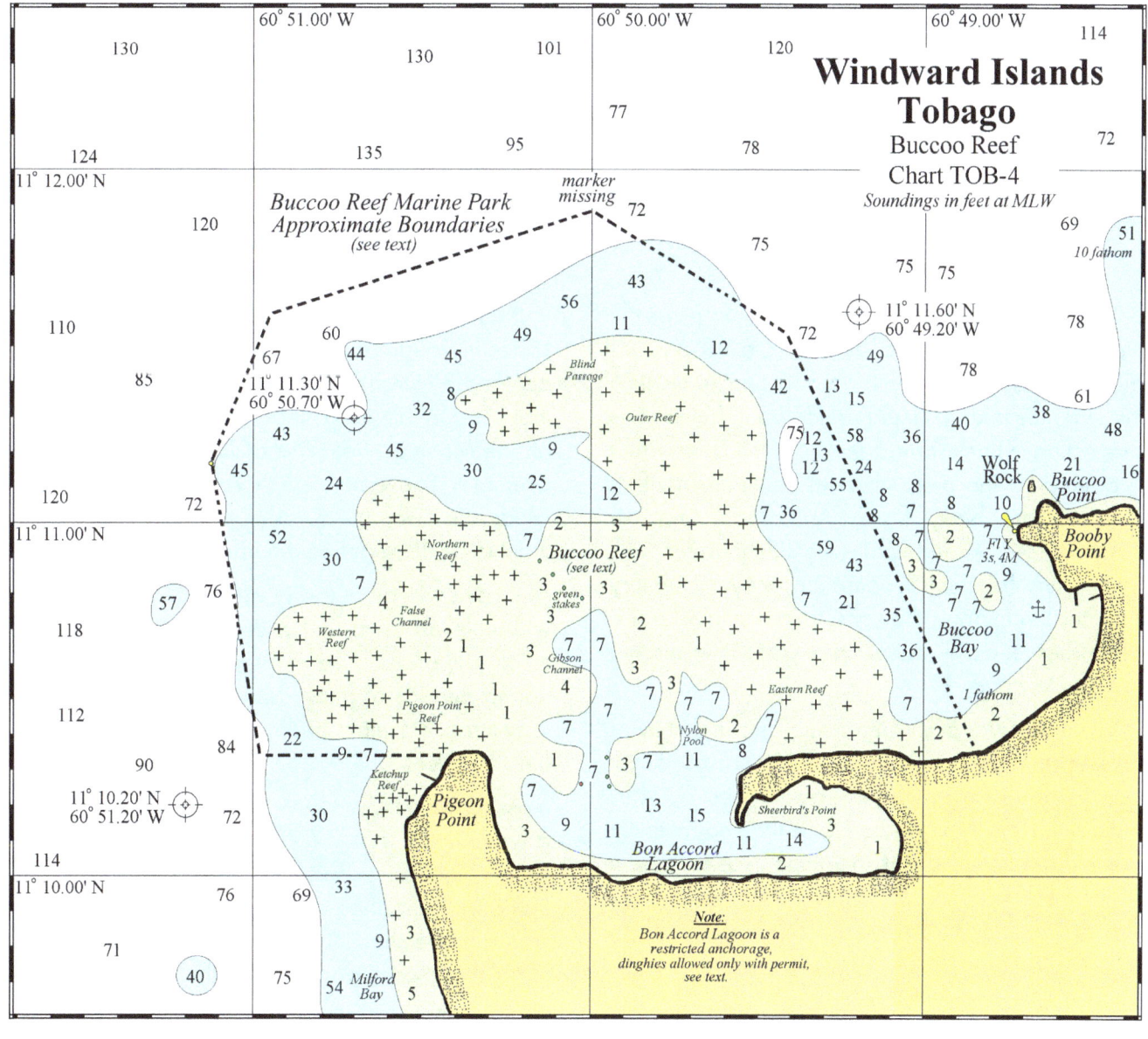

Chapter 17

The Northern Coast of Jamaica

ALTHOUGH PORT ANTONIO IS OFTEN TOUTED AS A GOOD HURRICANE hole by some locals, there is only one spot that you should even consider when seeking hurricane shelter on the northern coast of Jamaica, and that is at *Bogue Lagoon*, just south of *Montego Bay*. However, we shall discuss *West Harbour* at Port Antonio just in case you can't get to better protection at *Bogue Lagoon*.

Port Antonio: *West Harbour*

As shown on chart JAM-1 (see next page), a waypoint at 18° 11.75' N, 76° 26.80' W, will place you approximately ½ nautical mile north of the entrance channel into *East Harbour*, but the best protection will be found in *West Harbour* which is gained by the deep channel between Titchfield Peninsula and Navy Island. An easy landmark on the entrance from seaward is the lighthouse on Folly Point, which has been in operation for over a century. Another landmark is the conspicuous small stand of palm trees atop a hill about 1 mile east of the entrance to *East Harbour*.

From the waypoint head mid-channel staying between the red and green buoys. The best protection can be found in the southeastern end of *West Harbour* were Navy Island and the shoals to the west of the island break incoming seas. You can also check with the *Errol Flynn Marina* about a slip or hauling out for the duration of the storm.

Bogue Lagoon

If your goal is the fine shelter afforded by *Bogue Lagoon*, you must pass south of *Montego Bay* to access the entrance channel. Some folks might say that you can ride out a hurricane in the small cove just off the *Montego Bay Yacht Club* but I would never attempt it. I have been anchored there in moderate winds and it was like a washing machine, I would not want to be there during a hurricane.

Vessels with drafts to 6½' can enter *Bogue Lagoon* with little problem, but a draft of 7'-8' will need to play the tide. Use extreme caution in entering during periods of poor visibility and never attempt this entrance at night! Take it slow and easy through here and keep one eye on your depth sounder.

As shown on Chart JAM-10 (following next page), a waypoint at 18° 27.28' N, 77° 57.70' W, will place you approximately ¼ mile northwest of the unmarked entrance channel through the reefs as shown on the chart. If you are approaching from the north, from *Montego Bay*, give the reefs west of the

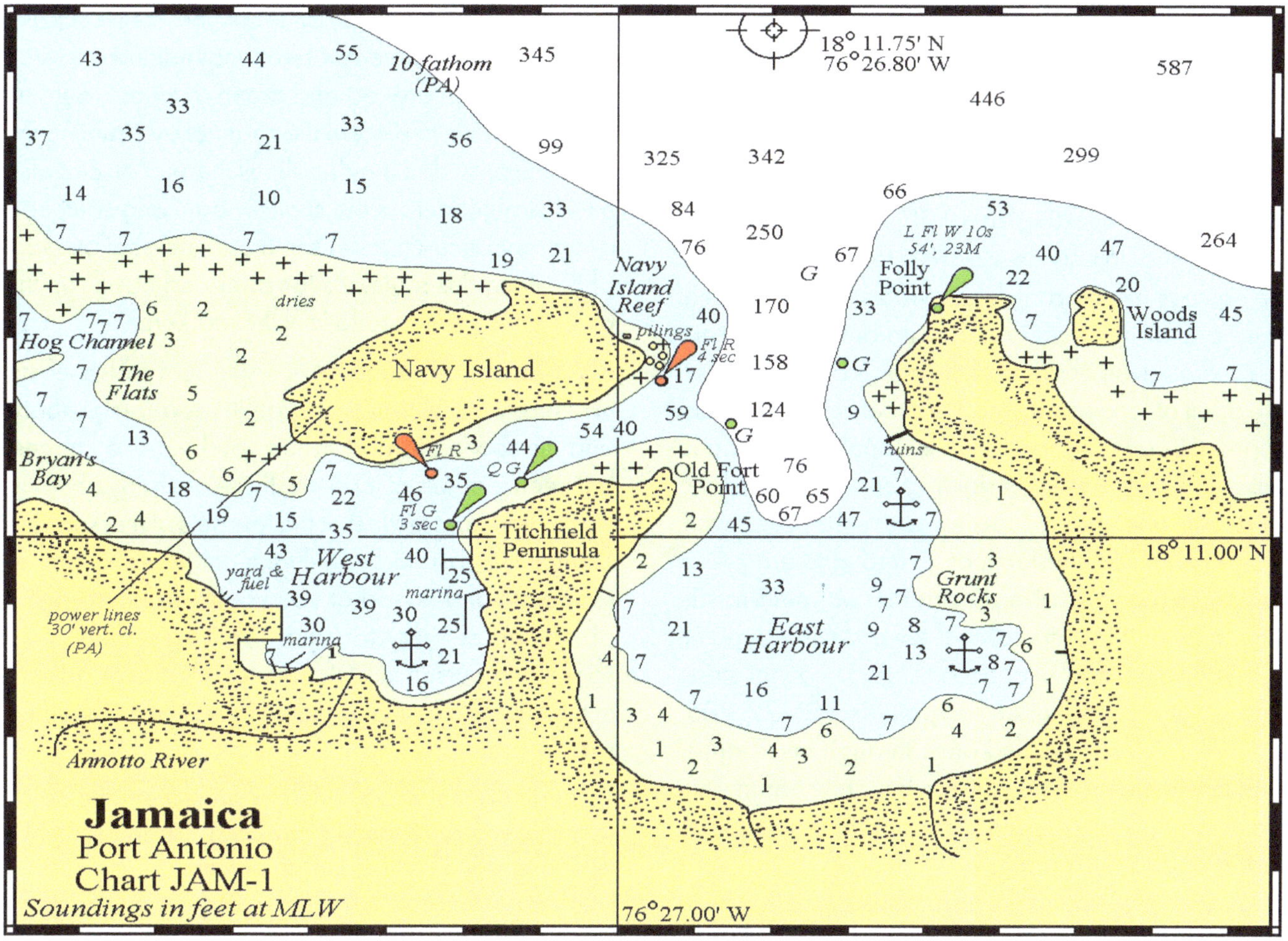

West Harbour, Port Antonio, with Navy Island in the background

THE NORTHERN COAST OF JAMAICA

Freeport Peninsula a wide berth. From the waypoint you'll easily see the reef on the southern side of the entrance channel, it dries in places at low water and is quite visible even at high water.

From the waypoint you can head in on an approximate heading of 113° magnetic, but this heading should only be used as a reference, the idea being you must keep the highly visible southern reef at least one boat length to starboard and steer toward the center of the opening of the mangrove island and the point of land that just northward from the southern shore of *Bogue Lagoon*. There is a small cone-shaped hill in the distance when sighting between the island and point and you can use that as a reference as well. There is a minimum depth of 7' through here. As a reference I can also give you an inner waypoint for the entrance, but don't try to steer toward this waypoint, use your eyes and pilot accordingly, I cannot stress that enough! The inner waypoint is 18° 27.205' N, 77° 57.548' W, and once through the channel and inside the reef you can turn to port to begin your passage through the markers and into the deeper water of *Bogue Lagoon*.

As you'll notice on the chart, there are several markers leading northeast and then southeast into the bay. These markers, although they're steel posts set in concrete, have not been maintained and you can't tell which are red and green anymore, but you should be able to discern the channel by counting the markers, provided they're all still there. You can also head straight across the shallow bar lying southeast of the entrance channel but there is a shallow spot of 3' along this route, it's far safer and easier on the blood pressure if you follow the markers.

As shown on the chart, as you turn to the east/ northeast, the first marker you'll keep to starboard and the second to port. Then you'll enter a channel between four sets of markers, these are fairly obvious. When you pass the last of the four sets of markers turn to the southeast and you'll take one more marker to starboard and ahead of you you'll see another pair of markers set very far apart, split these and you'll soon find yourself in 20' of water. Find a good spot in the mangroves along the perimeter of the bay and secure your vessel.

If you are not happy with the protection offered by Jamaica, you can head north to Cuba, Santiago lies 119 miles away and Portillas lies 80 distant.

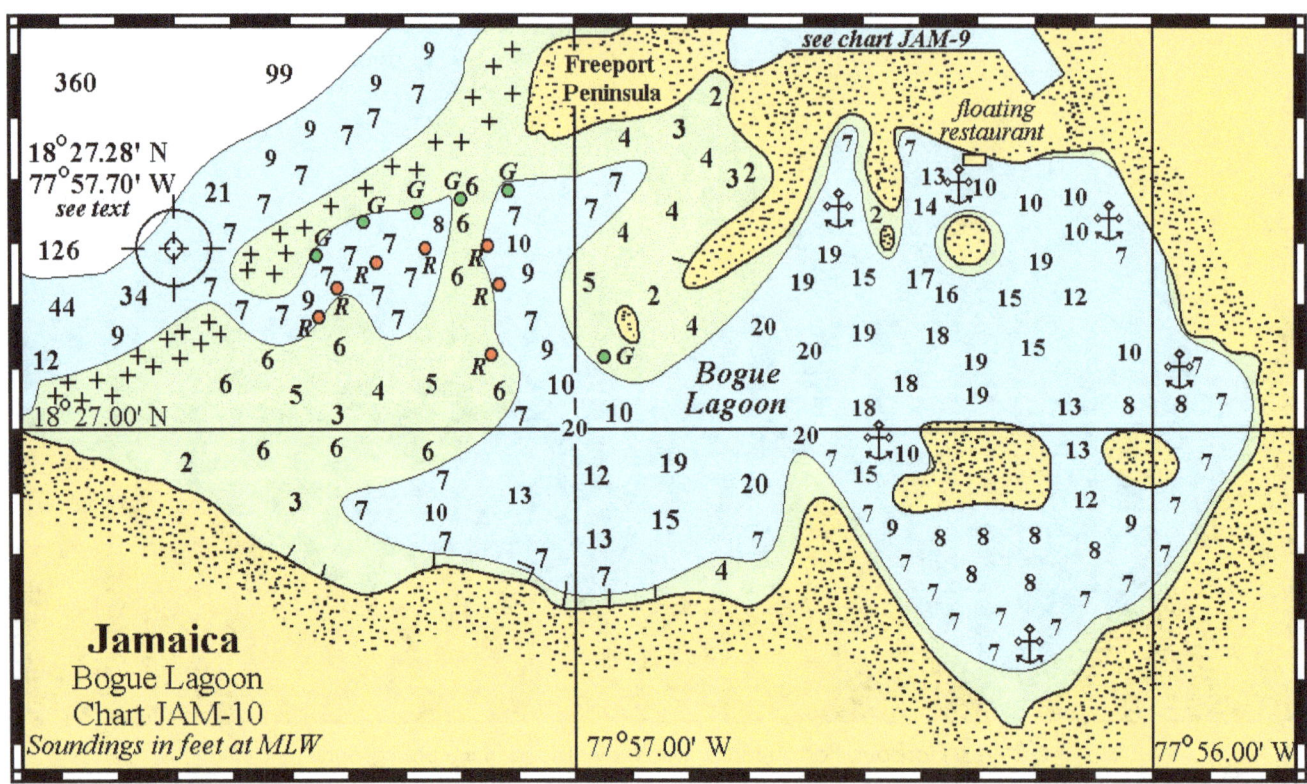

158 • THE CAPTAIN'S GUIDE TO HURRICANE HOLES

Chapter 19
Honduras and the Bay Islands

ON THE MAINLAND OF HONDURAS ONLY TWO PLACES CAN GIVE YOU any decent type of protection in the event of a hurricane, one being upriver of Puerto de Cabotaje at *La Ceiba Shipyard*, the other is *Laguna el Diamente* just southwest of Punta Sal. In La Ceiba you can count on the river at *Puerto de Cabotaje* to bring a lot of debris downstream as it did in Hurricane Mitch but in a pinch this shelter will suffice as you can haul out, get a slip, or anchor up-river, especially if there is not a strong storm surge. *Laguna el Diamente* offers very good protection but a strong storm surge will make this harbor untenable (as would a direct hit). If nothing else is available in a reasonable amount of time, *Laguna el Diamente* might be an option to consider as long as the hurricane is forecast to miss this area. In the Bay Islands you have a few choices, mostly on Roatán although small, shallow-draft vessels can work their way up the small canal west of Bonacca on Guanaja, and also into the lagoon at *East Harbour*, Utila.

If you plan to stay in any of the holes on Guanaja or Roatán for a hurricane remember that you'll likely be sharing your spot with a lot of local shrimpers. If you anchor behind some of the low, coastal islands off the southern shore of Roatán, remember that a strong storm surge could devastate these islands and the protection they offer. If you wish to anchor between Anthony's Cay and the northern shore of Roatán we cannot recommend this area. The small waterway is narrow, does not offer very good protection from the wind, and there are submerged power cables you would have to avoid. Coxen Hole has a small cove alongside the airport runway but it has a lot of jetty rocks on the bottom so it is not recommended.

Bay Islands: *Guanaja. El Bight*

El Bight is a semi-protected anchorage, not what I would truly call a "hurricane hole" by any means, but it does offer shelter and good holding from most weather and it is only mentioned here as a "last resort."

Located northeast of Bonacca as shown on Chart HON-5 (see top of next page), the entrance is easy, passing between Dunbar Rock and the point of land that is the southern shore of *El Bight* if you're approaching from *Savannah Bight*, or, if you are approaching from Bonacca, by passing west of Dunbar Rock and being careful to avoid a couple of shallow spots as shown on the chart, you can then pass north of the shoal lying north of Dunbar Rock to enter *El Bight*.

Use caution when rounding the point to the east of *El Bight*, the waters here are 7'-8' in depth several hundred yards off the point. The best shelter head is found to the eastern end of the cove, as far as you past the wrecks as you can get. The cove is open to the WSW-NW.

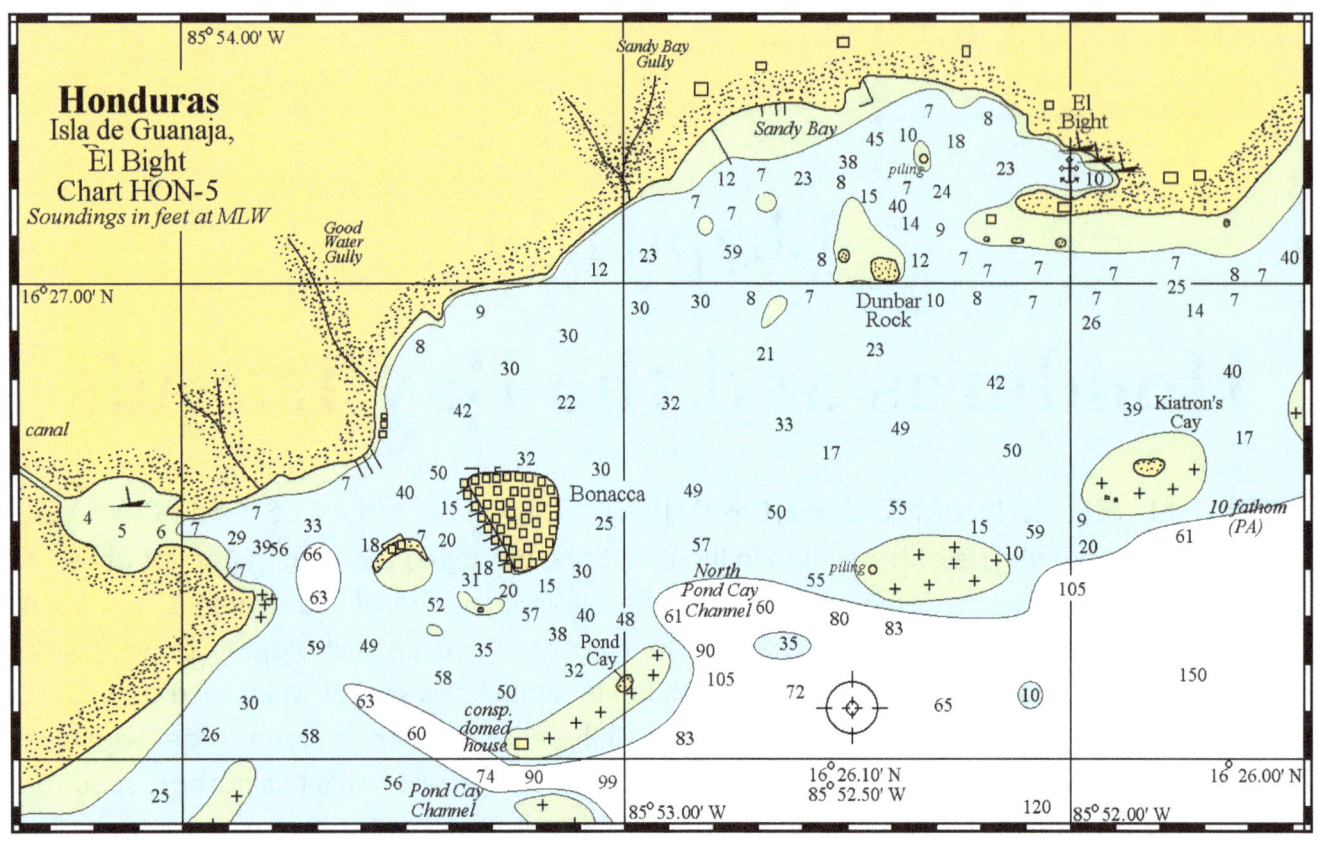

Bay Islands: *Guanaja. West of Bonacca*

West of Bonacca, often shown as Sheen Cay on some charts, is a small cove that is the eastern terminus of a canal that crosses Guanaja to the northern shore at *Pine Ridge Bight*. The canal, while too shallow in spots for most cruising boats (it was dredged to 9' in the 1980s but today the controlling depth is about 4'), it still offers good protection from a storm. As shown on Chart HON-5 (see above), when entering the bay from Bonacca stay to the north side of the channel and then stay south of the pair of wrecks in the inner bay. Shallow draft boats can work their way farther up into the safety of the narrow canal. You can be sure that local boats will fill up this canal quickly.

Bay Islands: *Roatán, Old Port Royal*

We will discuss the protection offered by the southern shore of Roatán from east to west (there is no true protection on the northern shore). Old Port Royal is the first stop as we travel along the southern shore of Roatán westward.

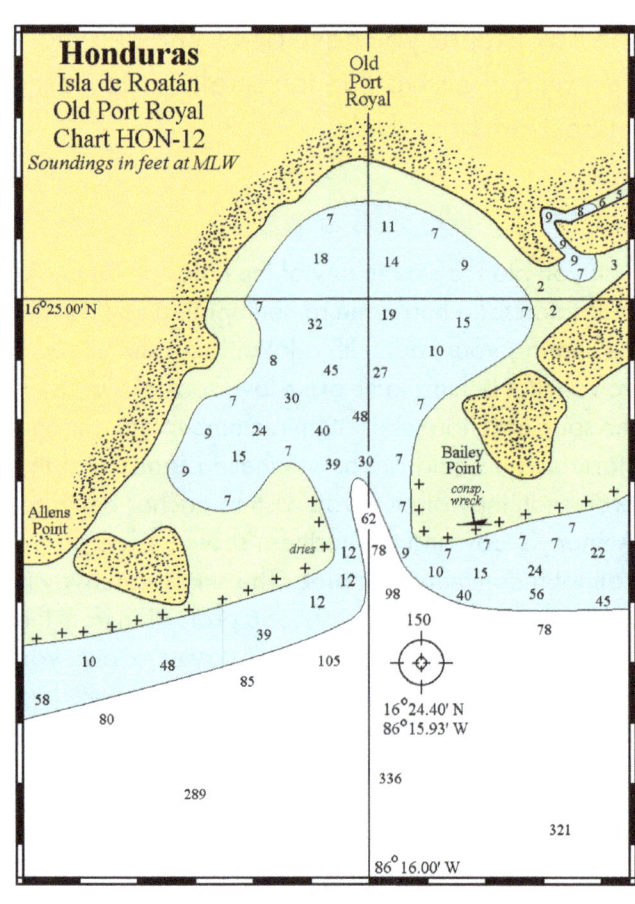

As shown on Chart HON-12, a waypoint at 16° 24.40' N, 86° 15.93' W, will place you approximately ¼ mile south/southeast of the channel. From the waypoint head a bit west of north into the deep channel between the unnamed cay off Allens Point and Bailey Point as shown on the chart. Once inside the harbor you can work your way to the NE corner where you will find the shallow entrance to a narrow creek that offers good protection. When entering, favor the north side of the channel and try to enter on a course of 90°.

Bay Islands: *Roatán, Calabash Bight*

Calabash Bight offers good protection if you tuck yourself well up into the northern extremity of the bay. Bear in mind that if you plan to stay in Calabash Bight for a hurricane that you'll be sharing this spot with a lot of local boats, many of them large, unattended, steel, shrimpers! Most folks here monitor VHF ch. 72 or 22.

A waypoint at 16° 23.00' N, 86° 20.00' W, will place you approximately ¼ mile south/southeast of the entrance channel as shown on Chart HON-14 (below). Line up and pass mid-channel between the reefs and the red and green buoys that mark the channel's edge. Once inside pass about ½ boat length to the west of the white buoy, this is a narrow channel between two shoals and caution must be exercised.

If you would rather get a slip instead of anchoring, on the west side of *Calabash Bight* is *Turtlegrass Marina* (http://www.turtlegrass.net/) 400' of

Southwest portion of *Oak Ridge Harbor*, Roatán

dockage (the marina can handle boats up to 48' LOA with a draft of 9' and monitors VHF ch. 16 and 72) and one slip that can accommodate one 30' wide catamaran.

On Chart HON-14 you will also notice *Fiddler's Bight* and *Oak Ridge Harbor*, both of which have power lines overhead near their northern ends. For this reason, we cannot recommend them as a hurricane hole for the simple reason it might be too easy (not to mention dangerous) to be blown into the power lines in a powerful storm.

An alternative is to anchor in the far western tip of the canal that leads southwest just inside the mouth of *Oak Ridge Harbor* (see photo above), but it will likely be full of local boats. A strong storm surge may also decimate the skinny arm of land that offers protection.

Bay Islands: *Roatán, Hog Pen Bight*

Hog Pen Bight offers little to the cruiser save good protection from all weather. As shown on Chart HON-15 (see top of next page), a waypoint at 16° 23.05' N, 86° 21.75' W, places you approximately ¼ mile SSE of the well-marked entrance channel. From the waypoint head just a bit west of north until you can pick out the red and green markers as shown on the chart, the green marked is backed by a white maker further inside. Split the markers and head up into the bay avoiding the shoal in the center, and although you can pass it on both sides (if you draw less than 7'), the deeper water is to the east of the shoal. You can anchor wherever your draft allows but the best protection is offered at the extreme northern end.

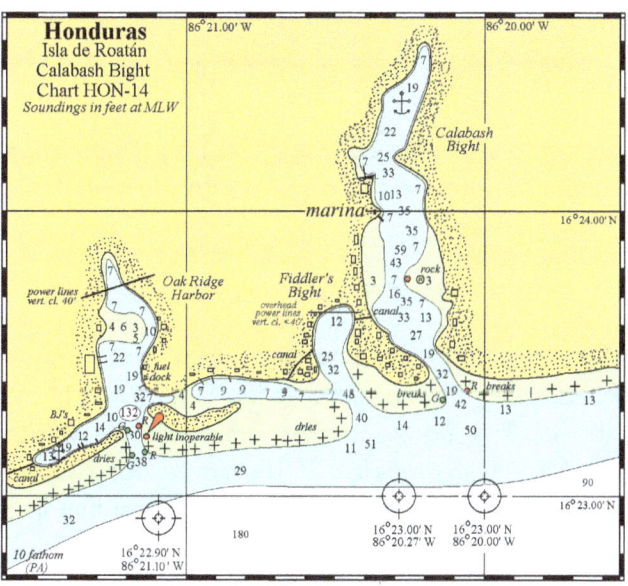

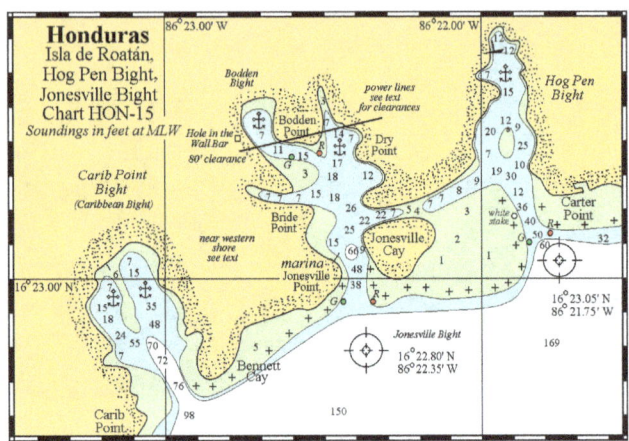

Parrot Tree Marina, Second Bight, Roatán

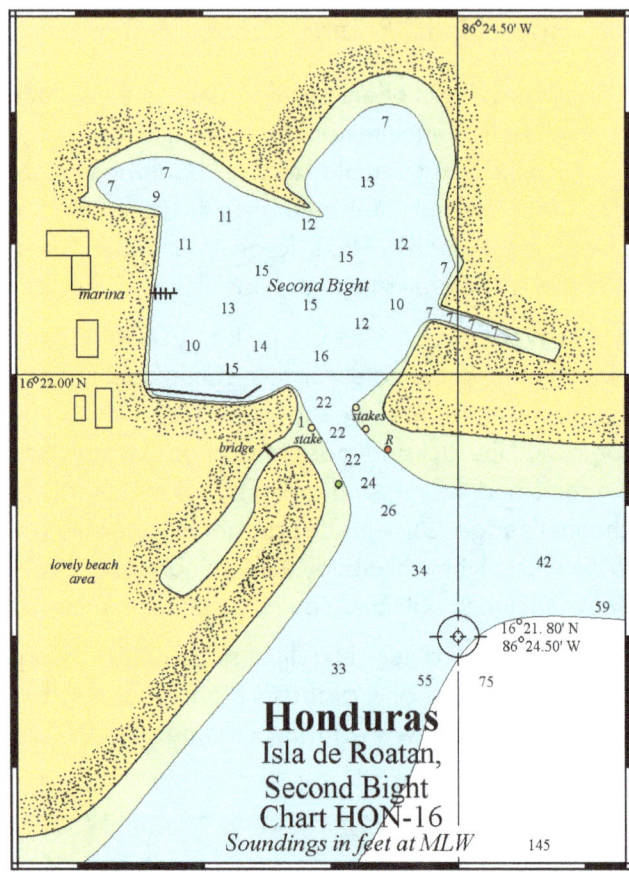

Bay Islands: *Roatán, Bodden Bight*

Bodden Bight is the small cove at the extreme NW corner of *Jonesville Bight* and although you must pass underneath a set of overhead power lines (again, this is one reason to consider NOT using this cove), the small bay offers good protection. There is a small marina to port as you enter *Jonesville Bight*, it's called *Jonesville Point Marina* (504-9967-3803) and has 4 side-tie slips available. The docks are low and fixed, and I cannot recommend riding out a hurricane here.

As shown on Chart HON-15 above, a waypoint at 16° 22.80′ N, 86° 22.35′ W, will place you ¼ mile south of the entrance channel into *Jonesville Bight*. From the waypoint head northward through the red and green markers as shown on the chart, past Jonesville Cay and into the harbor itself. Once inside you can anchor on the eastern side of the harbor or head northwest to anchor in *Bodden Bight*.

Those wishing to anchor in *Bodden Bight* will at first wonder how in the world did those sailboats already anchored there get under those power lines? The power lines will appear daunting at first but there is a way under them. To the east of *Bodden Bight*, across much of the bay, the clearance under the power lines is around 30′-40′, but on the western side of the harbor the clearance increases to almost 80′ along the western shoreline.

As you approach the northern part of *Jonesville Bight* you'll see a small red buoy in front of you as shown on the chart. Keep the red buoy to starboard and avoid the shoal to port as you pass the green buoy on your port side. You're heading for the peach-colored house on the shore. Stay close inshore as you parallel the shoreline into *Bodden Bight* and you'll be fine.

Bay Islands: *Roatán, Second Bight*

Second Bight is home to the largest marina on Roatán, the *Parrot Tree Plantation Marina* (see photo above). The marina can accommodate vessels to 170′ LOA with drafts to 13′. The marina has 20 fixed docks for vessels to 75′ LOA and several hundred feet of side-tie dockage. The marina (http://www.parrottree.com/) monitors VHF ch. 16 and 63'

As shown on Chart HON-16 above, a waypoint at 16° 21.80' N, 86° 24.50' W will place you ¼ mile south/southeast of the marked entrance channel. The channel is narrow, only 100' wide, but it carries 22' of depth. From the waypoint, heard towards the opening, you'll see a red marker, keep it to starboard. Past the red marker you'll see two steel rails to starboard and a bent one to port, pass between the two and the very visible shoals and you're inside *Second Bight*. The marina docks are directly in front of you. The marina suggests a heading of 328° as you enter the channel, but it is better to just keep your eyes on the shoals as the channel is only 100' wide.

You can anchor in the northeastern section of the bight, or perhaps in the man-made channel on the southeastern side of the bight.

Bay Islands: *Roatán, CoCo View Marina*

The *CoCo View Marina and Resort* has, in recent years, not been overly friendly to the cruising

CoCo View Marina, Roatán

community due to their facilities being abused by cruisers. However, north of the resort is a mangrove encircled cove (photo top of next column) that offers protection, but the waters are shallow and you'll have to pick your way in through some difficult to see shoals.

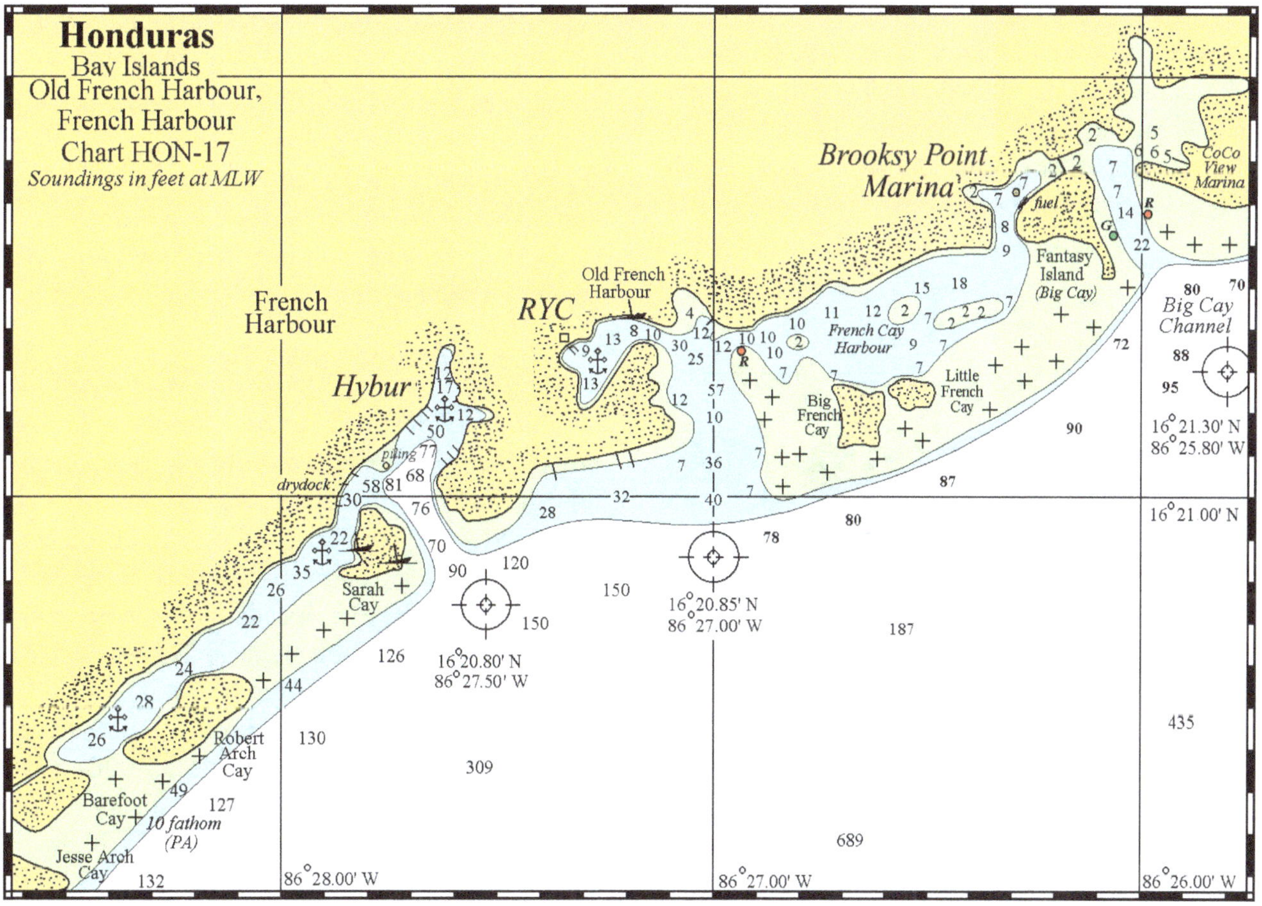

As shown on Chart HON-17), a waypoint at 16° 21.30' N, 86° 25.80' W, will place you ¼ mile south/southeast of *Big Cay Channel*. From the waypoint head in between the markers until you can round the point to work your way into the waters that sit north and northeast of the marina

Bay Islands: *Roatán, Brooksy Point Marina*

In a small, protected cove NW of Fantasy Island as shown on Chart HON-17 (previous page) sits *Brooksy Point Marina* (see photo at the bottom of this column). *Brooksy Point Marina* (http://brooksypointyachtclub.com/) (504-9455-2330), has 14 slips (bow or stern-to with floating docks) and a few mangroves where you might secure your vessel. The marina also has large concrete blocks on the bottom all chained together from each block up to a mooring ball.

A waypoint at 16° 20.85' N, 86° 27.00' W, is the gateway to *French Cay Harbour*, and *Brooksy Point Marina* as shown on the chart. If you wish to venture to *Brooksy Point Marina*, head north from the waypoint until you can take the red marker to starboard and turn towards the east. You will pass between the mainland of Roatán and Big French Cay as you work your way north of east to Fantasy Island keeping a close lookout for the shoals scattered about in *French Cay Harbour*.

As you approach Fantasy Island you'll want to work your way around to the northern shore of the island where the fuel dock is located. To port, WNW of Fantasy Island, you will find the small cove that is home to *Brooksy Point Marina*.

Old French Harbour, Roatán as seen from the northwest

Bay Islands: *Roatán, Old French Harbour*

Old French Harbour (see photo at top of next column) is a protected little cove with a small marina and is popular with the shrimpers during a hurricane.

As shown on Chart HON-17 (previous page), a waypoint at 16° 20.85' N, 86° 27.00' W, is the gateway to *Old French Harbour* (see photo above). From the waypoint you can head northward in the wide channel between Big French Cay and the unnamed point to port. To enter *Old French Harbour*, keep heading north until you can follow the channel as it curves to port into *Old French Harbour*. On the northern shore is the *Roatán Yacht Club* (504-9490-0042), formerly the *French Harbour Yacht Club*. The marina has 19 deep water slips. If you don't want a slip you can anchor but expect a few large, steel, shrimpers for neighbors.

Bay Islands: *Roatán, French Harbour*

West of *Old French Harbour* lies the entrance to *French Harbour*, home to one of the largest shrimping fleets in the Western Caribbean. *French Harbour* has a daily VHF radio net on ch. 74 at 0900 daily and ch. 72 is the hailing channel for *French Harbour*. Rest assured that this bay will be filled with shrimpers in the event of a storm, but you might find some room in the narrow channel between Robert Arch Cay and the mainland of Roatán. There are two places to haul out in *French Harbour*, Hybur (504-455-7590) and the *French Harbour Marine Railway*, but they both will likely be filled with shrimpers.

Brooksy Point Marina, Roatán

As shown on Chart HON-17, a waypoint at 16° 20.80' N, 86° 27.50' W, will place you approximately ¼ mile southeast of the entrance channel into *French Harbour*. From the waypoint head generally northwest into the deep wide channel until you can turn to port to head down the narrow channel between the mainland of Roatán and Sarah Cay. After entering the harbour, do not try to turn to starboard to anchor off town (not recommended as it's TOO BUSY with shrimping boats). Beware of the shoal on the N shore, it's marked by a piling.

Bay Islands: *Roatán, Brick Bay*

Brick Bay is home to two small marinas if you require a slip. Be forewarned that the upscale *Barefoot Cay Marina* is particularly susceptible to storm surge (see photo top of next column). You might be able to find a spot to anchor here in the canal that leads in from the east, from *French Harbour*, it is just east of the private shipyard.

As shown on Chart HON-18 (see bottom of next column), a waypoint at 16° 19.65' N, 86° 28.65' W, will place you approximately ¼ mile south/southeast of the marked entrance channel into *Brick Bay*. From the waypoint pass between the outer markers as shown on the chart and work your way around to port to enter the small cove where *Brick Bay Marina* is located, or, if you prefer, you can turn to starboard once clear of the reef lying west of Jesse Arch Cay to head into *Barefoot Cay Marina* if they have room for you.

Brick Bay Marina (see photo in next column) has 600' of dock space and can accommodate a vessel to 15' LOA with a draft to 9'. There is another small dock next to *Brick Bay Marina* that has six dock spaces and is really not a marina per se, it is more of a private site that rents dock space.

The Bay Islands: *Utila, Puerto Este*

As shown on Chart HON-26 (next page), on the western side of the harbor at Puerto Este (and in the photo above) is the large, shallow, *Blue Lagoon*, sometimes call *Lower Lagoon*. The mangrove surrounded lagoon offers fair protection for shallow draft vessels. The bay is shallow with a mixture of a soft

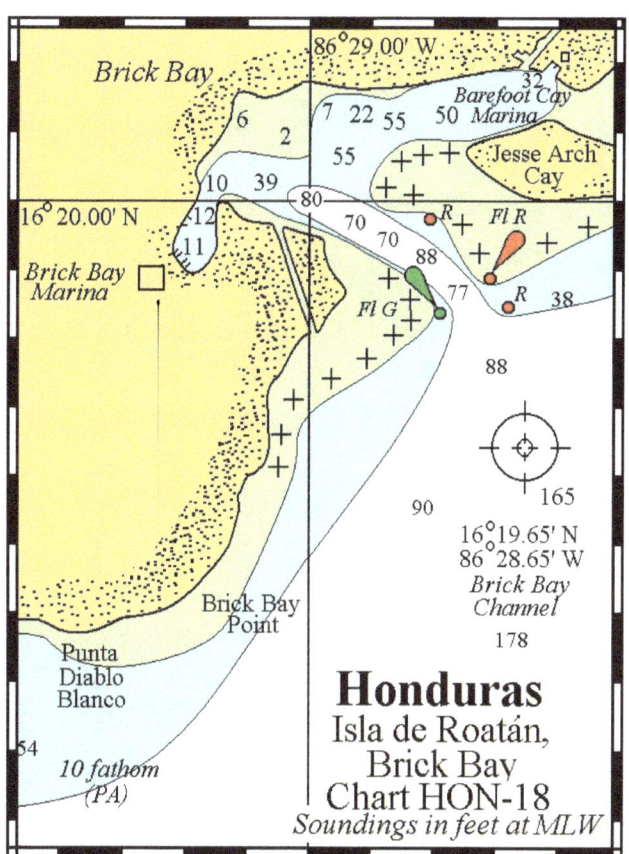

Barefoot Cay Marina, Brick Bay, Roatán

Brick Bay Marina, Brick Bay, Roatán

Puerto Este with *Blue Lagoon* center background, Utila

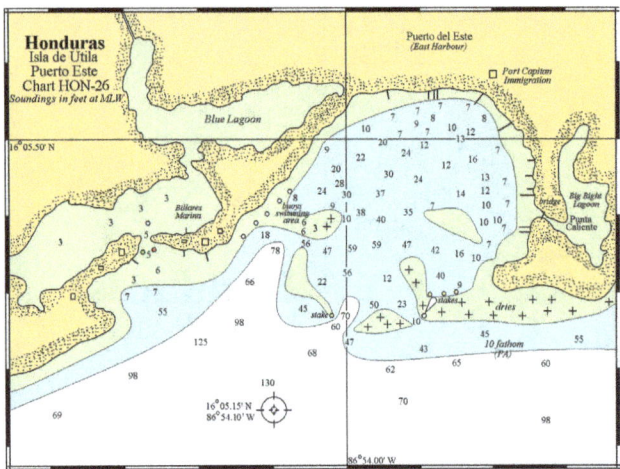

Puerto Cabotaje, shipyard at center, mainland Honduras

mud bottom sprinkled with lots of hard oyster beds, in fact, it is shown on some charts as *Oyster Bed Lagoon*.

To enter the dredged channel (almost 6' at MLW), pass between the red and green buoys (privately maintained so they may not be there at all) favoring the green, port side upon entering. Take the white buoy to port and you will see *Billares Marina* to starboard. The depths shallow to 4' quickly. You can anchor in the NE portion of the lagoon in 5'-6' at MLW. There is a 6' deep hole just inside the entrance, but a 4' draft can enter a good way into the lagoon. *Billares Marina* (504-425-3294) is primarily a private marina, you can check with the owners to see if they have a slip available. This is really NOT a recommended hole, but I wanted to discuss it here as there is no other place on Utila to seek shelter.

Mainland Honduras: *Puerto de Cabotaje*

Approximately two miles east of La Ceiba, at what is shown on older charts as Barro Boca Vieja, is Puerto de Cabotaje, a facility that was designed to accommodate the shrimping fleet and commercial vessels from the Bay Islands. Cruisers can find slips or a haul out at *La Ceiba Shipyard* and two narrow streams in which one could anchor (this is strongly NOT recommended-dangers consist of flotsam and jetsam during rains, local boats, and storm surge vulnerability. The shipyard facilities are located up a narrow river well inside the protective breakwater.

If you travel up the western stream you will come to the old *Lagoon Marina* which has been closed for a few years (although it appears open in the photo above). The two streams have 9'-15' of water.

As shown on Chart HON-31 next page), a waypoint at 15° 48.18' N, 86° 45.67' W, will place you approximately ¼ mile north/northwest of the entrance through the breakwaters. From this waypoint enter staying mid-channel between the jetties or a bit to the east of mid-channel, do not stray to the west of mid-channel, favor the eastern side of the channel. The water will shallow from 25'-30' to 9' in places, but don't worry, there's plenty of water in the channel (as long as you draw less than 9' and the tide is high). Once inside the jetties you can continue up river past the *muelle* (the town dock or wharf) to starboard, favoring the west side of the small harbor to avoid a shoal on the east side of the bay. Once past the *muelle* turn to port and then to starboard to go upriver to the marine facilities. A quarter mile or so upriver you'll come to a spot where two small rivers meet, and directly in front of you you'll see the haul out basin for the *La Ceiba Shipyard* (VHF ch. 69) which has side-tie docks on both small rivers.

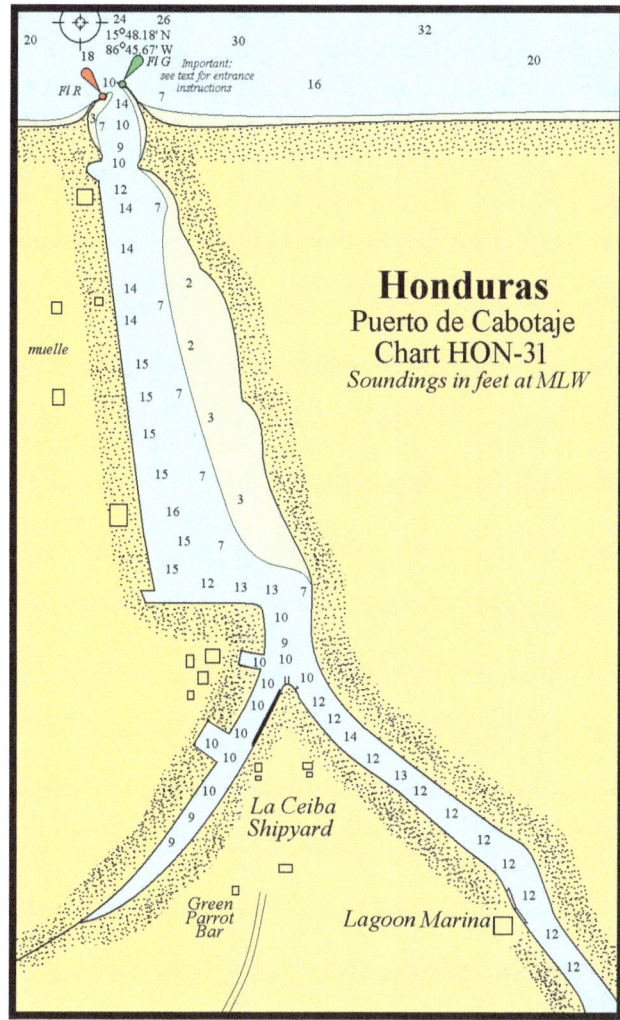

La Ceiba Shipyard is not a marina in the proper sense of the word, they do have side-tie dockage available. The yard has a 120-ton *Travelift* (they can haul vessels to 100' LOA or a multihull with a beam of up to 25' 4") and offers storage on the hard for cruisers who wish to leave their boat in a safe place.

La Ceiba Shipyard monitors VHF ch. 69, their phone number is 504-441-9426, and their email address is laceibashipyard@gmail.com.

Mainland Honduras: *Laguna el Diamante*

I'll begin this section with a warning. A couple of years ago a cruising boat was boarded here and the skipper murdered. Since then the number of cruisers visiting here have dropped off considerably. I would not anchor here alone, but with other boats and only if I needed to use the bay as a harbor of refuge.

As shown on Chart HON-33 (above right), a waypoint at 15° 54.00' N, 87° 39.40' W, will place

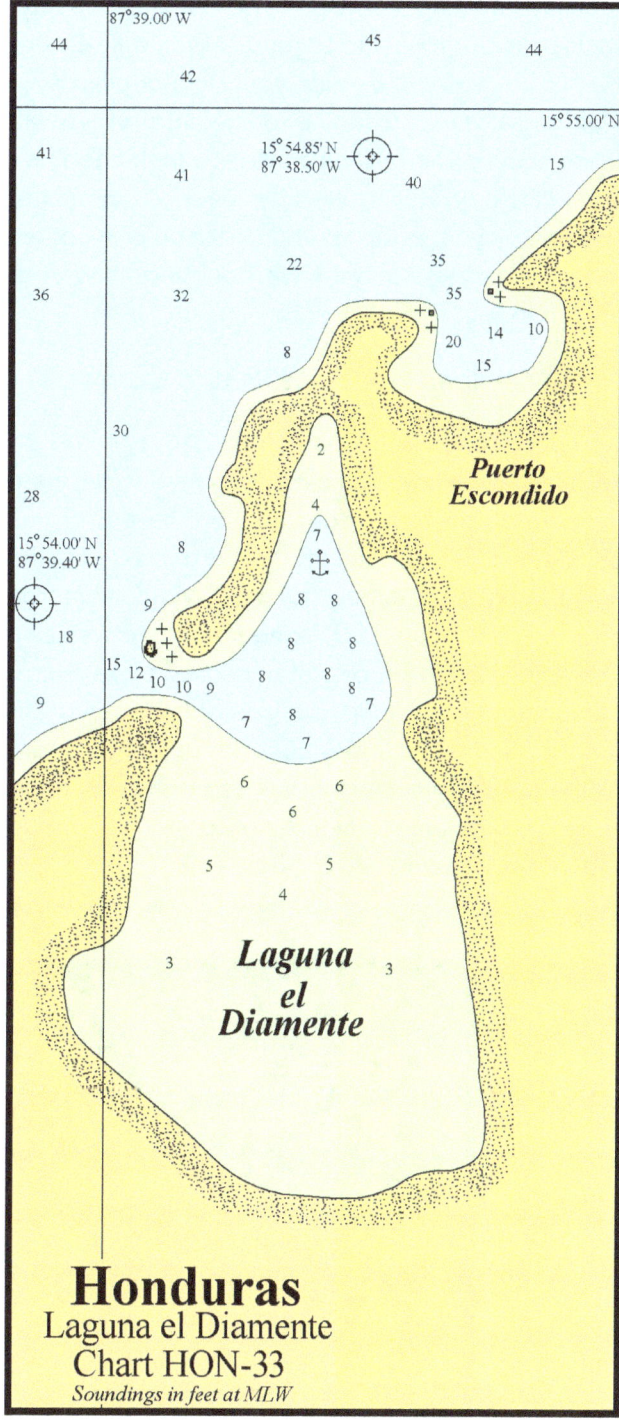

you ½ mile northwest of the entrance to *Laguna el Diamante*. The entrance to *Laguna el Diamante* can be a bit tricky if you've never been there before. The problem is that if you are approaching the entrance from a position too far to the east you'll see the channel between the rock that sits on the northern side of the channel and the mainland and you could easily mistake it for the entrance channel. IT IS NOT THE ENTRANCE CHANNEL. It is shoal and rocky and no place you would care to be. Make sure you keep that off-lying rock to port upon entering the bay! The entrance channel to *Laguna el Diamante* carries 10'-15' of water and inside you'll find depths of 7'-9' throughout except for the southern part where the water shoals. The best anchorage is in the northern part of the bay, well protected in all wind conditions. However, if you go too far north in the bay the bottom is soupy and you might have trouble getting your anchor to set.

Mainland Honduras: *Puerto Cortes*

Puerto Cortes (see Chart HON-34), is not to be considered a hurricane hole by any means, but there is a Honduran Navy Yard located there and you could arrange for a haul out (the yard has a 150-ton *Travelift* that is big enough for a 30' wide catamaran and routinely hauls 100' Honduran Naval vessels). The yard is in the process of becoming more cruiser friendly and the commercial part of the boatyard is run by civilians. The Navy base has an armed guard on the point by the river that lies just north of the base. The main problem is advance notice, you just have to show up and ask if they can haul you.

If you wish to avail yourself of the boatyard, do not tie up at the Navy pier or enter the haul-out area with your boat. These areas are guarded, and you could technically, be in violation of the law. It is best to dinghy in (and not to the Navy base) and walk up to the front gate of the base to make arrangements to haul your boat. English is not spoken here so please have an interpreter with you unless you have a good grasp of the Spanish language.

As shown on Chart HON-34 (below), a waypoint at 15° 51.20' N, 87° 58.60' W, will place you approximately ¾ mile west/northwest of Punta Caballos. Do not head straight for this waypoint if approaching from the east, you must first pass north of Punta Caballos. At this waypoint you will see a red buoy to port, that marks the western edge of the shipping channel, and unless you draw as much as a freighter you're okay passing west of it. From the waypoint you can head southeast passing between a pair of red and green lighted buoys to continue generally southeast towards the conspicuous water tower by the Honduran Navy base at the southeastern end of the bay.

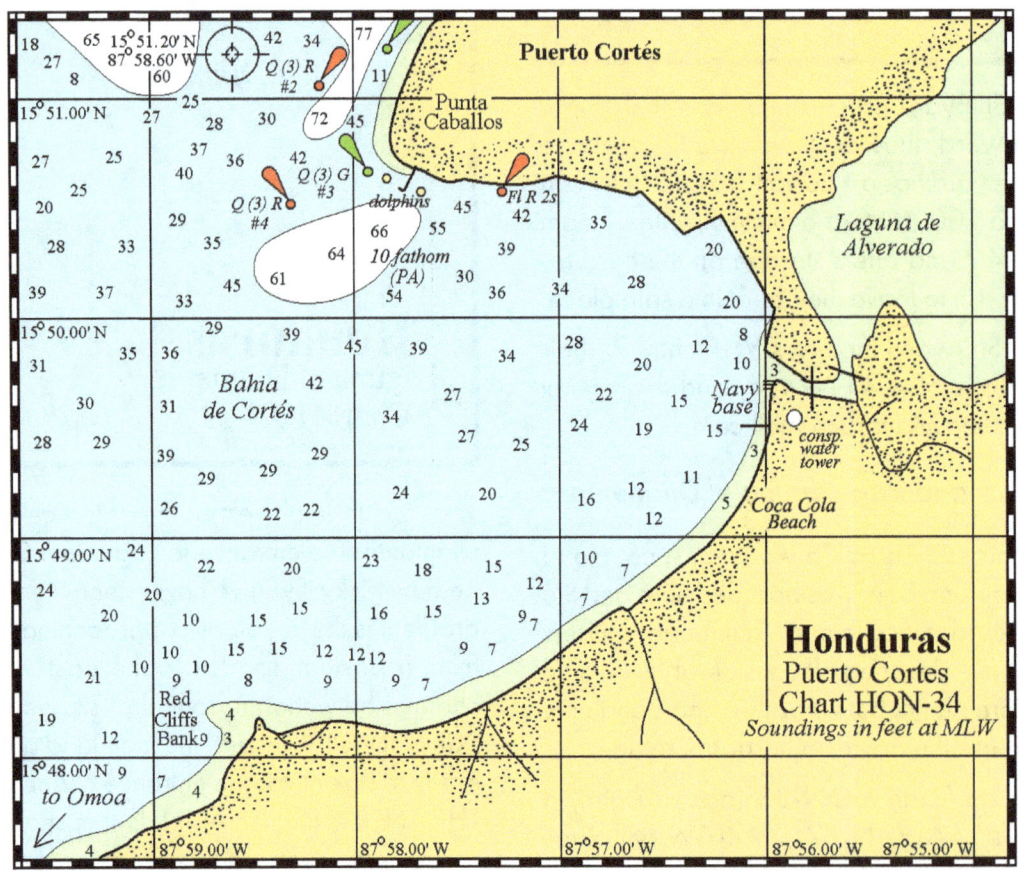

Chapter 18

The Cayman Islands

IF I WERE IN THE CAYMAN ISLANDS AND A HURRICANE WAS APPROACHING, THERE is only one place I would dare to ride out a hurricane, deep in the confines of *North Sound* on Grand Cayman. My first choice for protection, depending on where the wind is forecast to come from, would be tucked in the mangroves along the eastern shore of *North Sound*, past the *Kaibo Yacht Club* or in *Little Sound* (see Chart CAY-9). There are three entrances to *North Sound* as shown on Chart CAY-9, from the west they are *Stingray Channel, Main Channel*, and *Rum Point Channel*, and with the exception of *Stingray Channel*, they are well-marked. I do not advise entering *North Sound* at night unless you are familiar with the waters and entrance channels. I also don't advise using *Rum Point Channel*, it is shallow, about 7' at MLW, and it has a large head in the middle of the channel just south of the markers; you will have a much easier entry if you use *Main Channel*, it is well-marked, wide and easy to see in most conditions.

Grand Cayman: *North Sound*

When approaching the entrance channels to *North Sound* from the west, from *West Bay* and George Town, you must clear Northwest Point and Boatswain Point, both of which can be passed ¼ mile off, and then work your way eastward keeping clear of the reef that begins on the western side of Conch Point as shown on Chart CAY-9 (next page).

This reef breaks in most conditions and is fairly easy to see (a lot of it is above water). To avoid the reef, I suggest staying north of 19° 24.00' N as you work your way eastward. If you have good visibility and are not daunted by the reef on your starboard side, you can work your way eastward by paralleling the reef, keeping to the 5-fathom line a couple of hundred yards north of the reef. Either way you choose, exercise caution.

From the west you can head to a waypoint at 19° 24.00' N, 81° 21.00' W, which will be your turning waypoint. From this position you can head to the waypoints I am about to give you at the beginning of *Main Channel*, and *Stingray Channel*. From the turning waypoint the outer green marker for *Main Channel* is easily seen. You can head to a waypoint at 19° 22.90' N, 81° 19.72' W, which will place you at the northern end of the entrance channel. From this position you can look southward and see the inner markers, pass between those and you are in *North Sound*.

Now let's enter via *Stingray Channel*. From the above mentioned turning waypoint, you can head

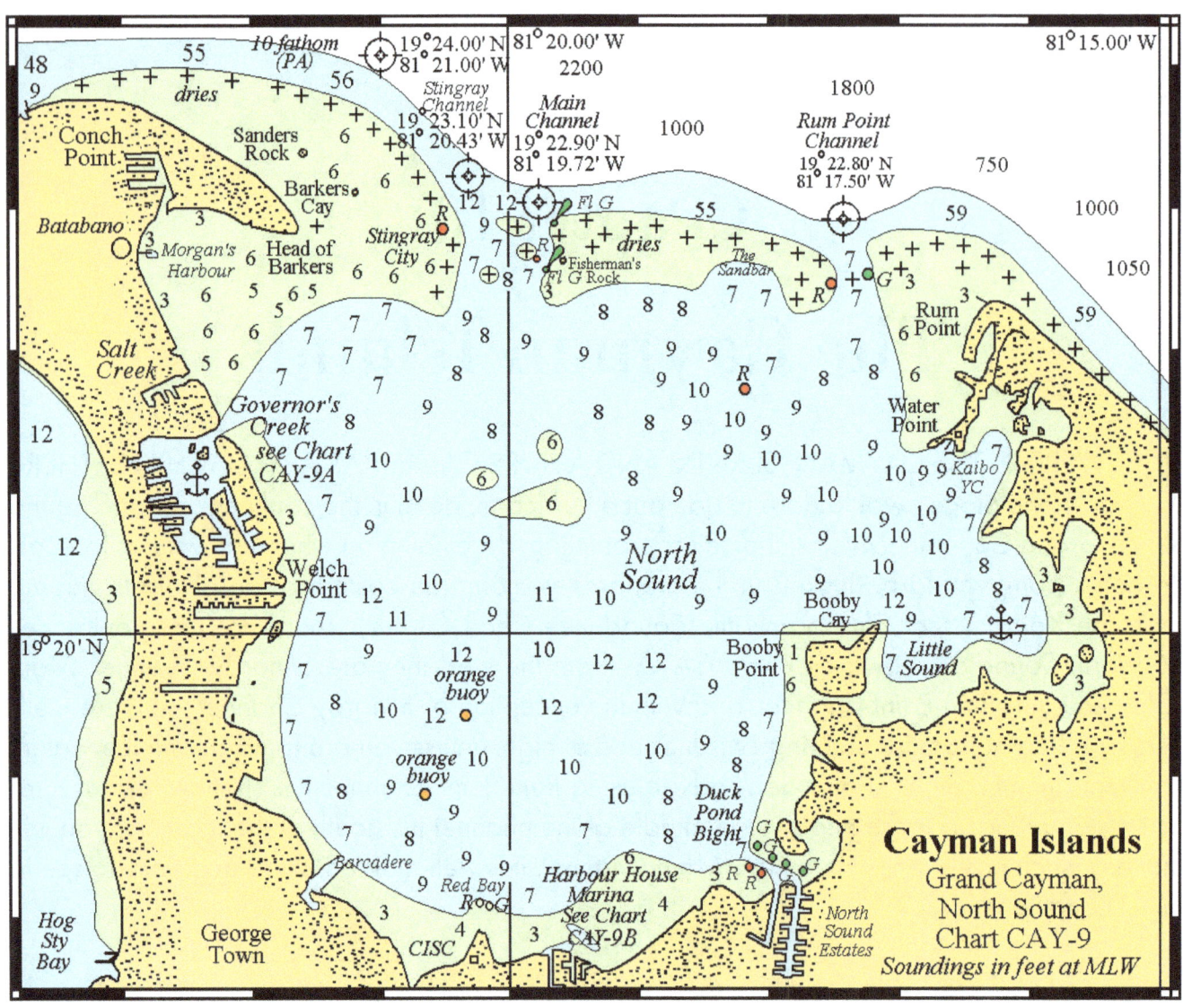

generally south/southeast to a waypoint at 19° 23.10' N, 81° 20.43' W, which puts you past the end of the reef on your starboard side and at the mouth of *Stingray Channel*.

From this position you can see a shoal well to port; head southward and then southeastward once past this shoal, paralleling the lie of the reef on your starboard side, and you are in *North Sound*.

Once inside *North Sound*, you can head in a SE direction to find shelter in the mangroves along the eastern shore past the *Kaibo Yacht Club in Little Sound*.

Grand Cayman: *Governor's Harbour*

A second choice for protection in *North Sound* would be to find shelter in *Governor's Harbour* on the western shore of the sound (enter using *Governor's Creek* as shown on Chart CAY-9 and in greater detail on Chart CAY-9A (see next page). It is very protected, but that area was hit hard by Hurricane Ivan several years ago. In all fairness, it must be mentioned that most of the docks of the *Cayman Islands Yacht Club* (*CIYC*) in *Governor's Harbour* were damaged or destroyed by Hurricane Ivan.

A waypoint at 19° 21.55' N, 81° 21.90' W, will place approximately ¼ mile east/northeast of the entrance into the *Governor's Creek* canals. From the waypoint head into the canal keeping the red buoy to starboard and keeping off the visible shoal to port. Once inside keep favoring the port side of the channel and wind your way back until you enter *Governor's Harbour*. A turn to starboard here will bring you to the *CIYC* and the anchorage just south of their seawall.

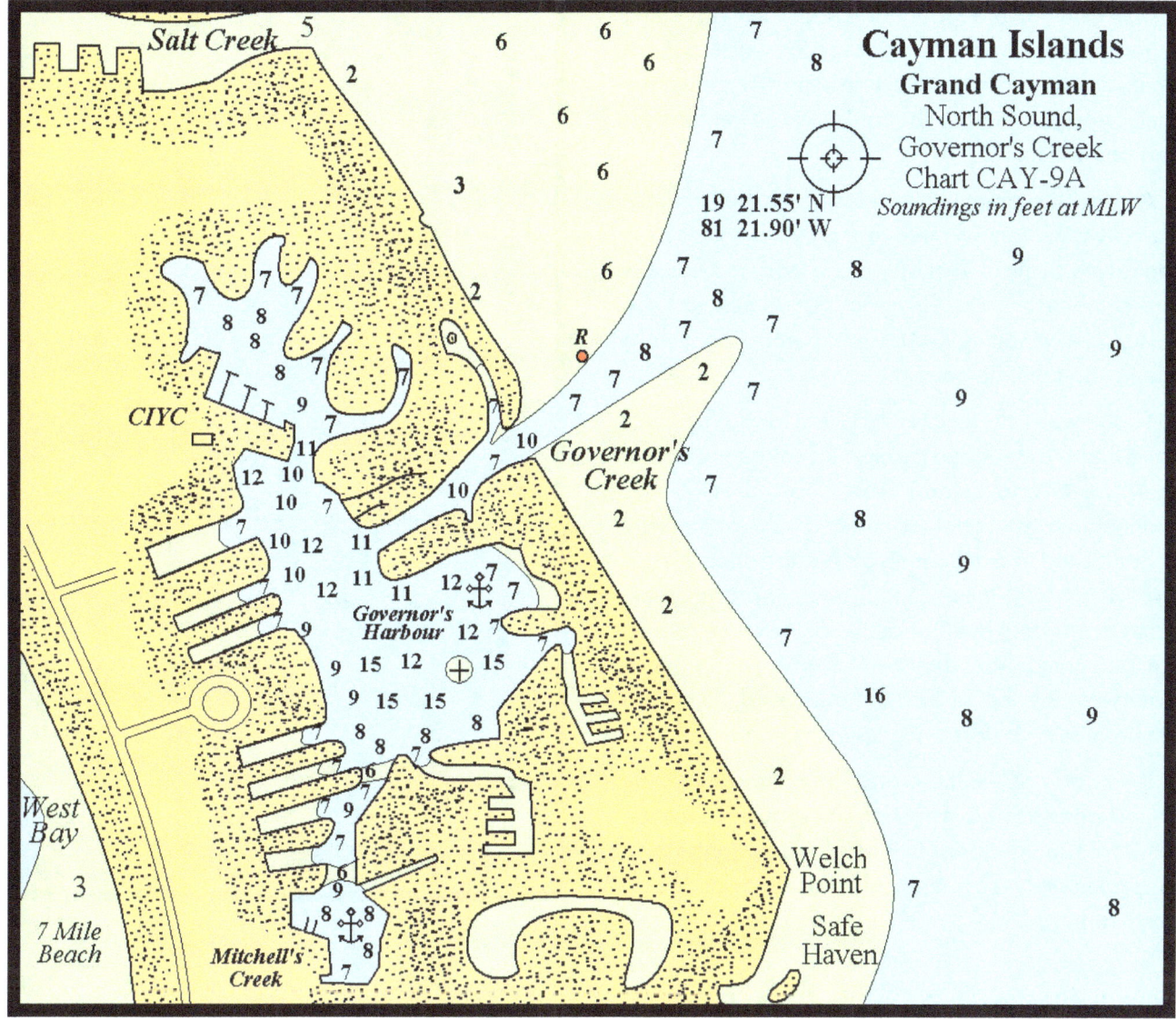

You can head further south from the CIYC and work your way through the canals to enter protected *Mitchell's Creek* where you can anchor in 7'-12' of water and a grassy bottom. You will have to thread your way through a couple of shallow areas, 6' at MLW, but keeping to mid-channel and going slow will get you through unless your draft is simply too deep.

Grand Cayman: *Harbour House Marina*

At the southern end of *North Sound* you will find the entrance channel to *Harbour House Marina and Boatyard*. The marina lies on a narrow, protected canal (highly vulnerable in north winds) and may or may not allow you to tie up inside, or be able to do a haul out. You should hail them first on VHF ch. 16 to see what their current hurricane policy is (http://www.harbourhousemarina.com/).

As shown on Chart CAY-9B (next page), a waypoint at 19° 18.10' N, 81° 19.50' W, will place you just off the entrance channel leading to the marina and boatyard. From the waypoint you will see several markers that define the channel which has a controlling depth of 5' at MLW in one spot right at the mouth of the channel. Remember that all markers inside *North Sound* are privately maintained so their configuration is subject to unannounced change.

From the waypoint head southward passing between the two red markers and one green marker. Once past the inner red marker, turn to port and head for the marina staying between the very visible shoal (visible only if there's good light) and the shoreline. When you are north of the canal off the western shore where the marina is located, turn to port and enter the canal, you will find a 6' bar at MLW but for the most part you will find 7'.

Around the point to the west of *Harbour House Marina* is a narrow, marked channel that has a controlling depth of 4' at MLW, and leads to the docks and clubhouse of the *Cayman Islands Sailing Club* (http://www.sailing.ky/). Seek permission before attempting to tie up here.

To the east of *Harbour House Marina*, at *Duck Pond Bight*, you will find the marked entrance channel to *North Sound Estates* (see Chart CAY-9) The channel and the canal will take 7' most of the way in; compared to other canals off *North Sound* this is a lot of water. With the exception of *Governor's Creek*, most canals only have about 4' 6" of water and so are used primarily by shallow draft powerboats. These waterways are full of private homes and it's best to get permission before tying up in the canals.

If you do not like the protection offered by *North Sound*, and if you have enough time, you can go north to Cuba, Cienfuegos or Casilda, for good protection (approximately 155 nautical miles north of Grand Cayman).

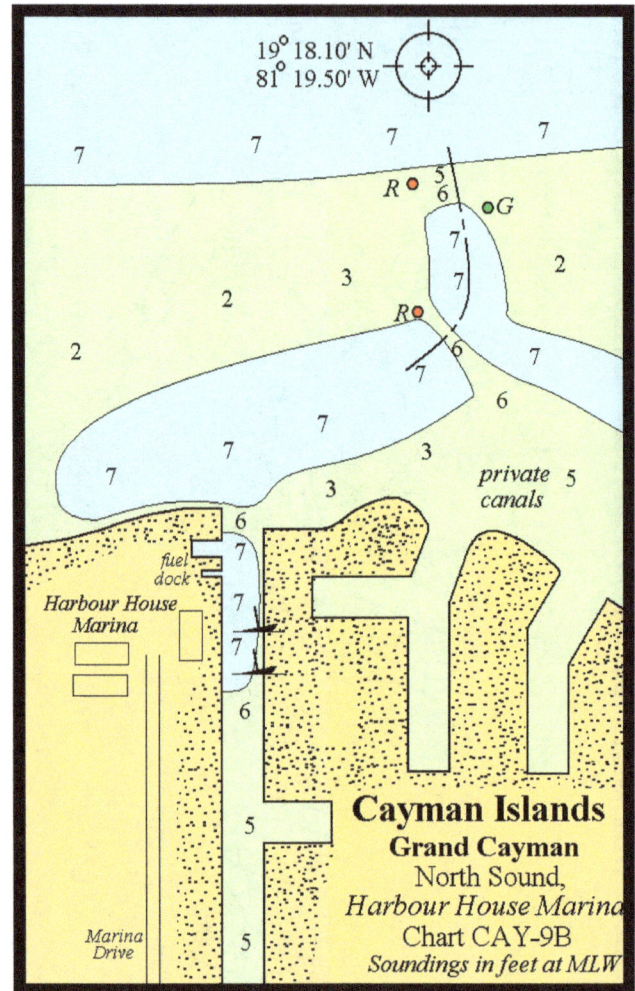

Chapter 20

Guatemala

AS FAR AS ACTUAL PROTECTION FROM HURRICANES FOR CRUISERS IN THE Northwestern Caribbean, in the broad area from Mexico to Honduras, the finest protection is up the *Río Dulce* in Guatemala. The *Río Dulce* offers excellent protection, economical prices, and an eclectic group of gregarious cruisers. The marinas are well upstream, miles from the coast and the worst of any hurricane surge, and the surrounding hills go a long way in lessening the strength of the wind. Arguably, the *Río Dulce* is probably the finest hole in the entire Caribbean. Why you may ask? It's not because it abounds in narrow mangrove-lined creeks in which to hide. In fact, if every boat on the *Río Dulce* sought shelter at the same time, the few good hurricane holes would be overflowing. The river's saving grace is its location. It is very difficult for a hurricane to make a direct hit on the river without crossing a good bit of mountainous land that would do nothing for the storm except weaken it. Even if a Category 5 hurricane were to make a direct hit at the mouth of the river, the storm surge at the marinas twenty miles upstream would be minimal, a third perhaps of the surge at the mouth of the river. If you are seeking hurricane shelter on the *Río* I advise you to use caution if you want to anchor in one of the smaller rivers that feed into the *Río Dulce*. These rivers can rapidly rise 4' or more with the rains from a hurricane, and all the water has no place to go except to rush downstream. If you'd like to look for yourself, take a dinghy ride up any river and you'll see logs strewn about here and there that on the next flood will be moving downstream and aiming at your vessel should you tie up for shelter in a questionable location. For hurricane protection on the *Río Dulce* there are several small coves you can access that offer protection from both wind and seas and we will highlight those areas here.

Bahía de Graciosa

Bahía de Graciosa lies on the eastern side of *Bahía de Amatique*, north of Puerto Barrios, and makes for a shallow, but well protected anchorage, not quite what I would call an ideal hurricane hole, but it would certainly do in a pinch. The entrance and interior of *Bahía la Graciosa* are home to several difficult to see shoals so use caution when entering.

As shown on Chart GTM-1 (see top of next page), a waypoint at 15° 52.50' N, 88° 34.50' W, will place you approximately ½ mile west/northwest of the small cays that lie north of Punta Manglar and the unmarked entrance to the bay. Watch out for local fisherman setting lines here.

From the waypoint head just a bit south of west to avoid the shoals on either side of you and enter the bay where you'll find depths of 7'-15' throughout most of it, anchor wherever your draft allows and be prepared for bugs with the setting sun. Keep an eye out for the sandbar that extends south of the entrance for about 200 yards. If your draft allows, you can work your way up into *Canal Lingles*, or the *Río San Fransisco* as it is sometimes shown.

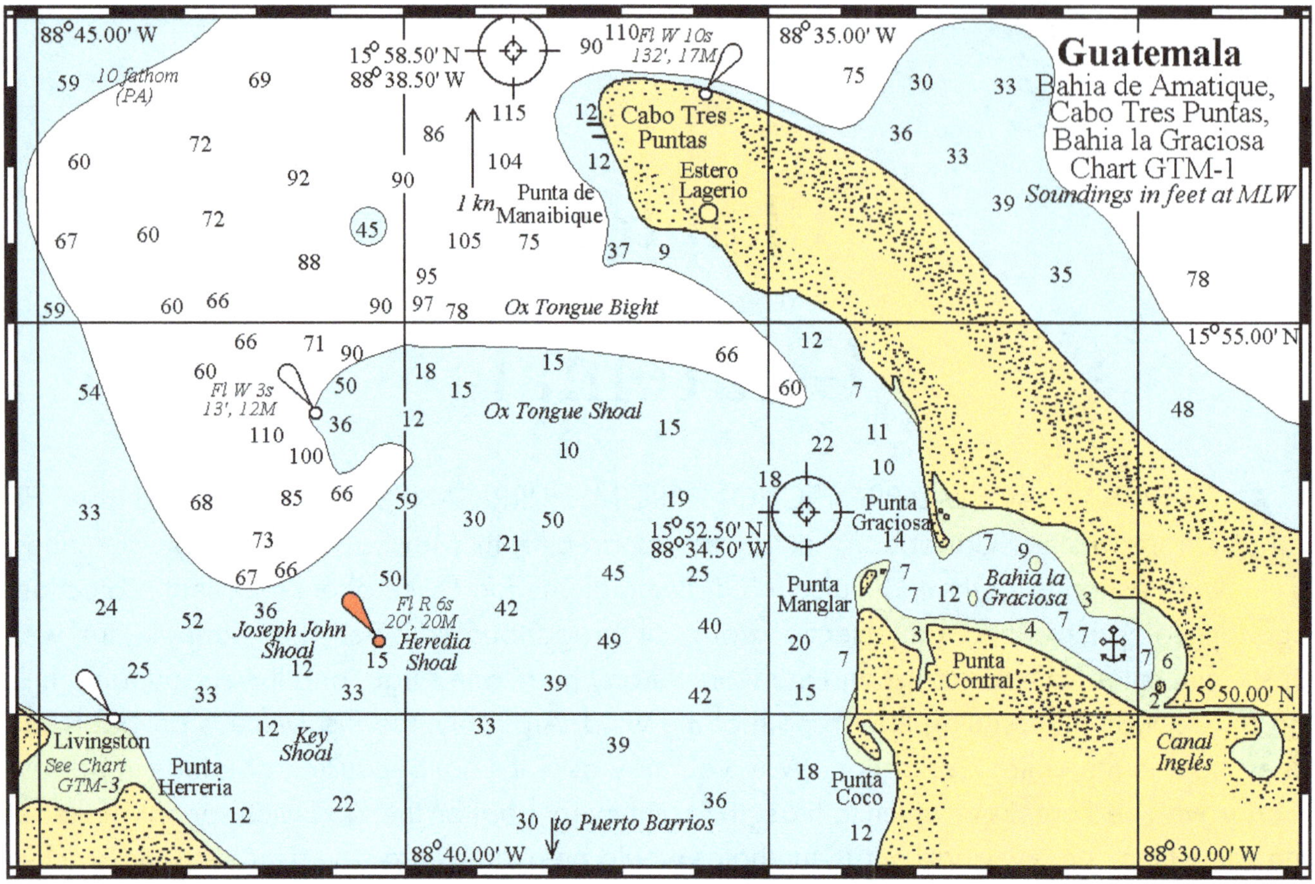

Puerto Barrios: *Ensenada San Carlos*

As shown on Chart GTM-2 (see next column), a waypoint at 15° 46.00' N, 88° 36.90' W, will place you approximately 1½ miles north of the *Bahía de Santo Tomás de Castillo*, shown on some charts as *Bahía de Gálvez*, and approximately ½ mile north/northeast of the light that marks *Bajo Villedo* (8'-12' depths). It is here, in the bay that is home to the port city of Puerto Barrios, that you can find a tiny bit of shelter in Ensenada San Carlos, but expect local boats to pack the cove before you can get there.

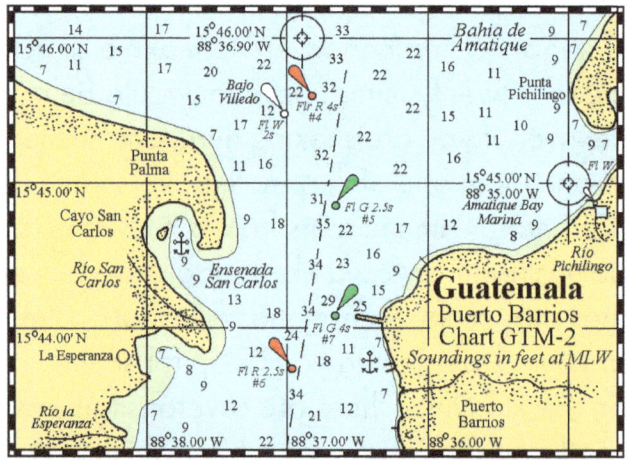

Amatique Bay Marina, Puerto Barrios

Puerto Barrios: *Amatique Bay Marina*

Located northeast of Puerto Barrios, in a resort complex built on the *Río Pichilingo*, is a well-protected marina, the *Amatique Bay Resort and Marina*, as shown on Chart GTM-2 (previous page with photo). The marina has a newly dredged channel (7' and soon to be 8') and depths at the dock range from 8'-10' consistently.

A waypoint at 15° 45.00' N, 88° 35.00' W, will place you approximately ¼ mile northeast of the entrance to the marina at the mouth of the *Río Pichilingo*. From the waypoint you'll see the resorts lighthouse and jetty on the northern side of the entrance, keep them to port and enter the narrow river and wind your way into the sheltered marina basin.

The marina has 25 slips (concrete piers) and has plans in the works to raise that number to 180 with dry storage for powerboats up to 30' You can contact the marina on VHF ch. 68, by phone at 502-7931-0000, or look them up on the internet at http://www.amatiquebay.net/.

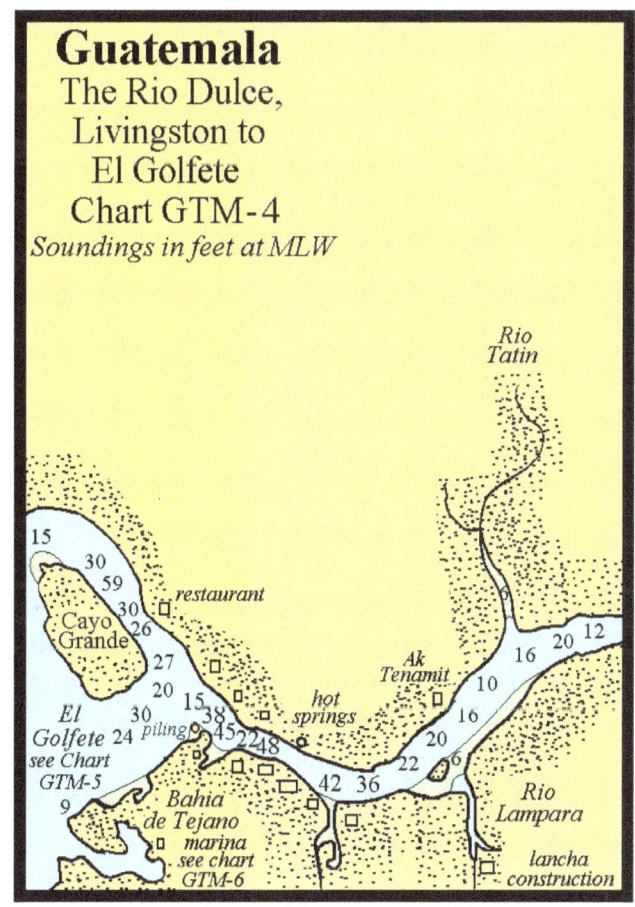

Rio Dulce: *Río Tatin, Río Lampara*

About a mile or so before reaching *El Golfete*, and just before *Ak Tenamit* as shown on Chart GTM-4 (see below).

To starboard, on the northern side of the river, is the *Río Tatin*. Vessels can enter the river and proceed a short distance upstream for shelter keeping in mind the danger of flotsam and jetsam during heavy rain.

On the opposite side of the river, the southern shore, is the mouth of the *Río Lampara* with a very visible shoal at its mouth. At the entrance, the deeper water lies to the east of the small as shown on the chart. The river itself is very busy, lots of local homes, a luancha builder, and a restaurant, but you may be able to find a spot to secure your vessel.

El Golfete: *Laguna Salvador, Laguna Calix*

A few miles upriver from Livingston *El Golfete* will open up in front of you as shown on Chart GTM-5 (see top of next page). It is a large lake approximately 2 miles wide and 6 miles long. As you enter *El Golfete* and pass Cayo Grande to starboard, you can turn to starboard, approximately NW where just west of the manatee reserve is a small river that leads to two very protected land-locked coves, *Laguna Salvador* and *Laguna Calix*, that are excellent places to hide in the event of a hurricane. Both coves offer good holding in sandy mud and lots of mangroves.

The river has 9'-11' at its mouth and you will find 7'-10' all the way into *Laguna Salvador*. Just before you enter *Laguna Salvador* you will find a small river that branches off to the southwest and leads into *Laguna Calix*. You will have 7' leading into this cove. There is a narrow river that leads away to the south just before you enter *Laguna Calix* that is only for dinghy exploration although you can enter and tie off in the narrow waterway, but you won't be able to get too far due to submerged debris.

If you don't enter *Laguna Calix* and opt to enter *Laguna Salvador*, you can proceed down the middle of the lagoon for approximately 300 yards until you see a small notch of a cove in the shoreline to port. There is a house on the hill to your starboard as you tuck into the notch. Drop bow and stern anchors as well as tying up to the mangroves and trees that line

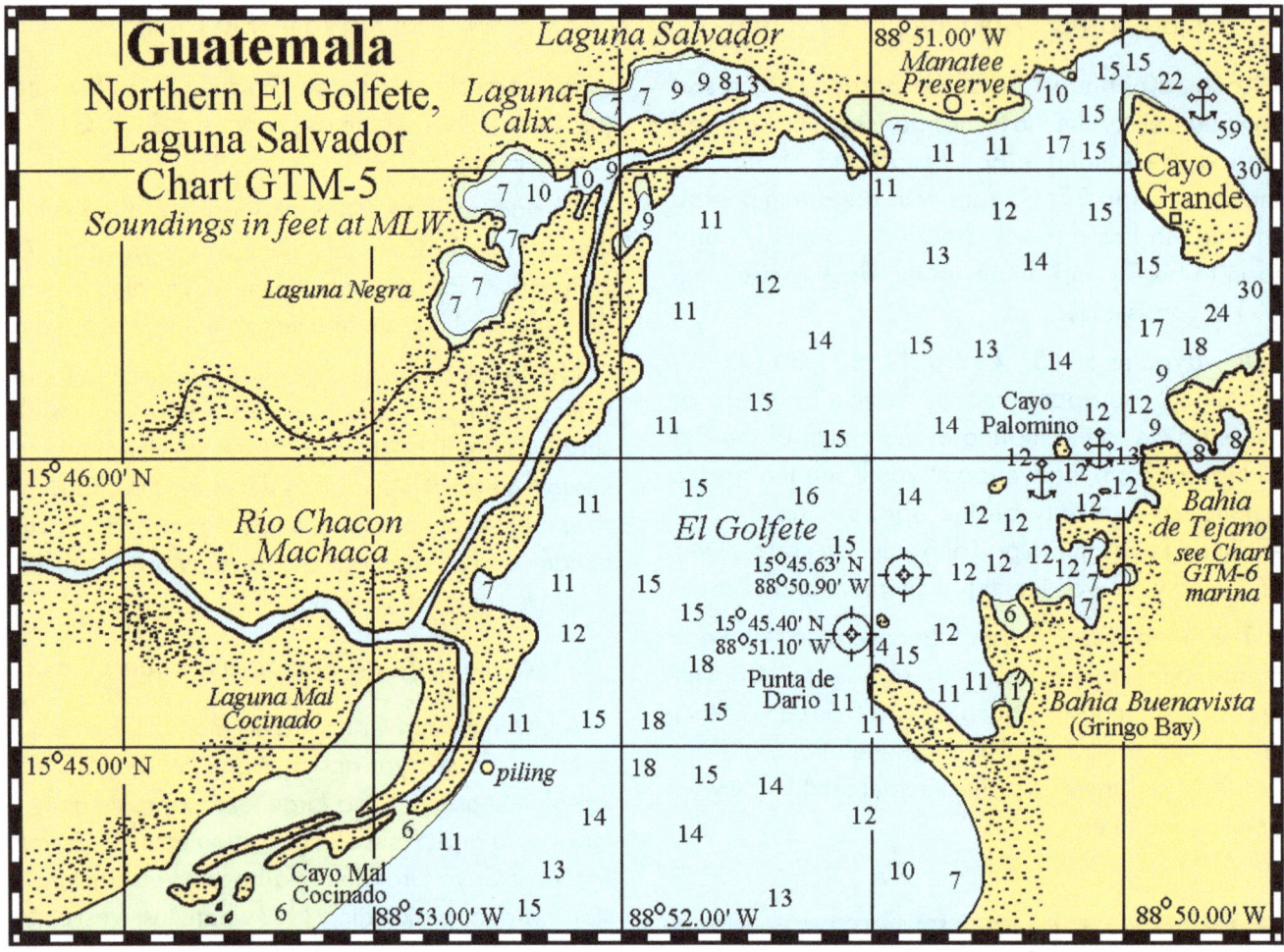

the lagoon. Placing anchors to starboard both fore and aft will be necessary since there is not much land to tie to on that side (perhaps a tree to the north) As an alternative, stay far enough towards the center, favoring the northern part, of *Laguna Salvador* to anchor but allow for swinging. There is enough room in the center of the lagoon for at least 10 boats but it probably will not be crowded.

El Golfete: *Bahía de Tejano (Texan Bay)*

Upon entering *El Golfete* and passing Cayo Grande to starboard, on our port side you will find *Bahía de Tejano* (*Texan Bay*, once shown on an old chart as *Bahía Durate*) also known as Burnt Key or Cayo Quemodo (the name of the nearby Mayan village). Here good protection can be found in the small mangrove encircled bay off the marina. Here you'll find good holding in mud with room for 3-4 boats. There are two small marinas within the bay but will probably be full during hurricane season.

As shown on Chart GTM-6 (top of next page), a waypoint at 15° 46.30' N, 88° 50.10' W, will place you approximately ¼ mile NW of the entrance into *Bahía de Tejano*. From the waypoint head into the bay as shown on the chart working your way to the inner lagoon and anchor wherever you choose.

El Golfete: *Unnamed Bay*

South of *Bahía de Tejano* and north of *Bahía Buenavista* as shown on Chart GTM-5 above, is a small bay with three smaller coves leading off to the north, east, and west. While these three small covers are open to seas and wind from the west, the still offer good protection from other wind directions and you can carry 7' of draft inside upon entry. There will likely be few other boats here and the holding is good in mud. Here you can set bow and stern anchors and tie off to the trees along the shoreline. There is one set of markers at the entrance.

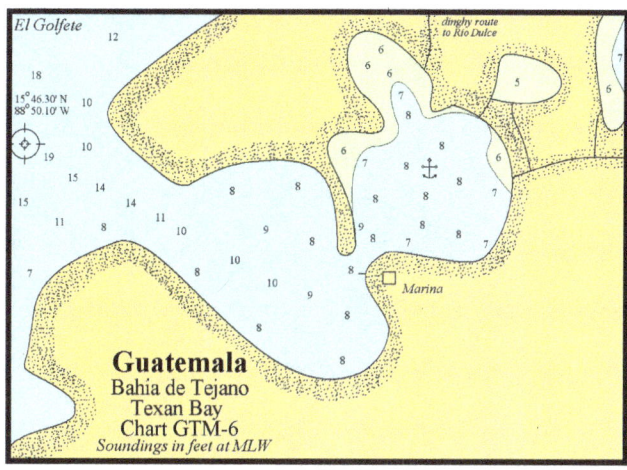

El Golfete: *Bahía Buenavista (Gringo Bay)*

Bahía Buenavista, often called *Gringo Bay* as shown on Chart GTM-5, offers good holding all over the bay so please don't anchor near the moorings to give the boats there plenty of swinging room.

There are actually two entrances to *Bahía Buenavista*, one for cruisers approaching from Fronteras, from upstream, and one for inbound cruisers, those approaching from downstream. Both entrances are straightforward and deep. The waypoints given (depending on your approach) are on either side of the small cay that lies off the entrance into the bay as shown on Chart GTM-5.

If you are inbound from Livingston (heading upriver), a waypoint at 15° 45.63' N, 88° 50.90' W, will place you approximately ¼ mile northwest of the cut between the off-lying island and the mainland; from the waypoint head generally southeast into *Bahía Buenavista* keeping the small island to starboard.

If you are outbound (heading downriver) from the town of *Río Dulce*, a waypoint at 15° 45.40' N, 88° 51.10' W, will place you approximately ¼ mile west of the entrance into the bay; from the waypoint head generally south/southeast into the bay keeping the small island to port. Both of these routes have plenty of water and offer no obstacles or hazards.

El Golfete: *Río Chacon Machaca*

Approximately 2.5 miles SW of the entrance to *Laguna Salvador* on the northern shore of *El Golfete*, is the entrance to the *Río Chacon Machacha*.

The entrance may be difficult to discern at first due to the tall brush sprouting from the deeper water. Eventually you will see a narrow gap leading into the river and you can begin to work your way into the river in 6' of water if you follow the flow of water coming out of the river (brown in color). You will be entering from a position slightly left of the entrance and once inside the depths will go to 20'.

Immediately upon entering, turn 90 degrees to your port and anchor in the 40' wide canal. Set bow and stern anchors and tie off to as many trees as you can.

Note: that there will be debris flowing downstream from the main river so proceed as far in the canal as you can, you'll have a maze of shallows, fallen trees, and small islands to work your way through.

El Golfete: *Laguna Quatro Cayos*

As shown on Chart GTM-7 below, in the northwestern corner of *El Golfete* is a small bay called *Laguna Quatro Cayos* lying north of Cayo Largo. This bay offers 360° protection when tucked into the starboard branch with 7' and good holding mud.

A bit to the east of *Laguna Quatro Cayos*, is a small bay lying north of Cayo Julio. You can anchor here gaining protection from the surrounding land as well as Cayo Julio to the south. This spot is open to the west but there probably won't be any other boats to contend with and the holding is good in mud. This unnamed bay can accommodate vessels with drafts to 7'.

Rio Dulce: *Laguna Escondida*

Approximately 1.5 miles upriver of *El Golfete* there is a small cove to port called *Laguna Escondida* as shown on Chart GTM-8 (top of next page). The cove is small, encircled by mangroves, but open to the north

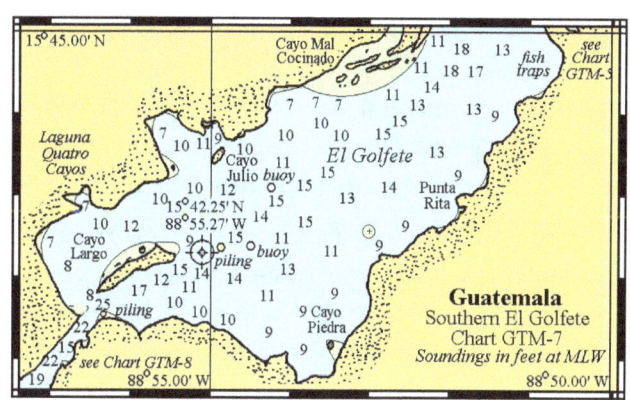

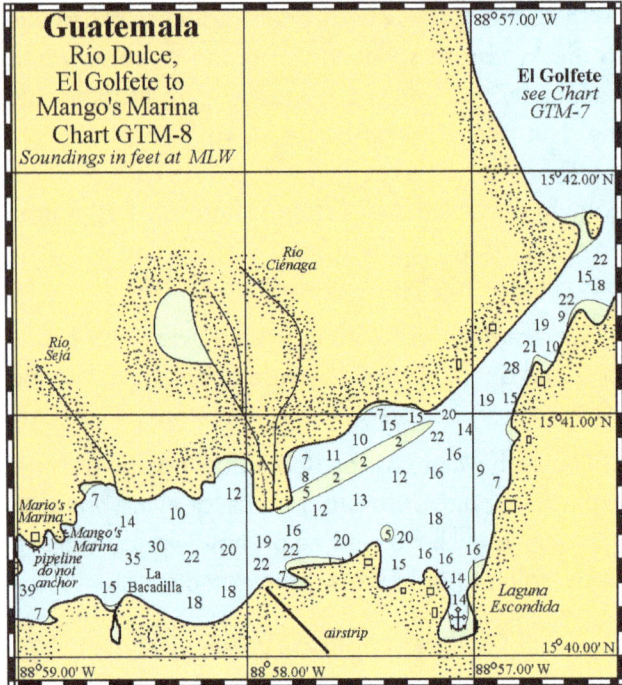
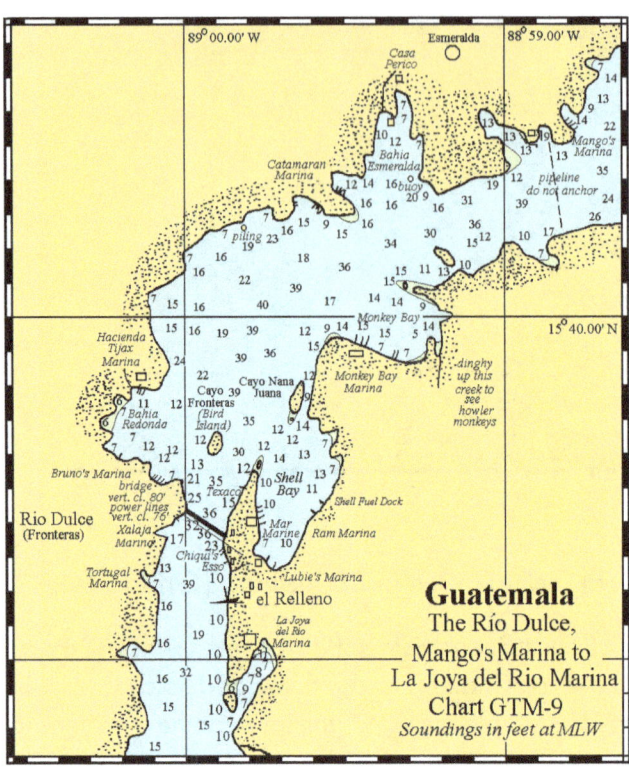

to check the forecast wind direction before trying to enter here. Enter through the narrow mouth and secure your vessel accordingly.

Rio Dulce: *Río Ciénaga*

As shown on Chart GTM-8 (at the top of this column), downstream from *Mango's Marina* and upriver of *Laguna Escondida*, you'll notice what appears to be two rivers emptying into the *Río Dulce* on the northern shore, their mouths converging at the same point of land just west of a long shoal in the river. It is actually one river, the *Río Ciénaga* with a small island at its mouth (at the time of this writing).

To enter the river, take the eastern most channel. Boats can navigate this river for several miles as depths of 6'-8' are good for quite a way upstream if you take the fork that leads to the east, the fork on your starboard side as you head upstream. But you'll have to keep a sharp eye out for submerged logs which may or may not make the river passable and may certainly be a hindrance if trying to exit the river after a storm and the accompanying rains.

If you take the stream that leads off to port you will find that ¼ mile upriver are some low-hanging power lines. Their vertical clearance appears to be less than 30' but nobody can seem to tell me the exact height so 30' is an estimate. Past the power lines the river leads into a small lagoon that offers protection to a vessel that can make it under the power lines.

Rio Dulce: *La Bacadilla*

As shown on Chart GTM-8 (above), across the river from the mouth of the *Río Ciénaga* and a bit upstream is a point of land called La Bacadilla. Just west of the point is a narrow creek that leads into a tiny, mangrove encircled cove that offers good protection for vessels with drafts of less than 4'.

Rio Dulce: *Monkey Bay Marina*

As you approach the bridge, to your port you will see *Monkey Bay* and *Monkey Bay Marina* (http://www.monkeybaymarina.com/) as shown on Chart GTM-9 (at the top of this column). If you wish to get a slip the marina has 22 side-tie slips for catamarans and mono-hulls, but the real protection is not to be found at the marina, or in *Monkey Bay* itself.

Looking at the marina from the river, to port, in the NE corner of *Monkey Bay*, is the entrance to a small river, *Laguneta Madre Vieja*. Many cruisers like to take their dinghies up this river to view howler monkeys. But this river offers very good protection and 7' of water at the entrance. Enter and round to starboard in the first branch of the channel where you can set bow and stern anchors and tie to the larger trees on shore. You can also head further upstream for a short distance but be careful not to pass the submerged trees that lie 1' below the surface. Water visibility will be poor.

Rio Dulce: *Fronteras*

The main community on the *Río Dulce* is Fronteras (often just called *Río Dulce*) and it is here that the many marinas and haul out yards are located.

If you plan to get a slip in any of the marinas in and around Fronteras for a hurricane be sure to plan on the forecast wind direction when making your choice. There is a list of marinas and contacts in the chapter "List of Marinas."

If you wish to haul out you have three choices. The first is *RAM Marina and Yacht Club* and *Nana Juana Marina*, both in *Shell Bay*, and *Astillero Magdelena*, better known as Don Abel's (Don Abel Ramirez's boatyard). *RAM Marine* has an 85-ton *Travelift* and an 8,000-lb. forklift for small powerboats. *RAM Marine* can handle large vessels. *Don Abel's* has two 150-ton marine railways (this means they can haul out large catamarans and trimarans-over 30' in width and boats up to 80 tons), a 75-ton marine railway, and an 85-ton *Travelift*. *Nana Juana* can haul catamarans up to 50' LOA.

There are numerous small creeks leading into the river in the area and you might wish to explore those with your dinghy as a possible hurricane refuge.

Rio Dulce: *La Joya del Rio Marina*

South of the bridge is the small cove where *Suzana's Laguna Marina* used to be located. It is the best protected cove on this stretch of the river and now home to *La Joya del Rio Marina* (502- 7930-5594 , VHF ch. 68, https://www.facebook.com/lajoyadelrio.marina).

The marina has 75 well-protected slips with fixed wooden docks and individual finger piers for each slip (unless you get a side-tie). As shown on Chart GTM-9 (top of previous page), head upriver, south of the bridge towards *Lago Izabal*, and the entrance to the small cove where the marina is located. To enter, pass south of the small mangrove island shown on the chart before turning to port to enter the cove on a NNE heading.

Lago Izabal: *Puerto Refugio*

One of the most protected anchorages on *Lago Izabal* (see Chart GTM-10 at the top of the next page) can be found in *Puerto Refugio*, also known as *Enseñada de Balandras*, located at the southwestern end of the lake. The anchorage is protected from all but heavy southwesterly and westerly winds in which case the long fetch could build up some dangerous waves.

As shown in greater detail on Chart GTM-11 (next page), a waypoint at 15° 24.65' N, 89° 16.30' W, will place you approximately ½ mile north of the western tip of Punta Chapin. From the waypoint round the shallows that lie west of the point and enter *Puerto Refugio* to anchor.

Lago Izabal: *Río Oscuro*

As shown on Chart GTM-10 (top of next page), just southwest of *Puerto Refugio* is the *Río Oscuro*, a pristine jungle river with a bar at its mouth that limits drafts to under 6' depending on the amount of recent rainfall. Depending on the amount of rainfall and debris, the *Río Oscuro* may be navigable for as much as 3 miles from its mouth. Remember our warning about flotsam and jetsam flowing downstream with heavy rains!

Lago Izabal: *Bocas de Bujajal*

The *Bocas de Bujajal* (see Chart GTM-10 on next page) are the mouths of the *Río Polochic*, a very busy river in times gone by, and still not dead yet. Although it was once possible to work your way some thirty miles upriver, it is not so today for a cruising boat due to debris from heavy rains. There are good times and bad times to enter the river, the mouth changes every year due to the buildup of debris, some years making entry impossible.

As shown in greater detail on Chart GTM-12 (bottom of next page), a waypoint at 15° 28.90' N, 89° 21.50' W, will place you approximately ½ mile north of the mouth of the *Río Polochic*. From this waypoint you can head south to anchor in the unnamed bay between *Enseñada el Padre* and the mouth of the *Río Polochic*. Here you can find good protection unless the winds go north/northeast.

Guatemala
Lago Izabal
Chart GTM-10
Soundings in feet at MLW
El Castillo
Finco Jocolo
Punta Dario
Finca Paraiso
(hot springs)
El Boqueron
El Estor
nickel mine
Punta del Muerto
Punta Murciélago
Punta Santa Cruz
Punta Caimanes
Punta Brava
Denny's Beach
Mariscos
Punta Gruesa
Punta Chapin
Puerto Refugio
See Chart GTM-11
Punta de Chile
Ensenada de las Minas
Laguneta el Amatillo
Bocas de Bujajal
Rio Polochic
see Chart GTM-12
Rio Oscuro
Rio Zarquita
piling
15°36.50′ N
89°03.00′ W
15°33.50′ N
89°12.50′ W
15°31.20′ N
89°20.30′ W
15°28.75′ N
89°01.41′ W
15°25.69′ N
89°05.97′ W
15°35.00′ N
15°30.00′ N
15°25.00′ N
89°20.00′ W
89°15.00′ W
89°10.00′ W
89°05.00′ W
89°00.00′ W

Guatemala
Lago Izabal,
Bocas de Bujajal
(Mouths of the Polochic)
Chart GTM-12
Soundings in feet at MLW
Lago Izabal
Ensenada Los Lagartos
Ensenada Boca Ancha
Ensenada El Padre
platform
Río Polochic
15°28.90′ N
89°21.50′ W
15°28.00′ N
89°25.00′ W

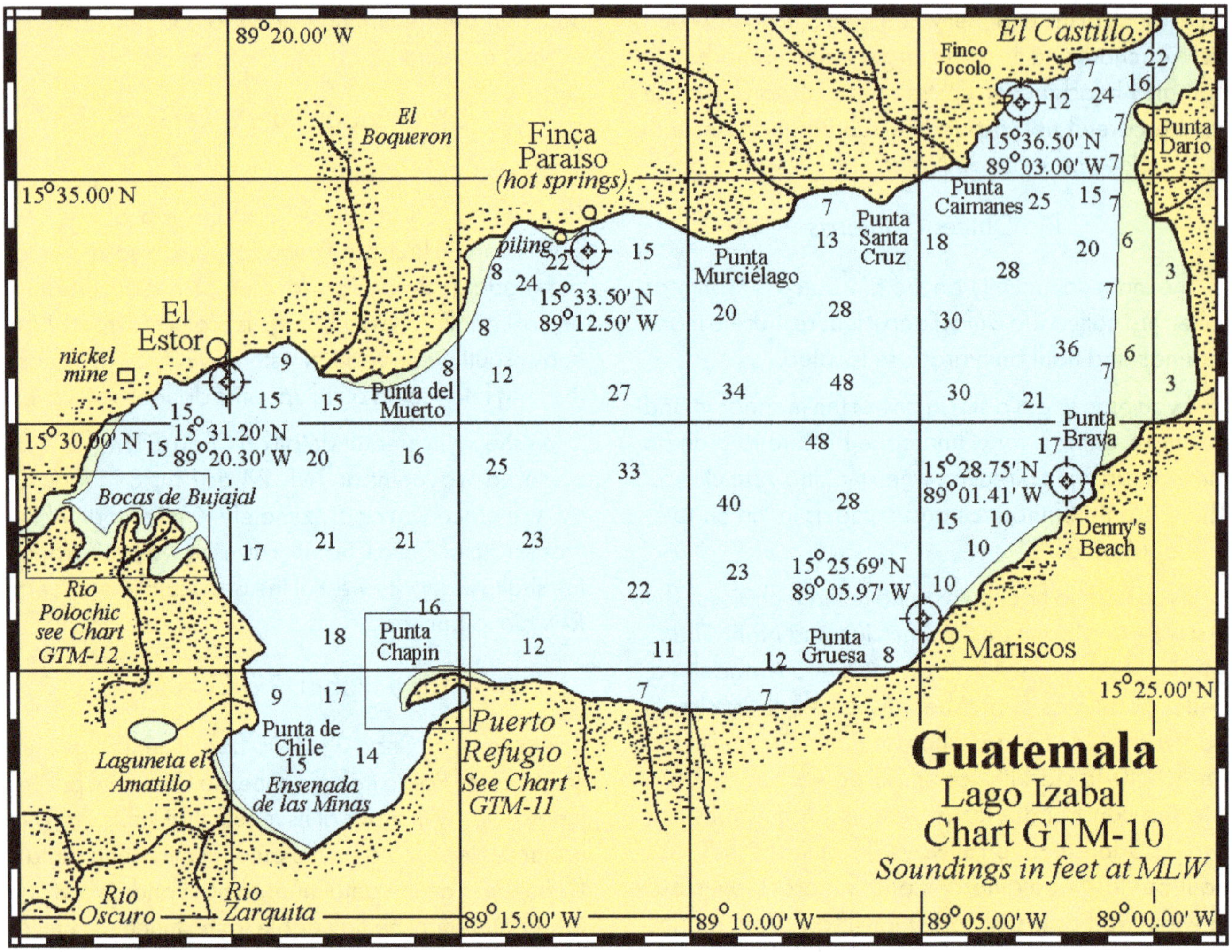

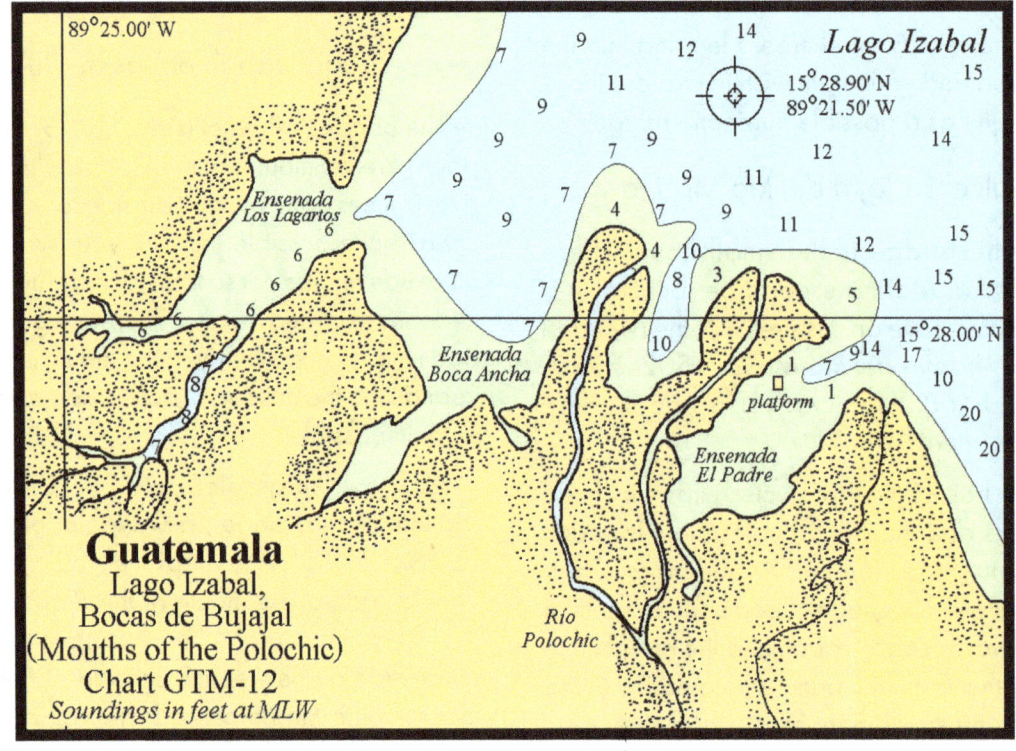

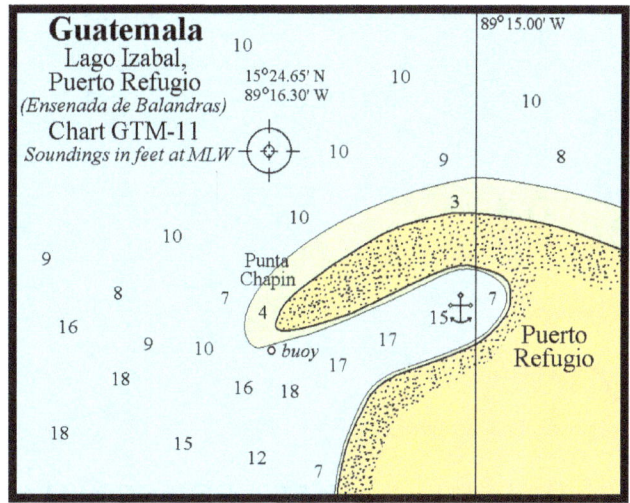

If you wish to enter the *Río Polochic*, the entrance can vary from easy to extremely tricky, and sometimes it is downright impassable. The problem lies with heavy rains and the flotsam and jetsam that they bring down the river. I've been to the *Río Polochic* when a muddy bar extended over ½ mile out and was laced with logs and all manner of debris, and the entrance depth was only a few feet. At other times, usually during the dry season, the river is much more clear and easier to enter. Another problem you'll need to be aware of is that the bottom is extremely silty and if you're churning up the bottom by dragging your keel across the bar that you stand a good chance of clogging your raw-water intake and overheating your engine.

You will notice that the *Río Polochic* appears to have two mouths, one on either side of the unnamed bay. The entrance between the unnamed bay and *Ensenada el Padre* is not passable, the only entrance is the mouth to the west of the unnamed bay, between the bay and *Ensenada Boca Ancha* as shown on the chart. Once over the bar the river deepens to 10'-30' in places and narrows to less than a hundred feet not far upriver. The trip upriver is spectacular, first the banks are low and flat but as you wind your way upstream you'll be surrounded by mountains that rise over 2,400'.

If the mouth of the river is open, vessels may also enter *Ensenada Los Lagartos*, sometimes shown as Ensenada Laguna, which lies northwest of *Ensenada Boca Ancha* and west/northwest of the waypoint. The entrance has shoaled, and drafts are limited to less than 6'.

Chapter 21

Belize

AS FOR A FIRST CHOICE AND IF I WERE WITHIN 60 MILES OF GUATEMALA, I would head to the *Río Dulce* in Guatemala. If there were other considerations and it was virtually impossible to make it in time, then I would stay in Belize where one can hide in the mangroves, get a slip, or haul out as one desires.

Cucumber Beach Marina

Located just to the southwest of Belize City, *Cucumber Beach Marina* (also known as *Old Belize Marina*) is relatively low to the water (see photo at top of next page) so bear that in mind before deciding whether to haul out here or not. The marina has a 20-ton lift with a depth of 6' at the lift slip and can accommodate a beam of 19'.

As shown on Chart BLZ-1 (next column), a waypoint at 17° 28.07' N, 87° 25.52' W, will place you approximately ¼ - ½ nm SSE of the entrance to the marina. From the waypoint take up a heading of 321° and enter the jettied channel leading in to the marina.

Cucumber Beach Marina can be reached by phone at 501-222-4129 or by email at marina@oldbelize.com.

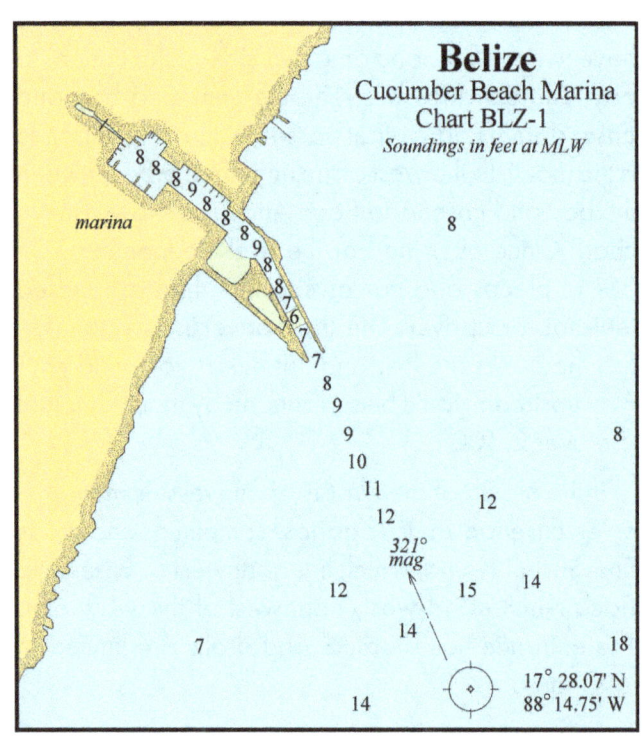

Cucumber Beach Marina, Belize

Sapodilla Lagoon

In the central coast of Belize, *Sapodilla Bay* offers good protection. It is 42 miles south of Belize City and the bay itself offers fine protection while the docks could probably be used in a storm but at your own risk.

As shown on Chart BLZ-2 (bottom of next column), a waypoint at 16° 45.93' N, 88° 17.92' W, will place you approximately ½ nm south of the entrance channel and the approach to the waypoint is without any hazards. From offshore, a good landmark is to approach on a heading of 280° aimed at 3,681' high Victoria Peak, the highest point in Belize. You will have 25' of water right up to the entrance of the lagoon where it shallows slightly to 13'. Note that the *Navionics* charts show an island at the entrance of the lagoon that does not exist.

There are seven sets of lighted red/green navigational markers that direct you through the 12' deep channel in the lagoon and into the marina area.

If you choose, you can anchor within the lagoon before you reach the marina area in several creeks that lie to starboard (northeast) of the channel. The first creek to starboard allows you to tie to the mangroves if you position your boat at the mouth of the creek in 6' of water. You can tuck further in the creek where there are 4'-5' depths. For the second creek you need to stay slightly starboard of the center at the entrance which allows for 4'-5' depths. Tie to the mangroves. For the third creek, again stay slightly starboard of center as you enter where you'll find 5'-6' depths and can tie to the mangroves.

Continuing into the marina area offers another layer of protection behind a peninsula, the channel to port (south) taking you to the marina and the branch to starboard takes you into a long canal. The marina, called *Sanctuary Belize*, currently not fully completed as of this writing but accepting boats, has well over 100 slips, catamaran slips, fuel, water, a small grocery, and electric at several side tie slips with 30/50-amp service. Unfortunately, none of the docks are floating which could cause problems if there were a significant storm surge. Juan Jimenez is in charge of the 24-hour security. The marina monitors VHF ch. 16 and ch. 68 and can also be reached through their web site at www.SanctuaryBelize.com, or by phone at 501-533-7565. No damage from Hurricane Earl was recorded within the marina.

At the starboard (north) branch of the channel opposite the marina entrance, there is a long, winding 60' wide canal with 12' depths. Once inside the main canal there is a smaller branch to port (west) that is 5' deep with a comfortably narrow 35' wide anchorage but you would need to set your anchors on the shore. There is a low bridge at the end of this branch that

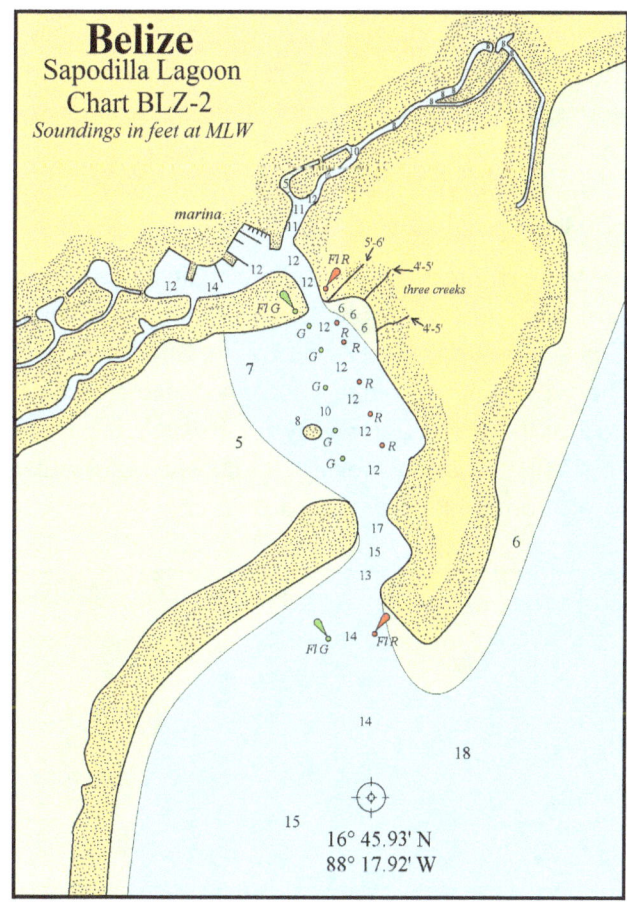

would cause problems should you break free. You could anchor anywhere along the main canal, tying to the mangroves on the east side but being aware of any other boat traffic trying to pass further into the canal. If you continue to the end of the main canal there is a secure anchorage for several boats where there are mangroves to tie to. Tall pine trees exist to the west of the canal which should help in blocking some of the west winds.

Hakim's Boatyard

As shown on Chart BLZ-3 (see below), Hakim's Boatyard is located just northwest of Belize City. The yard has a 150-ton lift capable of lifting catamarans with up to a 30' beam. The lift canal is 15' deep with the approach being 6-7' all the way from Belize City. There are several small coves northwest of Belize City, including one community built around their own canals. These are all private and entrance should be by invitation or with permission of the owners.

The only problem here is that the yard sits just 3' above sea level and when the Category 1 Hurricane Earl came through in 2016, there was 4' of water throughout the yard. Even though freight containers floated freely through the yard, no boat was reported damaged. All boats are strapped down during storms. You can reach the very helpful owner, Peter Teichroeb, and his friendly staff, Sherlene, by email at belizedivehavenmarina@gmail.com, or you can give them a call at 501-615-9341.

When approaching from Belize City, head north and pass between Moho Cay and North Drowned Cay to a waypoint at 17° 32.57' N, 88° 15.32' W. The waypoint will place you approximately ¼ nm east of the breakwater/jetty for *Hakim's Boatyard* so simply head west taking the curved breakwater to port and enter the boatyard complex. Needless to say, this route should be attempted on a half-tide rising at least depending on your draft.

Mango Creek

One of my first choices for hurricane protection in Belize would be to head into *Mango Creek* if my draft permitted it. *Mango Creek* is located within the confines of *Placencia Lagoon* and the entrance only carries 4.5' at low water, but it is very soft mud that can easily be plowed through in a catamaran or shallow draft mono-hull. Of course, if you play the tides you can pick up an extra 1.5' in depth.

As shown on Chart BLZ-4 (see below), a waypoint at 16° 32.00' N, 88° 23.00' W, places you just at the shallow mouth of the entrance into *Mango Creek*. Once inside the creek stay in the center and you'll have 10'-15' depths. Take the first branch to starboard (north) and then the next branch to port (west) for the best protection. *The Moorings* will bring their fleet of 25 boats into the cove that branches here to starboard so, as mentioned, take the branch to port. Proceed upstream to the next branch where straight ahead is an excellent mangrove lined canal roughly 30' wide. You won't be able to go in very far as the water shoals rapidly. Set bow and stern anchors, and at least one midship line both starboard and port, into the mangroves.

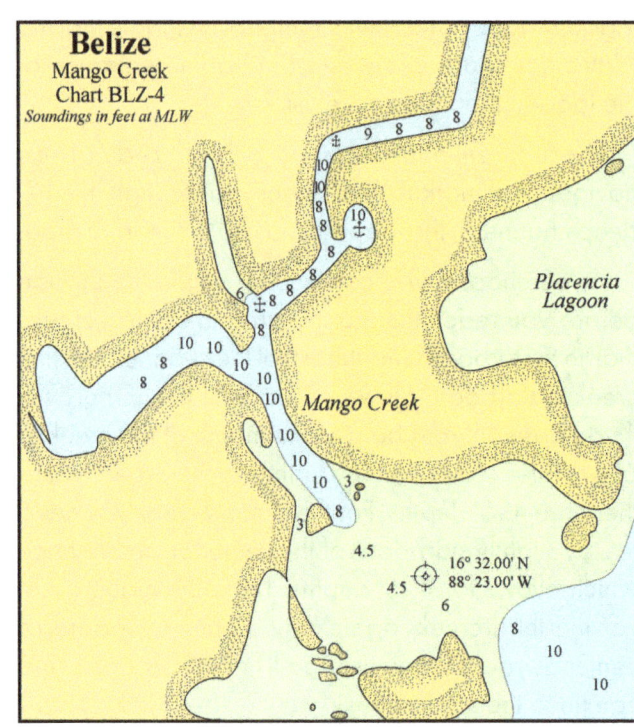

184 • THE CAPTAIN'S GUIDE TO HURRICANE HOLES

As an alternative, at that branch, continue to starboard (north) running right next to (within 5') the mangroves to starboard to avoid the shoal in the middle of the channel. Continue upriver until you find a comfortably narrow location for you to set a bow and stern anchor and tie at least one line from both starboard and port sides into the mangroves.

Ycacos Lagoon

South of Placencia is *Ycacos Lagoon*. The entrance carries 6' but is a little tricky. As shown on Chart BLZ-5 (see below), a waypoint at 16° 15.00' N, 88° 36.10' W, will place you approximately ¼ - ½ nm SSW of the entrance.

Upon entry, you should stay close to the peninsula that lies between *New Haven* and *Ycacos Lagoon* for the deepest water. Parallel the opening until half way across and then turn gradually to starboard, favoring the left side of the opening. Go slow as you will have to weave back and forth at the entrance to avoid the shifting bar. Water visibility will be poor. Once you are 50 yards inside the mouth, stay within 10' of the mangroves on your starboard side until you reach the lagoon. Continue in as far as your draft allows and either tie to mangroves or move towards the center to allow for the boat to swing.

Big Creek

A few commercial captains from as far away as Belize City like to go south to *Big Creek* during hurricanes, which is in the area of Placencia (about 80 miles south of Belize City and 2 miles west of Placencia). *Big Creek* is a commercial port for southern Belize that is utilized by large freighters making it off limits to cruisers except during hurricanes. There are some shallow 'brown bars' on the approach to the entrance channel but the lit markers are easy to discern. Once past the freighter docks the controlling depth is 6' and can be taken for quite a distance up in the mangroves. You will NOT be allowed to tie to the docks. We must mention that this is where a live-aboard dive boat flipped during Hurricane Iris with the result of 20 people drowning (details can be found in the chapter on *Hurricane History*). If you are unhappy with the protection that *Big Creek* offers, *Mango Creek* lies just a couple of miles to the north.

A waypoint at 16° 29.10' N, 88° 22.90' W, (not shown on Chart BLZ-6 below) will place you approximately 1/4 nm SE of the marked entrance channel that is shown on the chart. From the waypoint head in a northwesterly direction until you can see the markers for the beginning of the channel. Watch out for the small islands and shoals to the west. Enter the marked channel and head up *Big Creek* using the newly dredged channel that is not shown on any other charts except ours as of this writing. Best protection is well past the freighter docks in the winding, mangrove line creeks with a controlling depth of 6'.

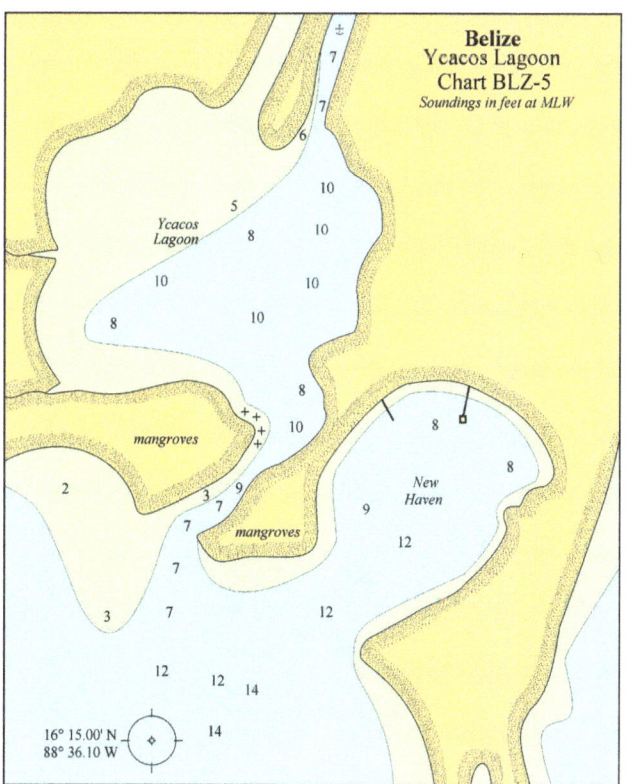

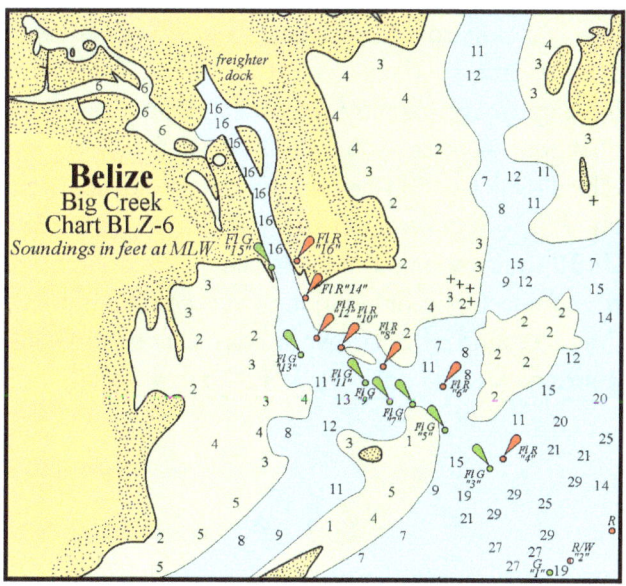

Other Marinas and Boatyards

If you are in the Placencia area and if you wish to haul-out, try Thunderbird's. They have a 15-ton lift that has lifted the 39' catamarans of *The Moorings* company. They have 10' of water to the haul out slip as well as 8 wet slips (they were full during Hurricane Earl). The yard does not strap down boats that are hauled out. For more information contact Rick Coh at 501-670-3737. The marina monitors VHF ch. 16 and ch. 68. If you need just a slip in Placencia, *Robert's Grove Marina* can accommodate 30 boats up to 75' LOA and with drafts to 8'. The marina can be reached by phone at 501-523-3565, or by email at info@robertsgrove.com.

The *Laru Beya Resort Marina* is located in Placencia, just south of *Roberts Grove* and is where *The Moorings* keep their fleet of vessels. There usually is space available and the marina can be reached by phone at 800-890-8010 or by email at info@larubeya.com.

Located up the *Sittee River* is the protected *Sittee River Marina* (in fact, the river has some mangroves to tie to in the event of a storm but overhead power lines a mile north of the marina restrict further passage upstream to powerboats). In fact, one may ignore this hole altogether since the well-protected canals at *Sapodilla Lagoon* lie just some 3.5 nm south.

The *Sittee River Marina's* floating plastic composite docks can accommodate vessels to 110' LOA with drafts to 19', but the river mouth restricts entry to 4.5' at low tide but once over the bar at the mouth (tricky, it is narrow and marked by pvc pipes, stay to the northern side of the entrance as it shoals rapidly to the south side) depths will deepen to 10'-25' and you will be fighting an approximate 1 knot current as you move upstream. A vessel with a draft of over 5' will have a difficult time crossing the bar. The marina monitors VHF ch. 9, and can be reached by phone at 501-533-7888, and by email at info@sitteerivermarina.com. If you need help entering the river, call the marina, they can send a boat down to help you negotiate the entrance (please note this is NOT a tow boat).

Chapter 22

Mexico's Caribbean Coast

THE CARIBBEAN COAST OF MEXICO, ALSO KNOWN AS QUINTANA ROO, DOES not offer a great number of protected anchorages although there are a couple of good marinas and boatyards. Moving from south to north, from *Bahía Del Espiritu Santo* to Cancún, we will show you a few places to hide from an approaching storm whether it be by anchoring out, getting a slip, or hauling out in a boatyard.

Bahía del Espiritu Santo

Bahía Del Espiritu Santo is a large, reef protected bay located approximately 90 nm north of San Pedro, Belize, offering a couple of layers of protection.

A waypoint at 19° 22.90' N, 87° 25.62' W, will place you just east of a 1.5-mile-wide break in the reef as shown on Chart MX-1 (top of next page). The break in the reef is wide but can be difficult to discern on calm or overcast days.

On a heading of 265° on the northwest point of Isla Chal (also called Owen island (and the point being Owen Point on some charts) there are 25'-30' depths between *Fupar Reef* and *Noja Reef* and 15'-17' can be carried until 2.5 nm from the west end of the village of la Victoria, located on the north shore of the bay. Take up a heading of 292° to the west end of the village. This allows a minimum of 6' depth to within .6 nm from the village. Turn to a heading of 172° to 300 yards off Isla Chal and you'll find up to 15' to Isla Chal. Round the NW point of Isla Chal and continue up the lee shore, staying only 100 yards off gives minimum of 12' depths.

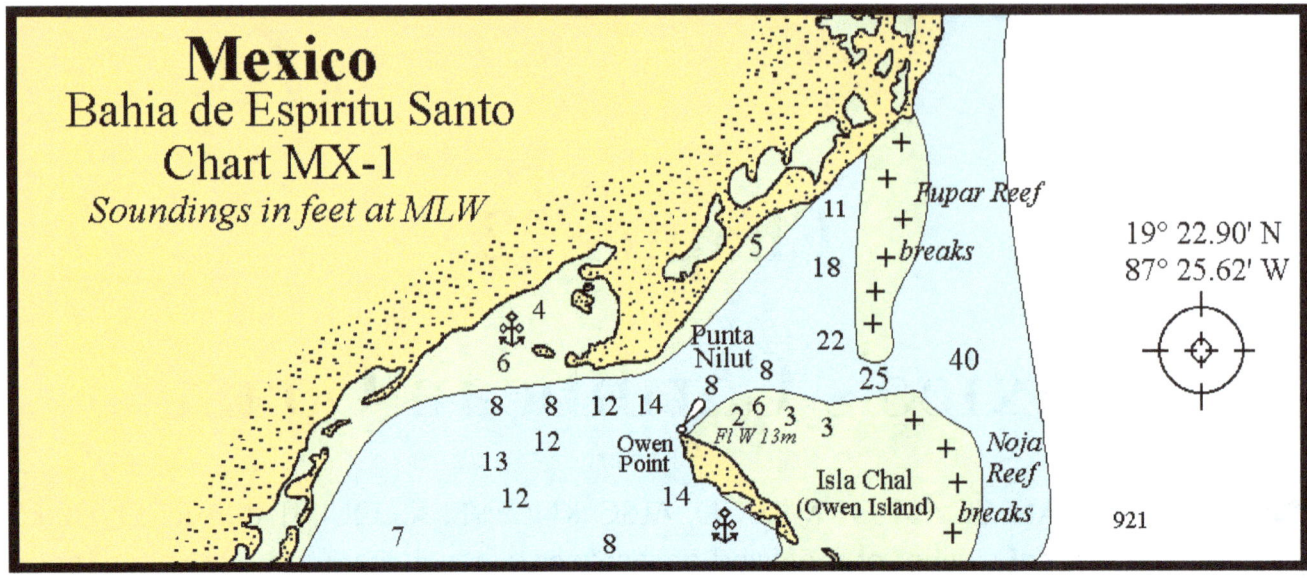

Approximately 1.9 nautical miles southeastward along the western shore of Isla Chal there is a mangrove lined pocket in the shoreline that is easily identified (see photo, below). Although open to the northwest for the fetch of 6 miles across *Bahía Del Espiritu Santo*, there are mangroves on three sides to tie to and a 5' depth at the entrance. Holding is very good in mud.

There will probably be no other boats to contend with here. If another boat has already claimed this spot, the best you'll be able to do at this point is drive your boat into the mangroves anywhere along the west shore of Isla Chal that your draft allows or try to gain entrance around Nilut Point on the north side of the bay. This offers 360° of protection but only for boats with less than a 4' draft unless a storm has given you additional depth.

Bahía Ascension

Bahía Ascension is located 24 nm north of *Bahía Del Espiritu Santo* and is also reef lined. A waypoint at 19° 45.63 N, 87° 25.52 W will place you at the entrance as shown on Chart MX-2 (see bottom of this column).

From the waypoint, take up a heading of 257° and you will have 20'-25' of water until reaching the large, very visible, white sand bore (4'-5') lying to starboard. Here you will be in 8' of water if you stay on the edge of the sand bore. Round the sand bore to gain access to the west side of the peninsula by taking up a northerly heading of 340° to place you between the mainland and Punta Allen lighthouse where the water deepens to 12' then shallows very gradually to 6'.

Anchor almost due west of the village of Punta Allen. This will give you protection from all but the southwest. Holding here is good in mud. Local boat guides told

Isla Chal, *Bahía Del Espiritu Santo*

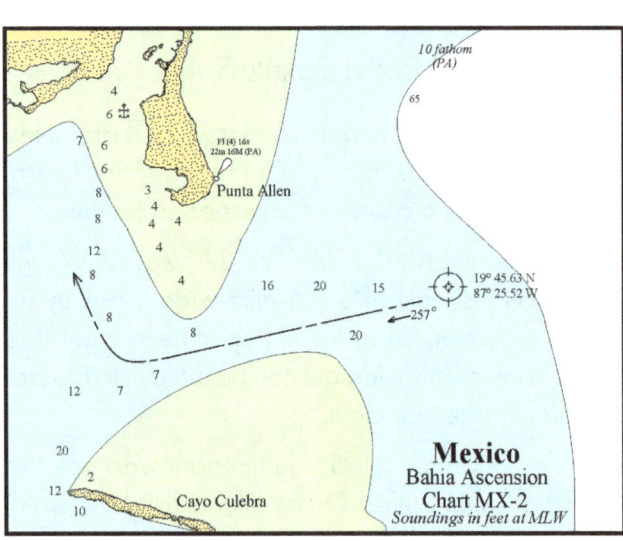

me they pull their skiffs out of the water and leave the area if a strong hurricane is eminent.

Puerto Aventuras Marina

Approximately 43 nautical miles north of *Bahía Ascension* is *Puerto Aventuras Marina* where the very helpful dockmaster, Gerardo Segrove, explained that they try and accommodate everyone in the event of a hurricane. Gerardo can be reached at 984-873-5108, by fax at 984-873-5008, and email at pamarinaf@puertoaventuras.com.mx. *Puerto Aventuras Marina* stands by on VHF ch. 16 and 79).

Puerto Aventuras Marina has 70 med-moor style slips) that can accommodate vessels from 10'-150' LOA with water depths of 7.5'-8' at low water. However, local ferry boats and fishing boats have first priority to the 10' high solid rock docks, but they have agreed to let others side tie to them in an emergency. There are numerous private canals that could also be used if available. The whole process operates on a first come, first serve basis.

The entrance to the marina is beset by a reef on each side and is subject to a dangerous side swell, especially with a south wind. In this case, you MUST arrive 3-4 days prior to any storm or the entrance will be impassable. There were approximately 20 days this occurred during the 2016 hurricane season.

As shown on Chart MX-3 (next column), a waypoint at 20° 29.70' N, 87° 13.50' W. approximately ¼ mile south of the entrance between the jetties. From the waypoint just a bit west of north to enter the channel and marina complex (see photo at the top of the next column). The entrance has a lit range consisting of two yellow posts (never attempt this entrance at night). The range appears to be a bit off to port upon entering so we tend to keep more to the starboard side of the range. The range is 349° upon entry.

Cancún: V&V Marina

The best protection in the area of Cancún is *V&V Marina* (just north of Cancún) where you will find haul out capabilities for vessels to 110-tons with a maximum beam of 19'. The yard has room for 18 boats and they have tie-downs anchored in concrete. The nearby high-rise condominiums offer good protection from strong east-northeast winds.

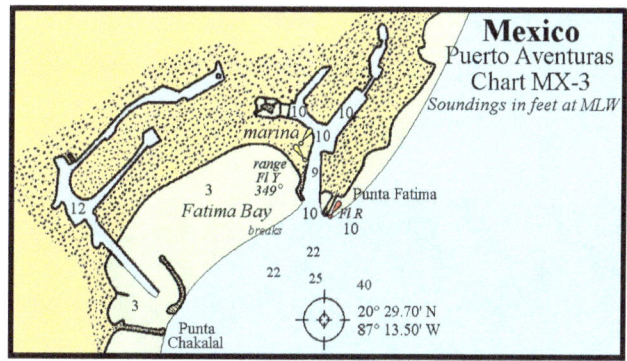

The entrance to Puerto Aventuras

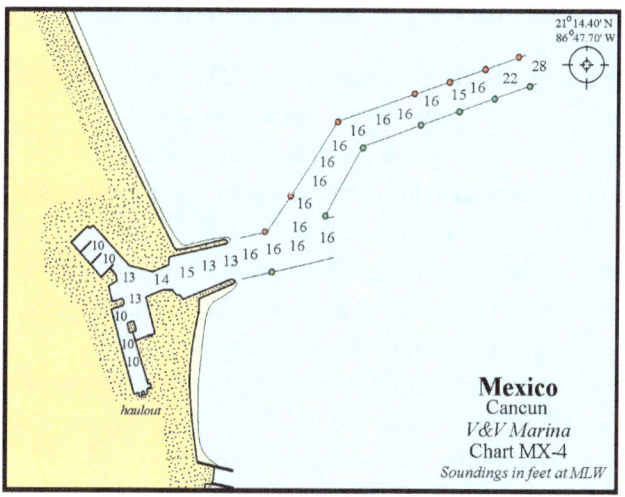

Unfortunately, the marina does not permit boats to stay at their docks during a hurricane (even though they have 176 slips with 8 that can accommodate vessels to 200' LOA). The marina monitors VHF ch. 12, and can be reached by phone at 998-234-0100, or at info@marinavv.com.

The entrance to the marina is on the Mexican mainland WSW of the anchorage at Isla Mujeres across *Bahía Mujeres*. As shown on Chart MEX-4

The last marker at the entrance to *V&V Marina*, Cancún

Entrance to Laguna Makax at Isla Mujeres

V&V Marina, Cancún, as seen from the north

The lift at Isla Mujeres

(previous page), a waypoint at 21° 14.80 N, 86° 47.30 W, will place you approximately 1 nm NE of the beginning of the dredged entrance channel into the marina basin. From the waypoint head in a general WSW direction and you will see the outer markers of the channel through the reef. Follow the channel markers and you will make a dogleg to port and then to starboard to enter the channel between the jetties (see photos above). Once inside turn to port just past the office/fuel dock and head south for the marina's haul out facility at the extreme southern end of the basin. The dredged, marked channel is approximately 75' wide between the markers.

Isla Mujeres

Isla Mujeres is the place that everyone from Morelos northward plans to go to in a hurricane. The island is 25 miles north of Morelos and 5 miles northeast of Cancún across the *Bahía Mujeres*.

On normal days all of the marinas are packed as well as the outer harbor anchorage. But the main place for hurricane protection in the inner harbor, *Laguna*

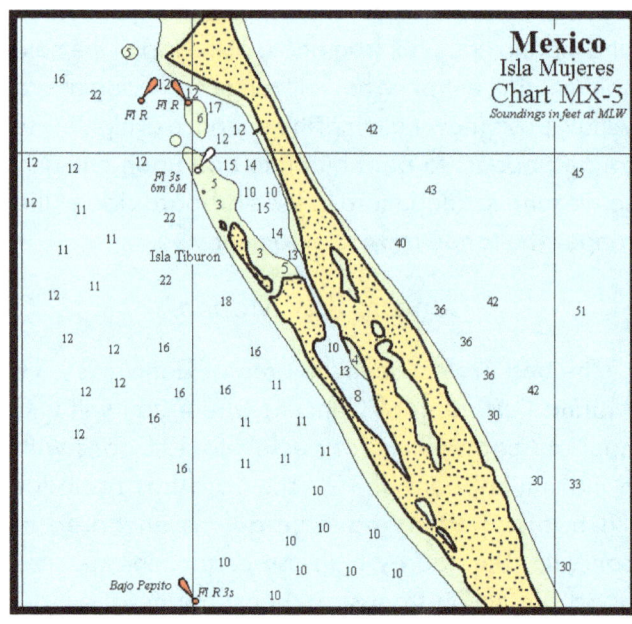

Makax (also shown as *Puerto de Abrigo*), that is extremely well-protected from 360°, encircled mostly with mangroves, but the holding is poor in mushy mud/grass. Everybody moves to this inside harbor when a norther threatens, and many proceed to drag anchors during even that small of a blow.

If you prefer a slip or haul out, the *Westin Marina* (also known as *Puerto Isla Mujeres*, 998-887-0330, VHF ch. 16 and 11, and their website is located at http://www.puertoislamujeres.com), is located just inside the cut through from the outer harbor to the inner harbor on the east side and is capable of accommodating very large boats in a well-protected part of the harbor. The marina has 64 slips for vessels to 80' LOA and 10 slips for vessels to 175' LOA with drafts to 10'. However, their slip fee goes up to $5 per foot in a storm. The yard has a 150-ton *Travelift* to haul boats (see photo previous page) but the locals seem to have priority, regardless of your prior reservation! The yard is relatively small, probably only capable of holding 20-25 average sized boats.

Just past the *Westin* on the same side (east), there exists a deeper water anchorage that carries 7' all the way up to the mangroves. This little pocket is protected by the *Westin Marina* to the north and would allow you to tie off to the mangroves to the east. There is little chance there would be any boats anchored to the west because that is the main channel and there is very shallow water on the other side of that. The only problem might be if your westerly set anchors didn't hold, you would wind up on the mangrove shore, which in this case consists of mostly rocks. There is a sunken ferro-cement sailboat up in the mangroves that could act as another anchor to the southeast.

There are a couple of other marinas on the eastern shore of *Laguna Makax*. *Oscar's Marina* (*Varedero deo Oscar*) and *Marina del Sol* has 22 slips but drafts over 5' may have to deal with the soft mud bottom. There are also a few marinas in the outer lagoon (and not as well-protected), *Marina Paraiso*, *El Milagro Marina*, and *Enrique's Marina* (22 slips). Bear in mind that these marinas are open to westerly winds.

Isla Blanca: *Laguna Chakmochuk*

Laguna Chakmochuk is a semi-protected cove on the western shore of Isla Blanca off the northern shore of the Yucatán peninsula just south of Isla Contoy. The only problem with Laguna Chakmochuk is that if the hurricane moves to the east of your position you will be getting hit with northerly winds and seas, so protect yourself accordingly.

As shown on Chart MX-6 (see below), if approaching from Isla Mujeres or V&V Marina proceed north from Isla Mujeres along the reef that runs almost continuously to Isla Contoy hugging the reef, at times only 30 yards to the west of it (see dashed line on chart). As you approach the NE tip of Isla Blanca you will find shifting sand bars northeast

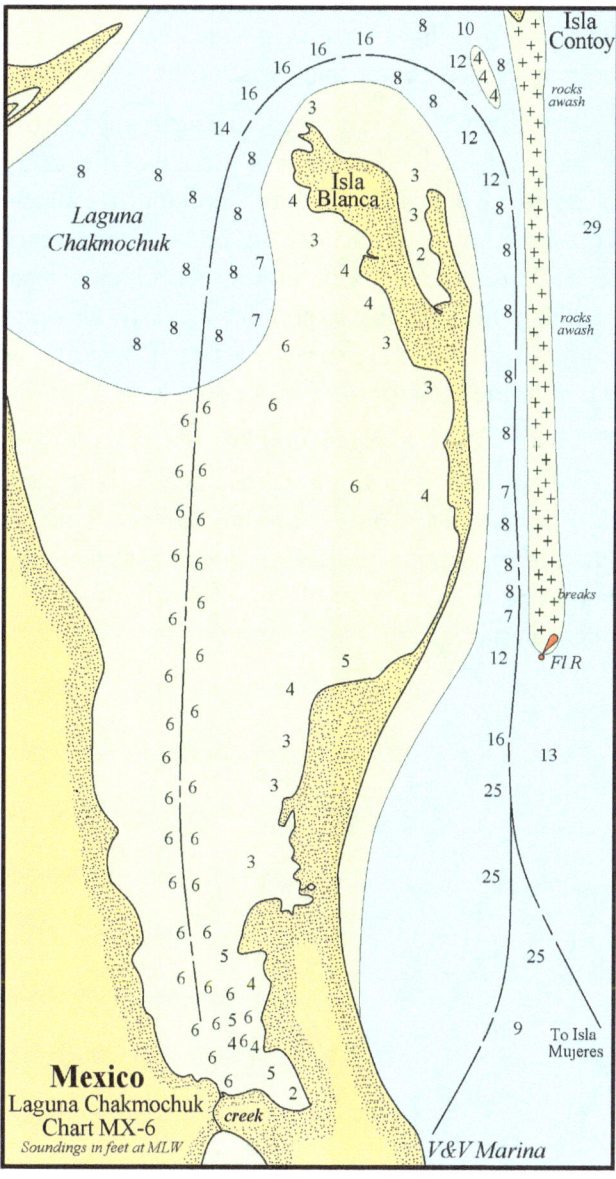

of the northern most tip of Isla Blanca. There is only 6' when abeam of Isla Blanca point.

At 21° 24.48 N, 86° 46.83 W, take a slight turn to port on a heading of approximately 303° to stay to the left of the extensive sand bar. You'll find 12'of water. Carry on into the wide lagoon entrance and, staying in the middle, you'll have 8-10' all the way to the end where it shallows to 6'. (Note that the island in the middle of the lagoon shown on *Navionics* charts does not exist).

There is a creek in the mangroves at the southern end of the lagoon at latitude 21° 14.92' N, 86° 50.340' W, with a 5'-6' approach. At the entrance to this creek there is a notch to your port side that is lined with mangroves on three sides. You'll either want to anchor facing the north and tie to the mangroves or pull your bow directly into the mangroves and set stern anchors. There is 8'-10' of water over a soft mud bottom. Watch out for small boat traffic in the creek.

Alternately, there is a cove to the east (not the very shallow one at the end of the lagoon) that can carry 5' draft to within 150 yards of the mangrove shore. You should see a small dock just north of this anchor spot. Good holding exists and good protection from anything from the east to north to south. This is open for the 1.3-mile fetch from the west. BEWARE! This water inside the lagoon is impossible to read as it is yellow/brown with 6" of visibility.

A final note on *Laguna Chakmochuk*. Northerly hurricane winds could make this bay untenable. Northerly winds would indicate the eye of the hurricane is to the east of you. If this happens act accordingly.

Chapter 23

Cuba

I WILL FIRST MENTION THAT, ACCORDING TO HURRICANE TRACK ARCHIVES, LOCAL boaters, fishermen, and dock masters, the east coast of Cuba, followed by the west coast, are much more likely to be in the path of hurricanes than the central region. Combining this with the high mountain range found in the center of Cuba, the central north coast would be my first choice for finding refuge if I were in the area of the largest island in the Caribbean. As a final note, there are many more Cuban hurricane holes that we have not shown in this publication because on the entire 800-mile-long northern coast of Cuba, the Government only permits non-resident boaters to enter 6 locations.

Northern Coast: *Canal de Barco*

Canal de Barco at lat 21°.55.508N long 84°.48.476W, only 6 nautical miles east of Punta Morros de Piedra, also known as Morona or Cabo San Antonio (where you'll find Customs and Immigration), provides 360-degree protection and is lined with mangroves. The entrance is straightforward from the east. The holding is good in a muddy bottom. It is just east of Cayo Pta Afuera which borders lagoon *Ensenada Bolondron*. There is not much elevation to the surrounding land, however, there will probably not be any other boat to share space with. Although this has good protection, it is relatively wide (400'), and is far enough west that it would not be my first choice if a storm was heading up the *Yucatán Channel*.

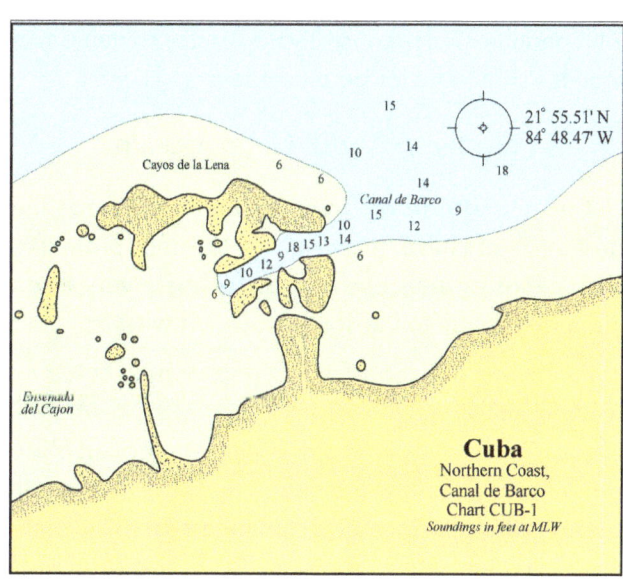

Cuba
Northern Coast,
Canal de Barco
Chart CUB-1
Soundings in feet at MLW

193

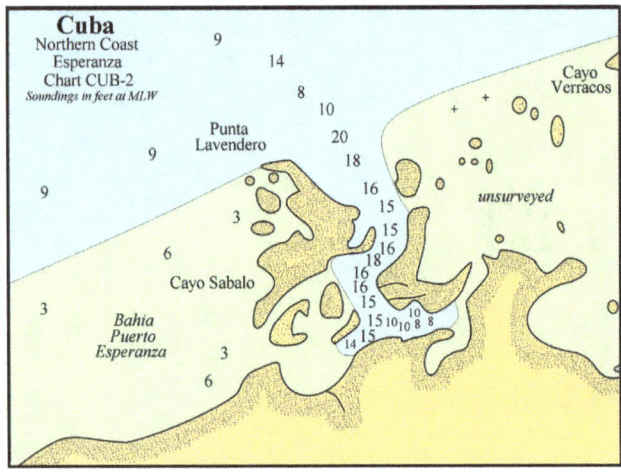

Northern Coast: *Esperanza*

About 35 nautical miles west of *Bahía Honda*, near the north coast town of Esperanza, is a wonderful hurricane hole. To enter you need to travel between the reef protecting the town of Esperanza at 22° 51.30' N, 83° 46.00' W, and continue on a heading of approximately 128° which will give you a minimum depth of 10'.

As shown on Chart CUB-2 (above), these numerous canals just east of Esperanza are a spiderweb of mangrove lined waterways providing 360° protection. Two of the many anchorages found are at 22° 47.39' N, 83° 41.53' W, and 22° 47.272' N, 83° 41.562' W. In either of these you can get close enough to the mangroves to tie off as well as anchor in good holding mud. The *Sierre de los Organos Mountains* offer protection from the south while a reef, several islands, and mangroves offer protection from the north, east, and west. There should be no other boats to contend with and the choices of anchorages are so numerous that this should never be a problem.

Northern Coast: *Cayo Morillo*

The anchorage at Cayo Morillo is perhaps the best in all of Cuba. It is centrally located, protected by a series of reefs offshore as well as a very large mountain range to the south, east and west.

As shown on Chart CUB-3 (top of next column), a waypoint at 22° 57.14' N, 83° 23.24' W, will place you at a wide break in the reef called *Pasa de la Mulata*. Follow the deep channel (30'+) south slowly turning to southeast and then east to tuck into the 60' mangrove lined canal south of Cayo Morillo as

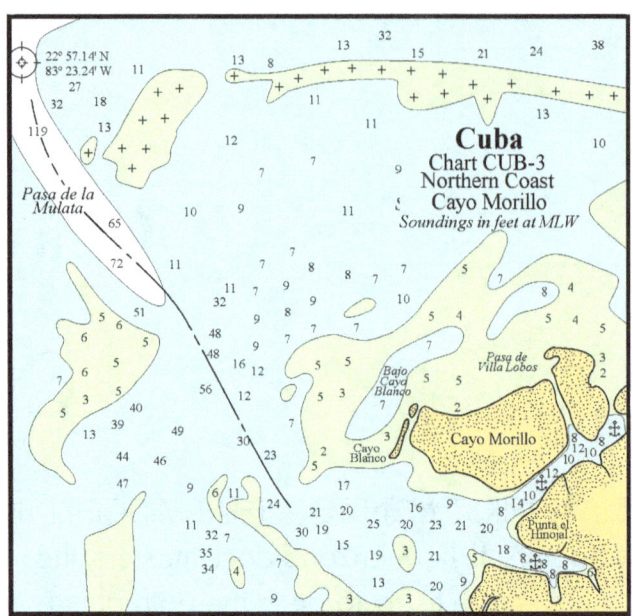

The mangroves at Cayo Morillo

shown on the chart and photo (above). Care must be taken to avoid the brown bars extending off the various points of land. There is 360° of protection in 8'-10' of water. Tie to the mangroves. Good anchorages can be found at 22° 54.759' N, 83° 20.609' W, at 22° 54.798' N, 83° 20.733' W, and at 22° 55.323' N, 83° 20.543' W, all in 8'-10' of water, 60'-75' wide, with good holding in mud.

Northern Coast: *Bahía Honda*

Bahía Honda is a "pocket bay" on the northwest shore that the Cuban government allows foreigners to enter. The entrance is straightforward through a marked channel which is overseen by a large old rusty shipwreck at its entrance.

Entrance, Bahía Honda

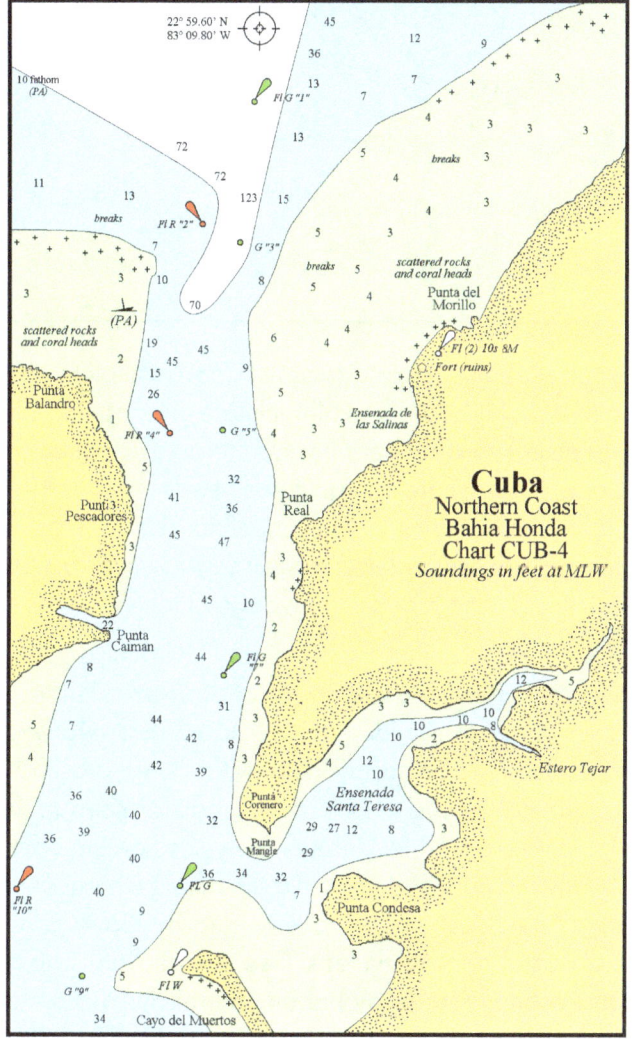

As shown on Chart CUB-4 (next page), a waypoint at 22° 59.60' N, 83° 09.80' W, will place you approximately 1 nm north of the entrance into the bay. From the waypoint head southward following the marked channel as per the chart (also see left).

As you enter, to your starboard (west) you will see the remains of quite a few other ships either waiting for their steel to be salvaged or simply sunk in the bay.

The area that is not off limits to foreigners is to your port side (east) after you pass the *Guarda* buildings and just past the first peninsula on the same side (port) and Punta Corenero (giving the shoal south of Punta Corenero a wide berth and the shoals south of Punta Mangle). The bay to the east and northeast here is *Ensenada Santa Teresa* and you will find a minimum depth of 12' upon entering with depths of 22' in the middle of *Ensenada Santa Teresa*.

There is a submerged cable running from Punta Corenero to the western tip of Cayo del Muertos and then south across *Bahía Honda*, do not anchor in its vicinity.

If you continue into *Ensenada Santa Teresa* you will see a small creek branching off to your starboard side (south) called *Estero Tejar*. This creek is mangrove lined, 45' wide, and has a depth of at least 8' with good holding in mud.

Ensenada Santa Teresa is reef protected, mangrove lined, and has a 1,000' mountain range offering protection from the south, east, and west. Further into the bay, you will see an oasis of tall palm trees. You can anchor anywhere your draft allows and tie to mangroves, but I would recommend continuing in slightly farther to stay away from the rock embankment on your port side. The bay continues to the east and northeast getting narrower and shallower.

Other boats should not be a problem here. The boat graveyard might result in some floating debris but tucked in as far as you can into the nooks and crannies to the east should nullify that problem.

Northern Coast: *Marina Hemingway*

Marina Hemingway is the most westerly located marina on the north shore of Cuba and was undamaged by Hurricane Irma in 2017. The entrance can be a bit choppy on a northwest wind or any wind above 30 knots and no place to enter if a hurricane is close and the wind and seas are up.

The marina consists of four long parallel canals, each approximately 80' wide. The surrounding land is rather low, and water has been known to come across

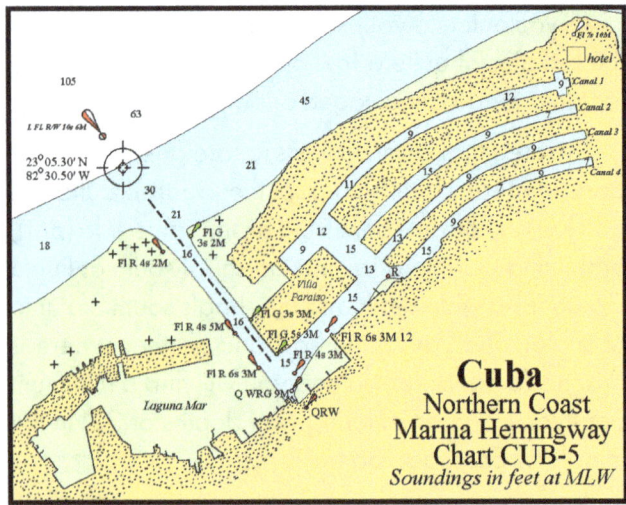

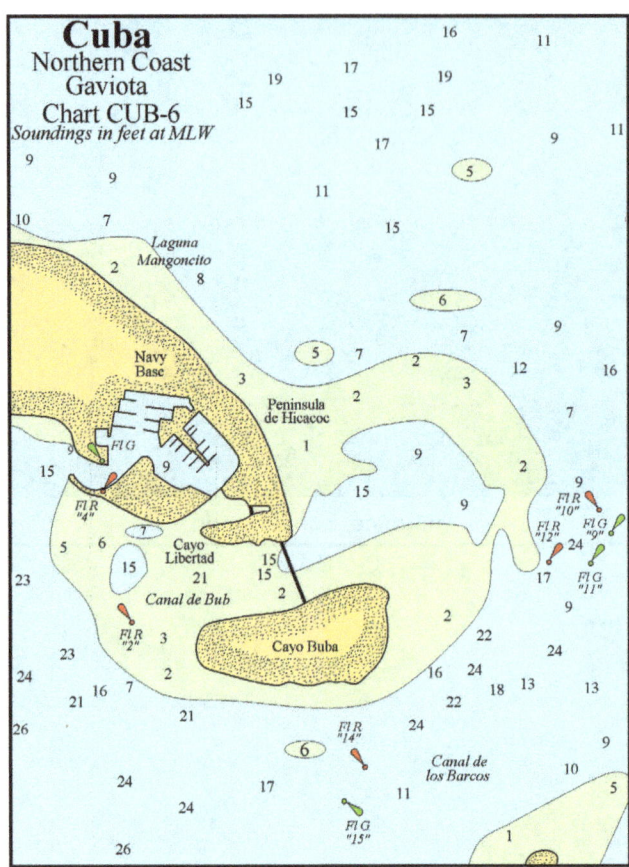

the outermost spit of land in a bad storm. However, during a hurricane, all boats are moved to the center of the inner 3rd and 4th canals and tied off to the low, fixed concrete sea walls. You would need some long dock lines to accomplish this but, reportedly, no boats have been damaged in the past 30+ years by following this technique.

There is a hotel within the marina complex that would be a good place to get a room for the duration of the storm. The marina can be reached by phone at 537 - 204-6848 / 209-7928 / 209-7270, or through their website at http://www.hemingwaycuba.com/marina-hemingway-cuba.html, and they also monitor VHF Ch. 16 and Ch. 77. As a final note, remember that even though the marina has security, it would be a good idea to lock your vessel when leaving it. Co-author Capt. Dave Underill left his boat unlocked for 10 days here and nothing was missing thanks to the wonderful security and the people of Cuba.

As shown on Chart CUB-5 (see above), a waypoint at 23° 05.30' N, 82° 30.50' W, will place you approximately 1 nm NW of the entrance channel. From the waypoint head down the well-marked entrance channel following the dashed line on the chart (approximately 140°). Do not turn to port, to the east, to enter the canal system until immediately after passing Villa Pariso, do not go further along the range, just turn to the east.

Northern Coast: *Varadero, Marina Gaviota*

Marina Gaviota is located at the east end of peninsula Hicacos (where Varadero is located) and is correctly named the *Cancún of Cuba*, complete with numerous high rises and all-inclusive resorts lining the perfect white sand beaches and turquoise water. In Cuba, this is where you go if you need to haul out for a hurricane. *Marina Gaviota* monitors VHF ch. 16 and can be reached by telephone at 534-566-7755, or by email at dir_marina@delvar.gav.tur.cu.

Gaviota is a fine marina with 1,400 slips (able to accommodate drafts to 9') and an excellent dry storage yard that holds over 200 boats. Those wanting to haul out will be happy to know that there is a 100-ton lift in the yard and that the boats are strapped down onto concrete. Unfortunately, the floating dock marina will not let you stay in a slip during a hurricane, so all the local boats move to nearby Cayo Cruz Del Padre, about 5 nm east,(see next section). This is best as we learned by the amount of destruction to the yard and marina by Hurricane Irma in 2017 (for more information on Irma's effect on the Caribbean see the chapter Irma and Maria).

Although not shown on Chart CUB-6 (above), a waypoint at 23° 16.00' N, 81° 06.50' W, will place you approximately 1/2-1 nm NW of the marked entrance channel into *Marina Gaviota*. From the waypoint head in a generally southeast direction and you will pick up the markers to lead you in as per the chart.

Northern Coast: *Cayo Cruz del Padre*

The anchorage at Cayo Cruz Del Padre (see Chart CUB-7 below) consists of a series of natural canals intertwined between mature mangroves. There are a multitude of anchorages available although the over two dozen 80' catamarans, fishing, and dive boats from *Marina Gaviota* will all be here.

Although not shown on Chart CUB-7, a waypoint at 23° 15.74' N, 80° 57.12' W will place you off the entrance channel where you will find a series of mangrove line canals with approximately 8' of water at MLW.

A few of the more ideal anchorages with little fetch and deep water up to the edge of the mangroves are at latitude 23° 14.706' N, 80° 55.138' W, at 23°15.567'N, 80°.56.430'W, and at 23° 15.514' N, 80° 56.435' W. Holding is in thick turtle grass so a few days to set the anchor is advised. You will also be tying to the mangroves on all sides.

Northern Coast: *Bahía de Vita*

Bahía de Vita, another of the small "pocket bays" that the Cuban government allows foreigners to enter, lies approximately 111 nm west of the eastern tip of Cuba and gives you the geographic advantage of the protection of the high mountains to the south.

Although it is not shown on Chart CUB-8 (below), a waypoint at 21° 06.03' N, 75° 57.92' W, will place you approximately ½ nm NW of the marked entrance channel. Proceed down the marked channel and beware of a reported .6 knot current. There is a 10-meter lighthouse on the point.

In the bay there is a small marina and officials to clear a boat into the country. *Marina Internacional Puerto de Vita* monitors VHF ch. 16 and 13 and can be reached by phone at 532-430-4751. The marina's 38 fixed concrete slips and stern-to mooring

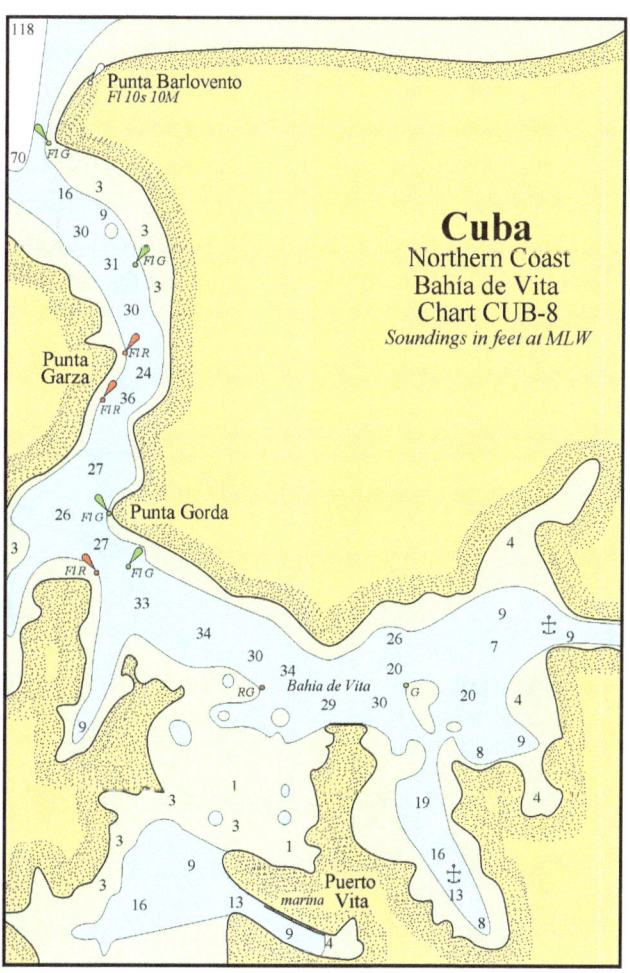

can accommodate vessels to 180' LOA with drafts to 18'. If you are having trouble negotiating the entrance channel, call the marina on VHF and they will send a boat to guide you into the bay. However, you might find that you will be better protected here tied off into the mangroves instead of stern-to a concrete dock.

The best location for protection with enough water depth to reach the mangroves would be to the east and south after you enter the bay. Tuck in as far as your draft allows.

Southern Coast: *Bahía de Baitiquiri*

Baitiquiri has a narrow opening (50' wide) through a marked channel that opens up into a 360-degree protected bay. It is the first such anchorage on the southeast portion of Cuba that you can get in to as you approach from the east. There are mangroves to tie to located on both port and starboard sides as you enter the bay. There is excellent holding in mud. Theft may be an issue if tied to the mangroves and abandoning your vessel. The center of the bay has good holding in 15'-20' of water.

As shown on Chart CUB-9 (below), a waypoint at 20° 01.00' N, 74° 50.80' W, will place you approximately ½ nm SSE of the entrance channel. From the waypoint head in a generally NNW direction and enter the channel into the small bay avoiding the shoals at the narrow entrance (see chart). There is a light (Fl 6s 8m) on the point on the eastern side of the channel.

Southern Coast: *Santiago*

Santiago is located on the southern shore west of Guantanamo. It has a well-marked channel leading through a winding, relatively narrow (200') channel graced by a Spanish fort, *Morro Castle*, on your starboard side upon entry. As shown on Chart CUB-10 (see below), a waypoint at 19° 57.80' N, 75° 52.40' W, will place you approximately 1 nm SSW of the marked entrance channel into the bay. From the waypoint, enter between the outer lights and follow the well-marked channel.

Once inside, the harbor opens up to a considerable width. There is a marina (*Marina Marlin Santiago De Cuba*) with a very helpful English speaking manager located at Punta Gorda as shown on the Chart. The marina's small docks are dilapidated and probably not reliable in any strong winds. The marina monitors VHF ch. 16 and sometimes ch. 72, and can be reached by phone at 532-269-1446, or by email at admin@marlin.seu.tur.cu.

The main harbor has plenty of fetch due to its relatively large size. There are numerous waterway branches reaching like fingers outward from the main harbor offering good protection with little crowding

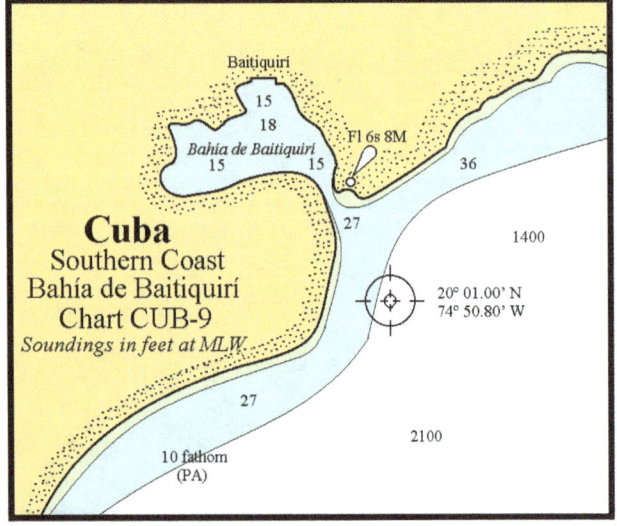

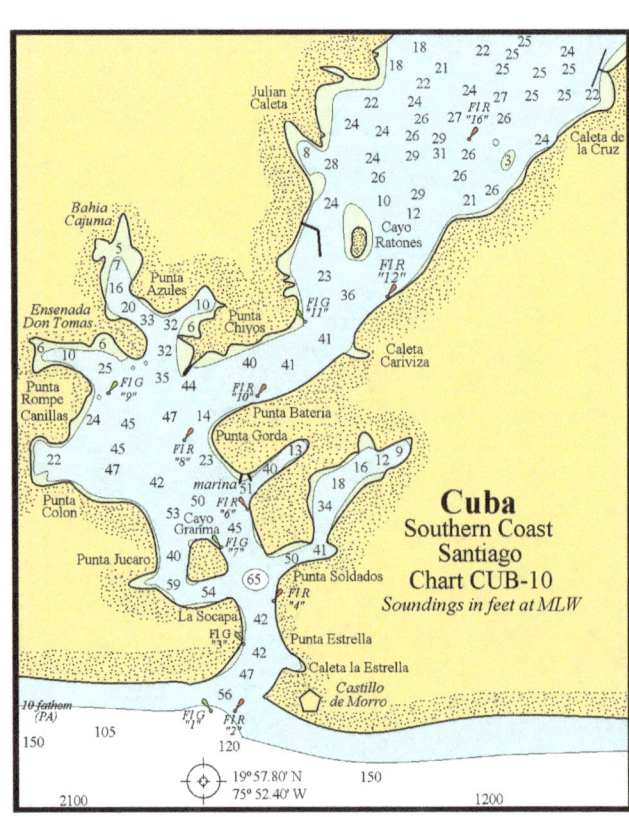

from other boats. It is up to the *Guardafronteras* where they will let you anchor.

Southern Coast: *Puerto de Casilda*

Puerto de Casilda is the commercial and fishing port for Trinidad. The port lies approximately 30 nautical miles southeast of Cienfuegos, and the entrance to Puerto de Casilda is located on the southern shore of Cuba almost in the middle of the island. Here you will find a small marina (*Marina Cayo Blanco*) in a mangrove surrounded hole for anchoring on the Peninsula de Ancon. You will be located close to the coast with only a 5-story hotel and the mangroves to lessen any significant wind.

Not shown on Chart CUB-11, a waypoint at 21°38.17' N, 79° 52.28' W, will place you just south of the well-marked entrance channel into this busy commercial port about 6 miles to the southeast of the main harbour.

From the outer waypoint, head north into the channel, passing Cayo Blanco and leaving it to your port side. The entrance is well-marked with good hurricane protection nestled up in the mangroves. To reach the narrow channel into the marina as shown on Chart CUB-11 (top of next column), you must first navigate through an area with controlling depth of 4.5'-5 as shown on Chart CUB-11. Once inside the mangroves you can anchor or continue to the small marina. There are moorings in the small bay, but I don't believe I would trust one if I were to ride out a hurricane here They are only good for vessels to 42' LOA. The marina monitors VHF ch. 16 and can be reached by phone at 534-199-6205, and by email at marinastdad@enet.cu.

Southern Coast: *Cienfuegos*

Cienfuegos, located on the southern shore, has a well-marked entrance channel, complete with communist graffiti painted on the overlooking walls to greet you (along with the military guard).

AS shown on Chart CUB-12, a waypoint at 22° 01.80' N, 80° 27.20' W, will place you approximately 1.5 nautical miles south of the entrance channel. From the waypoint head in a generally north direction, between the points of land, and you will pick up the well-marked channel entering the bay.

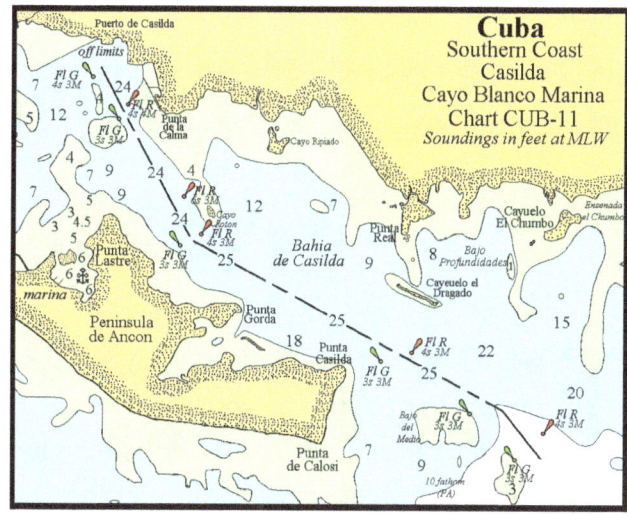

The winding channel opens up into a very large and deep harbor, both commercial and recreational, which would mean plenty of fetch in the main part of the harbor. There is a marina at Punta Gorda, but space is very limited as local charter sailboats have taken up most of the dock space (5343-255-1241, VHF ch. 16, 19/71, or rrpp@nautica.cfg.tur.cu) The marina has a 6.5' controlling depth.

There are a number of small bays branching off the main harbor that offer better protection with little crowding from other boats. Again, there is a question of where exactly the *Guadafronteras* will allow you to anchor.

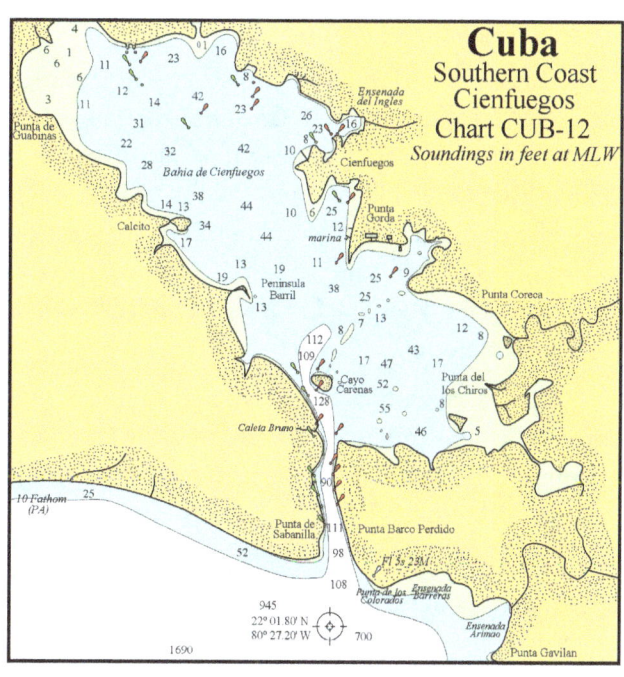

Chapter 24
Weather Broadcasts

THE ONE THING ALL MARINERS NEED DURING HURRICANE SEASON, IS access to accurate, timely weather reports for the areas in which they are cruising. Skippers will find that there are various forms of weather broadcasts available in The Bahamas and Caribbean from local radio weather reports to on-line and HF weather reporting. In this chapter we will endeavor to give you as complete a listing as we can for weather access in the islands. Remember to use weather reports with an eye on moving early if you need to move!

TV and Internet

You can access the *National Hurricane Center* at www.nhc.noaa.gov, and the NOAA weather site at www.nws.noaa.gov. *The Hydrometeorological Prediction Center* offers a broad overview of weather patterns at www.hpc.ncep.noaa.gov/.

For Bahamian weather information on the Internet, you can visit the website of *The Bahamas Department of Meteorology* at www.bahamasweather.org. The site features forecasts, tide tables, and satellite images. For Nassau-based radar coverage of Abaco visit www.bahamasweather.org.

If you are at a dock with cable TV, you can pick up *CNN* or the *Weather Channel* for the latest weather information every hour. For info online visit the *National Hurricane Center* website at http://wwwnhc.noaa.gov, the *Weather Channel* website at http://weather.com, at http://www.caribwx.com, and at the *WeatherCaribe* site at http://www.weathercarib.com/.

If you are located in St. Martin and Anguilla, you can pick up the *Weather Channel* during the hurricane system on the *Anguilla Community Broadcastings* daily broadcasts on TV Ch. 9.

In Belize you can access weather forecasts at www.hydromet.gov.bzl. You can pick up Cuban weather at https://www.yr.no/place/Cuba/.

HF Weather

If you have ham radio capabilities you can pick up the *Bahamas Weather Net* every morning at 0720 on 3.696 MHz, lower sideband. The net begins with the local weather forecast and tides from the *Nassau Meteorological Office*. Next, hams from all over The Bahamas check in with their local conditions which are later forwarded to the *Nassau Meteorological Office* to assist in their forecasting. If you are interested in the approach of a front you can listen in and learn what conditions hams in the path of the front have experienced. The local weather reports follow a specific order so listen in and give your conditions in the order indicated. If you have marine SSB capabilities you can pick up *BASRA's* weather broadcasts every morning at 0700 on 4003 KHz, USB.

At 0745 on 7.268 MHz you can pick up the *Waterway Net*. Organized and maintained by the *Waterway Radio and Cruising Club* (http://www.waterwayradio.net/), this dedicated band of amateur radio operators begin the net with a synopsis of the weather for South Florida and then proceed to weather for The Bahamas (with tides), the southwest north Atlantic, the Caribbean Sea, and the Gulf of Mexico.

The *United States Coast Guard* in Portsmouth, Virginia (*NMN*) weather broadcasts can be received on your SSB on 4428.7 KHz (ch. 409), 6506.4 KHz (ch. 601), 8765.4 (ch. 816), 13113.2 KHz (ch. 1205), and 17307.3 (ch. 1625). Times are 0600, 0800, 1400, and 2200.

Chris Parker conducts weather nets 6 days a week, Monday through Saturday, and also Sundays when Tropical or other severe weather threatens. Chris' summer schedule, April to October, begins at 1000 UTC with coverage of the Eastern Caribbean on 4.045 USB and 8.137 USB, moving to The Bahamas at 1030 UTC. Chris then moves up to 8.137 USB and 12.350 to cover the U.S. East Coast at 1130 UTC, the eastern Caribbean again at 1230 UTC, the western Caribbean at 1300 UTC, and then checks for other stray traffic to the east and northeast.

Chris' winter schedule, November through March, begins at 1100 UTC with coverage of the Eastern Caribbean on 4.045 USB and 8.137 USB, followed by coverage of The Bahamas at 1130 USB. At 1230 UTC, Chris moves up to 8.137 USB and 12.350 to cover the U.S. eastern Atlantic Coast, followed by the Western Caribbean at 1330 UTC.

When severe weather or tropical weather systems threaten Chris will also transmit in the evenings, usually on 8.104 MHz at 1900 AST/1800 EST and Chris will usually announce this on the morning net. Chris begins the net with a 24-48-hour wind and sea summary followed by a synoptic analysis and tropical conditions during hurricane season. After this, Chris repeats the weather for those needing fills and finally he takes check-ins reporting local conditions from sponsoring vessels (those vessels that have paid an annual fee for this service). Those who seek more information about weather, weather patterns, and the forecasting of weather, should pick up a copy of Chris Parker's excellent publication: *Coastal and Offshore Weather, The Essential Handbook*. You can pick up a copy of Chris Parker's book at his web site: http://www.mwxc.com.

George Cline, KP2G, can be found on the *Caribbean Maritime Mobile Net* located at 7.241 MHz, lower sideband at 0715 AST, 15 minutes into the net. George gives an overview of the current Caribbean weather beginning in Trinidad and working his way up the chain to Puerto Rico. At 0730 AST, George moves to 7.086, lower sideband for further Caribbean weather information. The same weather information is then transmitted in a weatherfax format. George returns to the airwaves at 1630 AST, 15 on the afternoon cocktail net at 7.086 LSB.

Virgin Islands Radio broadcasts on the following Marine SSB frequencies: 2506 KHz, 4357.4 (ch. 401), 4382.2 KHz (ch. 409), 6.515.7 KHz (ch. 604), 8728.2 (ch. 804), and 13100.8 (ch. 1201).

In the Northwest Caribbean, your primary source of weather on the SSB will likely be the *Northwest Caribbean Cruiser's Net* on 6.209MHz or 6.212MHz at 1400 Zulu. The net covers the entire Northwest Caribbean from the Yucatán to San Andres Island and I've picked them up as far away as Jamaica and the Caymans. The *Northwest Caribbean Cruisers Net* is a directed net and a weather report is given and check-ins are requested from boaters in Belize, Guatemala, Honduras, and Mexico, as well as anywhere offshore.

In Mexico, *XFP* in Chetumal stands by on 2.182 MHz, XFC in Cozumel stands by on 2.182 MHz. The *Belize Pilot Station* stands by on 2.182 MHz, while *Belize Customs* stands by on 2.750 MHz,

In Honduras, *La Vox Evangelica* is a religious station that broadcasts in English from 0300-0500 on Mondays at 810 KHz, 1310KHz, and 1390KHz, and on 4.8202 MHz.

Radio France offers full marine forecasts for the *Atlantic Ocean* and the Leeward Islands at 0739 on 15.300 MHz and 15.530 Mhz.

Local Weather Broadcasts

One good source of local weather information is often the local marinas who will sometimes print out the daily weather forecasts for all to see.

The Bahamas

Staying in touch with weather broadcasts presents little problem in The Bahamas. From Nassau you can receive the local Bahamian radio station *ZNS I* at 1540 KHz (with weather at 0735 and 0755), which broadcasts simultaneously on FM at 107.1 MHz. *ZNS II* on 1240 KHz and *ZNS III* at 810 KHz can usually be picked up in the northern Exumas. *WGBS* also from

Miami at 710 KHz has weather four times an hour 24 hours a day. In the New Providence area you will be able to pick up *BASRA* Coral Harbour giving the weather and tides at 0715 every morning. *BASRA* will place a call on VHF ch.16 and then move to ch. 72 for weather information. Skippers can contact the *Nassau Marine Operator* on VHF ch. 27 and ask for the latest weather report from the *Nassau Meteorological Office*. In Abaco, you can pick up the weather broadcasts of Silbert Mills on *Radio Abaco*, FM 93.5, at 0700, and again between 1800-1830 during the evening news. Vessels planning a Gulf Stream crossing can hail *Blue Dolphin* on VHF ch. 73 at 0730 for the latest on the Stream.

The Turks and Caicos Islands

On Grand Turk, *Flagstaff* comes on VHF ch. 16 at 8:00 a.m. local time and informs those who want to hear the latest southwest North Atlantic weather forecast to shift to VHF ch. 13. You can frequently pick up *Flagstaff*'s weather transmissions as far away as South Caicos. When *Flagstaff* is off the island Brian Riggs (of the *National Museum*), handles the weather broadcasts using the call *Bluewater*.

Puerto Rico and the Spanish Virgin Islands

In San Juan you can pick up hourly weather reports on AM radio. Tune in to *WOSO*, 1030 AM, and you'll receive the weather every hour after the news. If you are in the waters off the eastern shore of Puerto Rico or the Spanish Virgin Islands you should be able to pick up Virgin Islands Radio from St. Thomas, USVI. Virgin Islands Radio, *WXM*, broadcasts 24/7 on VHF Wx Ch. 3.

The United States and British Virgin Islands

You can pick up local weather on *WIVI*, 99.5 FM at 0730, 0830, 1530, and 1630, on *WVWI*, 1000 AM hourly from 1000, and on *WSTA*, 1340 AM and *Radio Antilles* on 830 AM. Tortola hosts *ZBVI Radio*, 780 AM with weather forecasts at 0730 and 0805, Monday-Friday, and at 0945 on Sundays; weather is also available on the half-hour from 0730-2130. In St. Croix you can receive weather on *WSTX* at 970 AM. In Puerto Rico you can pick up *WOSO* at 1030 AM with weather reports in English at 6 minutes after each hour, just after the news. You can also pick up local weather forecasts in the Virgin Islands on VHF weather ch. 03 and ch. 04. *Virgin Islands Radio* broadcast weather forecasts at 0600, 1400, and 2000 on VHF channels 28 and 85 (sometimes you can receive these broadcasts as far away as Fajardo).

The Leeward Islands

On VHF weather ch. 01 (162.55 MHz), as well as VHF ch. 12, you can pick up a 24-hour recorded weather forecast for St. Martin and vicinity. St. Martin also has a daily VHF net on ch. 14 at 0730.

Also on St. Martin, *PJD2 Radio* broadcasts a daily marine forecast on 1300 AM at 0830 and 102.7 FM at 0930. During hurricane season these forecasts are repeated once more each day. *Radio 91.9 FM* is new on the scene in St. Martin and broadcasts programs aimed at the visiting cruising community including marine weather and news. On Sundays mornings at 0900, tune back in to *FM 102.7* (PJD2), the *Voice of St. Martin*, for an hour-long nautical program specifically for mariners.

In Antigua you can get weather forecasts at 0750 on the FM band at 90.5 or on the AM band at 650. In English Harbour you can pick up *English Harbour Radio*, VHF ch. 06, at 0900 Monday through Friday, for their local and Leeward Islands forecasts. In Guadeloupe, the *MRCC* in Fort de France monitors VHF ch. 16 24-hours a day and if you hail them they will be happy to give you an English version of the current weather. In Dominica you can pick up the weather on the hour from Gem Radio at 93.3 FM and marine weather forecasts daily at 0703 and 0930 with marine news following the weather on Wednesdays. In St. Kitts you can pick up weather broadcasts on *Radio ZIZ*, 555 AM and 90.1 FM, or Radio Paradise at 825 AM. In Nevis, listen to the *Voice of Nevis* at 895 AM. In Anguilla tune in to the *Anguilla Broadcasting Service*, AM 650, for weather at 0750 daily.

The Windward Islands

In southern St. Vincent, *Sam's Taxi* gives weather forecasts at 0900 and 1730 daily on VHF ch. 06, and in Martinique, *COSMA* gives weather forecasts in French at 0730 and 1830 on VHF ch. 11. In

Rodney Bay, St. Lucia, there is a daily VHF net on ch. 68 at 0750, M-F, where you can pick up the latest weather forecasts. There is a cruiser's net in Prickly Bay, Grenada, at 0730 on VHF ch. 68.

In Martinique you can pick up *Radio Caraibes* at 89.5 on the FM band, but be forewarned that it is in French. In St. Lucia, you can pick up weather forecasts on the FM bands thanks to *WAVES* at 93.5 and 93.7 at 0730 and 1630. St. Vincent boasts *St. Vincent Radio* at 705 on the AM band and 100.5 on the FM band, (weather at 0745) and *Sound of the Nation* (weather after the 0700 news) at 89.7/90.7/107.5 on the FM band.

In Barbados you can listen to the *Barbados Broadcasting Company* on 900 AM for weather after the 0715 news show. In Carriacou you can pick up the weather on *Radio Kayak* at 106 FM at 0725 and 0915, while in Grenada, *Radio Grenada* gives weather forecasts after the 0700 news on 535/540 AM and *Sun Radio* will give you weather at 0700, 1200, and 1800 at 87.9/98.5/105.5 on the FM band.

Trinidad and Tobago

In Trinidad, *North Post Radio* gives weather reports at 0940 and 1640 daily on VHF ch. 27. In Chaguaramas tune in to the daily Cruiser's Net on VHF ch. 68, every morning at 0800. Trinidad and Tobago have a wealth of AM and FM stations where you can receive weather periodically during the day. On the AM band you can tune in to *NBS Radio* at 610, and *Radio Trinidad* at 730 on your dial. On the FM band, the popular *Hott 93* (93.5) has a local forecast every morning at 0730. Other FM stations with periodic weather broadcasts are *Central Radio* (90.5), Radio ICN (91.1), Love (94.1), *The Rock* (95.1), WEFM (96), *Music Radio 97* (97.1), Yes FM (98.9), *NBS Radio* (100), Power 102 (102.5), WABC (103), *Radio 104* (104), *Radio Tempo* (105), *Classic Radio* (106).

Mexico

In Mexico, the *San Miguel de Cozumel Pilots* can be contacted on VHF ch. 16 (in Spanish). On the AM and FM bands there are hundreds of stations up and down the dial. The *Isla Mujeres Net* broadcasts weather daily on VHF ch. 13 at 0730 and 0830.

Belize

In Belize you can pick up news and weather in English on *Radio Belize* at 830 KHz, 910 KHz, 930 KHz, and on the FM band at 88.9 MHz and 91.1 MHz at 0100, 0300, 1300, 1500, 1700, 1830, 2100, and 2300 UTC. *British Forces Broadcast Service* can be found on the FM band at 93.1 MHz and 99.1 MHz while the *Voice of America* relay station is on 1530 KHz and 1580 KHz, and the *Belize City Pilots* are found on VHF ch. 16. *Belize City Radio* has weather broadcasts at 0700, 1230, and 1900 daily on 834 AM and 93 FM. *The Moorings Base* in Placencia gives a local daily forecast on VHF ch. 74 between 0900-0930.

Cuba

You can receive weather from the Port Captains at *Marina Gaviota* and at *Marina Hemingway*.

Guatemala

In Guatemala try *Radio Cultural* at 730 KHz which broadcasts in English daily from 0300-0430 UTC and on Sundays from 2345-0430 UTC. *Unión Radio* broadcasts in English daily from 0200-0400 UTC on 1330 KHz. The daily cruisers net on the *Rio Dulce* offers weather as well as lots of other good information.

Honduras

In Honduras you can contact the *Puerto Cortés Pilots* on VFH ch. 6 or ch. 16 for weather info.

Jamaica

In Jamaica you can receive weather forecasts from *Radio Kingston* on 2.738 MHz, USB at 0830 and 1330 local time. *Radio Kingston* also broadcasts on VHF ch. 13 at 0930, 1430, and 2030 local time.

Chapter 25

List of Charts

CAUTION:

<u>All</u> charts are to be used in conjunction <u>with</u> the text in this guide. All soundings are in feet at <u>Mean Low Water</u>. All courses are magnetic. Projection is Transverse Mercator. Datum is WGS84. North is always "up" on these charts. The charts are designed strictly for orientation, they are NOT to be used for navigational purposes.

The prudent navigator will not rely solely on any single aid to navigation, particularly on floating aids.

Differences in latitude and longitude may exist between these charts and other charts of the area; therefore, the transfer of positions from one chart to another should be done by bearings and distances from common features.

The authors and publisher take no responsibility for errors, omissions, or the misuse of these charts. No warranties are either expressed or implied as to the usability of the information contained herein. Always keep a good lookout when piloting in these waters.

Chart #	Chart Description	Page #
THE NORTHERN BAHAMAS		
Grand Bahama		
Chart GB-1A	West End, Marina Entrance	25
Chart GB-2	Freeport Harbour	26
Chart GB-4	Xanadu Beach to Madioca Point	27
Chart GB-5	Silver Cove, Ocean Reef Marina	28
Chart GB-7	Bell Channel, Port Lucaya Marina	28
Chart GB-8	Fortune Point to The Grand Lucayan Waterway	30
Chart GB-9	Grand Lucayan Waterway, Southern Entrance	30
Chart GB-10	The Grand Lucayan Waterway, Northern Entrance	31
Chart GB-12	South Riding Point Harbour	31
The Abacos		
Chart AB-BI-5	Randall's Creek to Basin Harbour Cay	32
Chart AB-BI-13	Mores Island	33
Chart AB-4	Grand Cays	34
Chart AB-5	Double Breasted Cays	34

Chart #	Chart Description	Page #
Chart AB-7	Carter Cays	35
Chart AB-12	Umbrella Cay, Allan's-Pensacola Cay	36
Chart AB-19	Green Turtle Cay	37
Chart AB-19A	Green Turtle Cay, Black Sound	37
Chart AB-19B	Green Turtle Cay, White Sound	38
Chart AB-21B	Mariposa Project	38
Chart AB-21A	Treasure Cay Marina	38
Chart AB-22	Great Guana Cay	39
Chart AB-24	Leisure Lee	40
Chart AB-25	Marsh Harbour	40
Chart AB-27	Matt Lowe's Cay to Marsh Harbour	42
Chart AB-28A	Man-O-War Harbour	42
Chart AB-29	Hope Town Harbour	43
Chart AB-30	White Sound	44
Chart AB-32	Tilloo Pond	45
Chart AB-33	Snake Cay	46
Chart AB-35	Little Harbour	46
THE CENTRAL BAHAMAS		
The Biminis		
Chart BI-3	North Bimini, Harbour Entrance	49
Chart BI-5	South Bimini, Nixon's Harbour	49
The Berry Islands		
Chart BR-2A	Great Harbour Cay Marina	50
Chart BR-7	Little Harbour Cay	51
Chart BR-13	Chub Cay Marina	52
Andros		
Chart AN-5	Kamalame Cay Marina	52
Chart AN-7	Fresh Creek	53
Chart AN-7A	Fresh Creek Basin	53
New Providence		
Chart NP-7A	Lyford Cay Marina	54
Chart NP-8	Coral Harbour	55
Chart NP-10	Palm Cay Marina	56
Chart NP-11	Rose Island	57
Eleuthera		
Chart EL-2	Davis Harbour	57
Chart EL-3	Powell Point	58
Chart EL-13	Hatchet Bay Pond	59

Chart #	Chart Description	Page #
Chart EL-19	Spanish Wells	60
Chart EL-20	Royal Island	61
Cat Island		
Chart CT-1B	Flamingo Hills Marina	62
Chart CT-1A	Springfield Bay	62
Chart CT-1	Hawk's Nest Creek	62
Chart CT-6A	Bennett's Harbour	63
Chart CT-7	Arthur's Town, Orange Creek	64
THE EXUMA CAYS		
Chart EX-6A	Highborne Cay Marina	65
Chart EX-8	Norman's Cay	67
Chart EX-17	Warderick Wells, South Mooring Field	68
Chart EX-23	Pipe Creek	69
Chart EX-25	Staniel Cay	71
Chart EX-28B	Cave Cay	72
Chart EX-29	Musha Cay to Darby Island	73
Chart EX-41A	Emerald Bay Marina	74
Chart EX-45	The Holes at Stocking Island	75
Chart EX-48	Crab Cay, Red Shanks	76
THE SOUTHERN BAHAMAS		
Long Island		
Chart LI-1	Calabash Bay, Joe Sound	78
Chart LI-2	Stella Maris	79
Chart LI-5	Dollar Harbour	80
Chart LI-5A	Deadman's Cays	81
Chart LI-7	Little Harbour	81
The Jumentos		
Chart JU-14	Ragged Island Harbour, Duncan Town	82
Chart JU-15	Duncan Town to Little Ragged Island	83
THE TURKS AND CAICOS ISLANDS		
Chart TCI-C10	Providenciales, Sellar's Cut, Turtle Cove Marina	85
Chart TCI-C13	The Caicos Cays, Pine Cay to Parrot Cay, Ft. George Cut	86
Chart TCI-C8	Providenciales, Cooper Jack Bight, Discovery Bay	87
Chart TCI-C9	Providenciales, Juba Point Creek, Caicos Marina & Shipyard	88
Chart TCI-C14A	West Caicos Marina	89
Chart TCI-T2	Grand Turk	90
Chart TCI-T2A	Grand Turk, North Creek Entrance	91

Chart #	Chart Description	Page #
THE DOMINICAN REPUBLIC		
Chart DR-1	Luperón	94
Chart DR-2	Ocean World Marina	94
Chart DR-8	Puerto Bahía Marina	94
Chart DR-9	Bahía de San Lorenzo	95
Chart DR-10	Barahona	95
Chart DR-12	Puerto de Haina	96
Chart DR-13	Santo Domingo	96
Chart DR-17	Cap Cana Marina	97
Chart DR-15	Casa de Campo Marina	97
PUERTO RICO		
Chart PRW-5	Western Coast, Puerto Real	99
Chart PRW-6	Western Coast, Bahía de Boqueron	99
Chart PRS-2	Southern Coast, Arrecife Margarita to Caleta Salinas, La Parguera	100
Chart PRS-3	Southern Coast, Bahía de Guanica	101
Chart PRS-5	Southern Coast, Bahía de Guayanilla	102
Chart PRS-11	Southern Coast, Playa de Salinas	103
Chart PRS 12	Southern Coast, The Hurricane Holes at Jobos	104
Chart PRE-2	Eastern Coast, Palmas del Mar	104
Chart PRE-7	Eastern Coast, Puerto del Rey Marina	105
Chart PRE-8	Eastern Coast, Cayo Obispo Marina	105
Chart PRN-5	Northern Coast, San Juan Harbo	106
THE SPANISH VIRGIN ISLANDS		
Chart SVI-6	Culebra, Ensenada Honda	108
Chart SVI-13	Vieques, Ensenada Honda	109
Chart SVI-14	Vieques, Bahía Mosquito to Bahía Tapon	110
THE UNITED STATES VIRGIN ISLANDS		
Chart USVI-MB	St. Thomas, Mandahl Bay	112
Chart USVI-5	St. Thomas, Druif Bay to Flamingo Ba	112
Chart USVI-4	St. Thomas, Krum Bay	112
Chart USVI-10	St. Thomas, Sapphire Bay Marina	113
Chart USVI-6	St. Thomas, Benner Bay	113
Chart USVI-BB	St. Thomas, Brewer's Bay	114
Chart USVI-12	St. John, Frank Bay to Great Cruz Bay	114
Chart USVI-15	St. John, Coral Bay	115
Chart USVI-16	St. John, Coral Harbor	116
Chart USVI-23	St. Croix, Christiansted	117

Chart #	Chart Description	Page #
Chart USVI-26	St. Croix, Green Cay Marina	117
Chart USVI-27	St. Croix, Salt River Bay	118
THE BRITISH VIRGIN ISLANDS		
Chart BVI-1	Tortola, Road Town	120
Chart BVI-3	Tortola, Paraquita Bay, Maya Cove	121
Chart BVI-7	Tortola, Trellis Bay	122
Chart BVI-5	Tortola, The Camanoe Passages	123
Chart BVI-12	Tortola, Nanny Cay Marina, Sea Cow Bay	123
Chart BVI-20	Virgin Gorda, Virgin Gorda Yacht Harbou	124
Chart BVI-23	Virgin Gorda, North Sound	124
THE LEEWARD ISLANDS		
St. Martin/Sint Maarten		
Chart STM-2A	Anse Marcel, Radisson Marina	126
Chart STM-9	Oyster Pond	126
Chart STM-7	Simpson Bay Lagoon	127
Chart STM-6	Cole Baii	127
Chart STM-4	Port la Royale	128
Antigua		
Chart ANT-5	Jolly Harbour	128
Chart ANT-7	Falmouth Harbour	129
Chart ANT-8	English Harbour	129
Chart ANT-9	Indian Creek, Mamora Bay	130
Chart ANT-11	Nonsuch Bay, Green Island	131
Chart ANT-13	Judge Bay Point to North Sound	132
St. Kitts		
Chart STK-1A	St. Kitts Marine Works	132
Chart STK-3	Southern St. Kitts	133
Guadeloupe		
Chart GUA-5	Marina de Rivière Sens	133
Chart GUA-7	Point-à-Pitre	134
Chart GUA-8	Marina Bas du Fort	134
Chart GUA-9	Rivière Salée	135
Chart GUA-10	Grande Cul-de-Sac Marin	136
Chart GUA-15	St.-François	136
St. Barths		
Chart STB-4	Gustavia	137

Chart #	Chart Description	Page #
THE WINDWARD ISLANDS		
Martinique		
Chart MAR-7	Marina de Port Cohé	139
Chart MAR-5	Baie de Fort-de-France	139
Chart MAR-17	Cul-de-Sac du Marin	140
Chart MAR-18	Le Marin	141
Chart MAR-22	Havre du Robert	142
Barbados		
Chart BAR-2	Port St. Charles	142
Chart BAR-5	Bridgetown	143
Chart BAR-6	The Carenage	143
St. Lucia		
Chart STL-3	Rodney Bay Lagoon	144
Chart STL-5	Marigot Bay	144
St. Vincent and The Grenadines		
Chart STV-8	Ottley Hall Marina	145
Chart GND-7	Canouan, Glossy Bay Marina	146
Carriacou		
Chart CAR-3	Tyrrel Bay	146
Grenada		
Chart GRE 5	St. George's Harbour	146
Chart GRE-7	True Blue Bay	147
Chart GRE-8	Prickly Bay	148
Chart GRE-9	Mt. Hartman Bay, Hog Island	148
Chart GRE-10	Hog Island to Petite Bacaye	149
Chart GRE-11	Petite Bacaye to St. David's Point	151
TRINIDAD AND TOBAGO		
Chart TRI-4	Scotland Bay	152
Chart TRI-8	Gaspar Grande Island, Winn's Bay (Corsair Bay)	153
Chart TRI-10	Carenage Anchorage, TTSA	153
Chart TRI-12	Port of Spain	154
Chart TOB-4	Buccoo Reef	155
THE NORTHERN COAST OF JAMAICA		
Chart JAM-1	Port Antonio	157
Chart JAM-10	Bogue Lagoon	158
HONDURAS		
Chart HON-5	Isla de Guanaja, El Bight	160

Chart #	Chart Description	Page #
Chart HON-12	Isla de Roatán, Old Port Royal	160
Chart HON-14	Isla de Roatán, Calabash Bight	161
Chart HON-15	Isla de Roatán, Hog Pen Bight, Jonesville Bight	162
Chart HON-16	Isla de Roatán, Second Bight	162
Chart HON-17	Bay Islands, Old French Harbour, French Harbour	163
Chart HON-18	Isla de Roatán, Brick Bay	165
Chart HON-26	Isla de Utila, Puerto Este	166
Chart HON-31	Puerto de Cabotaje	167
Chart HON-33	Laguna el Diamente	167
Chart HON-34	Puerto Cortes	168
	THE CAYMAN ISLANDS	
Chart CAY-9	Grand Cayman, North Sound	170
Chart CAY-9A	Grand Cayman, North Sound, Governor's Creek	171
Chart CAY-9B	Grand Cayman, North Sound, Harbour House Marina	172
	GUATEMALA	
Chart GTM-1	Bahía de Amatique, Cabo Tres Puntas, Bahía la Graciosa	174
Chart GTM-2	Puerto Barrios	174
Chart GTM-4	Río Dulce, Livingston to El Golfete	175
Chart GTM-5	Northern El Golfete, Laguna Salvador	176
Chart GTM-6	Bahía de Tejano, Texan Bay	177
Chart GTM-7	Southern El Golfete	177
Chart GTM-8	Río Dulce, El Golfete to Mango's Marina	178
Chart GTM-9	Río Dulce, Mango's Marina to La Joya del Rio Marina	178
Chart GTM-12	Lago Izabal, Bocas de Bujajal	180
Chart GTM-10	Lago Izabal	180
Chart GTM-11	Lago Izabal, Puerto Refugio	181
	BELIZE	
Chart BLZ 1	Cucumber Beach Marina	182
Chart BLZ-2	Sapodilla Lagoon	183
Chart BLZ 3	Hakim's Boatyard	184
Chart BLZ-4	Mango Creek	184
Chart BLZ-5	Ycacos Lagoon	185
Chart BLZ 6	BigCreek	185
	MEXICO	
Chart MX-1	Bahía de la Espiritu Santo	188
Chart MX-2	Bahía Ascension	188
Chart MX-3	Puerto Aventuras	189

Chart #	Chart Description	Page #
Chart MX-4	Cancún: V&V Marina	189
Chart MX-5	Isla Mujeres	190
Chart MX-6	Laguna Chakmochuk	191
CUBA		
Chart CUB-1	Northern Coast, Canal de Barco	193
Chart CUB-2	Northern Coast, Esperanza	194
Chart CUB-3	Northern Coast, Cayo Morillo	194
Chart CUB-4	Northern Coast, Bahía Honda	195
Chart CUB-5	Northern Coast, Marina Hemingway	196
Chart CUB-6	Northern Coast, Gaviota	196
Chart CUB-7	Northern Coast, Cayo Cruz del Padre	197
Chart CUB-8	Northern Coast, Bahía de Vita	197
Chart CUB-9	Southern Coast, Bahía de Baitiquirí	198
Chart CUB-10	Southern Coast, Santiago	198
Chart CUB-11	Southern Coast, Casilda, Cayo Blanco Marina	199
Chart CUB-12	Southern Coast, Cienfuegos	199

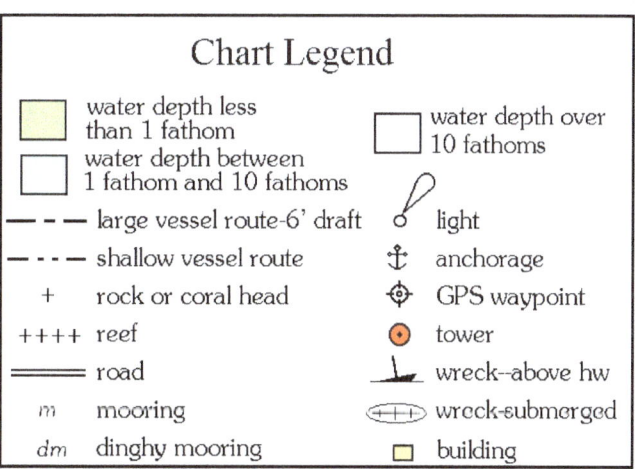

Chapter 26
List of Haul Out Yards

Bahamas

Northern Bahamas

FACILITY	LOCATION	TELEPHONE #	WEB OR EMAIL ADDRESS
Abacos			
Abaco Yacht Services	Green Turtle Cay	242-365-4033	
Edwin's Boat Yard	Man-O-War Cay	242-365-6006	edwinsboatyard.com
Lighthouse Marina	Hope Town	242-366-0154	www.htlighthousemarina.com
Marsh Harbour Boat	Marsh Harbour	242-367-5025	mhby.com/
Grand Bahama			
Bradford	Freeport	242-352-7711	bradford-marine.com
Knowles Marine	Grand Bahama	242-351-2769	www.knowles-marine.com

Central Bahamas

FACILITY	LOCATION	TELEPHONE #	WEB OR EMAIL ADDRESS
Eleuthera			
R & B Boat Yard	Spanish Wells	242-333-4462	rbboatyard.com
New Providence			
Bayshore Marina	Nassau	242-393-8232	bayshoremarina@hotmail.com
Brown's Boat Basin	Nassau	242-393-3331	rmb43@hotmail.com
Harbour Central	Nassau	242-321-2172	harbourcentralmarina.com
Exumas			
George Town Marina	George Town	242-345-5116	
Master Harbour Marina	George Town	242-345-5116	

Southern Bahamas

FACILITY	LOCATION	TELEPHONE #	WEB OR EMAIL ADDRESS
Long Island			
Stella Maris	Stella Maris, Long Island	242-338-2050	www.stellamarisresort.com
Turks and Caicos Islands			
Caicos Marina	Providenciales, Caicos	649-946-5600	caicosmarinashp@tciway.tc

Dominican Republic

FACILITY	LOCATION	TELEPHONE #	WEB OR EMAIL ADDRESS
Casa de Campo Marina	La Romana	809-523-2111	www.marinacasadecampo.com
Club Náutico	Boca Chica	809-683-2582	www.clubnautico.com.do

FACILITY	LOCATION	TELEPHONE #	WEB OR EMAIL ADDRESS
Club Náutico	Haina		
Ocean World Marina	Cofresi	809-291-1000	www.oceanworldmarina.com
Marina Zar-Par	Boca Chica	809-523-5858	www.marinazarpar.com
Tropical Marine	Luperón	809-440-9926	

Puerto Rico

FACILITY	LOCATION	TELEPHONE #	WEB OR EMAIL ADDRESS
Cayo Obispo (Isletta)	Cayo Obispo	787-863-0370	
Island Marine	Cayo Obispo	787-382-3051	
Palmas Del Mar YC	Humacao	787-656-7300	palmasdelmaryachtclub.com
Ponce Yacht	Ponce	787-842-9003	ponceyachtandfishingclub.com
Puerto Del Rey Marina	Ceiba	787-860-1000	www.puertodelrey.com
San Juan Bay Marina	San Juan	787-235-9633	www.sjbaymarina.com
Veradero Fajardo	Las Croabas	787-656-7300	
Veradero Puerto V.	La Parguera	787-899-5588	
Villa Marina	Fajardo	787-863-5131	www.villamarinapr.com
Villa Pesquera	Puerto Real	787-458-4332	

United States Virgin Islands

FACILITY	LOCATION	TELEPHONE #	WEB OR EMAIL ADDRESS
Caneel Bay Shipyard	Cruz Bay, St. John	340-693-8771	
East End Boat Park	Benner Bay, St. Thomas	For vessels under 25' LOA	
Haulover Marine	Subbase, S. Thomas	340-776-2078	www.subbasedrydock.com
Independent Boatyard	Benner Bay, St. Thomas	340-776-0466	ibyvi.com/
St. Croix Marine	Gallows Bay, St. Croix	340-773-0289	Stcroixmarinecenter.com

British Virgin Islands

FACILITY	LOCATION	TELEPHONE #	WEB OR EMAIL ADDRESS
Nanny Cay Marina	Nanny Cay	284-494-2512	chandlery@nannycay.com
Soper's Hole	Soper's Hole	284-494 2983	www.sopersholemarina.com
Tony's Refinishing	Nanny Cay	284-499-4189	
Virgin Gorda Yacht	Virgin Gorda Y. H.	284-495-5500	www.virgingordayachtharbour.com
VI Shipwrights	Soper's Hole	284-495-4496	
Workbench	Virgin Gorda Y. H.	284-495-5310	workbenchbvi@gmail.com

Leeward Islands

Antigua

FACILITY	LOCATION	TELEPHONE #	WEB OR EMAIL ADDRESS
A1 Marine Services	Jolly Harbour	268-462-6042	www.jolly-harbour-marina.com
Antigua Rigging	Falmouth Harbour	268-562-1294	www.antiguarigging.com
Antigua Slipway	English Harbour	268-460-1056	www.antiguaslipway.com
Antigua Yacht Paint	Falmouth Harbour	268-774-1461	Antigua-Yacht-Painting.com
Bailey's Boatyard	Falmouth Harbour	268-460-6054	Antigua-Marina.com
Falmouth Harbour	Falmouth Harbour	268-463-8081	
Harris Boat Works	Jolly Harbour	268-462-5333	harrisboatworks@actol.net
North Sound Marina	Parham	268-562-3499	Northsoundmarine.com
Shell Beach Marina	Parham Harbour	268-562-0185	shellbeachmarinaantigua.com

Guadeloupe

FACILITY	LOCATION	TELEPHONE #	WEB OR EMAIL ADDRESS
Cap Sud Chantier	Pointe-à-Pitre	0590-90 76 70	info@capsud.net
Chantier Naval	Le Bourg, Îles des Saintes	0590-90 34 47	
Chantier Pineau	Goyave	0590-95 84 41	

FACILITY	LOCATION	TELEPHONE #	WEB OR EMAIL ADDRESS
Karuplast	Capesterre Belle Eau	0590-86 82 53	
Lemaire Marine	Pointe-à-Pitre	0590-90 34 47	lemairemarineservices.com
Marina Bas du Fort	Pointe-à-Pitre	0590-93 66 20	www.marinaguadeloupe.com/en
Seminole Marine	Pointe-à-Pitre	0590-23 18 60	y.kihel@ool.fr
Top Gun Marine	Pointe-à-Pitre	0590-91 10 11	topgunmarine@antilladoo.com

Isle des Saintes

Roche à Move	Baie de Marigot	0590-99 53 15	

St. Kitts

St. Kitts Marine	Half Way Tree, St. Kitts	869-662-8930	www.skmw.net/

St. Barthélémy

St. Barth Boatyard	Gustavia	0590-29 00 03	www.2swedes.com/

St. Martin/Sint Maarten

Bobby's Marina	Airport Rd., Sint Maarten	599-545-2890	www.bobbysmarina.com/
Bobby's Marina	Philipsburg, Sint Maarten	599-542-2366	www.bobbysmarina.com/
Crown Boatyard	Philipsurg, Sint Maarten	721-542-6890	info@crownboatyard.com
Geminga	Marigot, St. Martin	0590-29 35 52	geminga@domaccess.com
Island Water World	Cole Bay, Sint Maarten	599-544-5310	sales@islandwaterworld.com
JMC Boatyard	Marigot, St. Martin	0590-77 10 05	
Polypat Caribes	Marigot, St. Martin	0590-87 12 01	saintmartinboatyard.com
Sint Maarten Ship	Simpson Bay Lagoon	721-545-3740	www.stmaartenshipyard.com/
TOBY	Marigot, St. Martin	0590-52 02 88	timeoutboat@hotmail.com

Windward Islands

Barbados

Willie's Marine Service	Bridgetown	246-424-1808	mikie@spiceisle.com

Martinique

Carenantilles	Fort-de-France	0596-63 76 74	www.carenantilles.com/
Carenantilles	Le Marin	0596-74 77 70	www.carenantilles.com/
Chantier Naval	Le Marin	0596-76 79 39	
Martinique Dry Dock	Quai Quest	0596-72 69 40	
Multicap Caraibes	Quai Quest	0596-71 41 81	MCM@multicaparaibes.com

St. Lucia

Rodney Bay Marina	Rodney Bay	758-452-0324	www.igy-rodneybay.com/

St. Vincent

Howard's Marine	Kingstown	784 457 1806	kpmarineltd.com
Ottley Hall Marina	Kingstown	784 457 2178	ottleyhall@vincysurf.com

Carriacou

Carriacou Marine	Tyrrel Bay	473-443-8878	carriacoumarine.com

Grenada

Grenada Marine	St. David's	473-443-1667	info@grenadamarine.com
Spice Island Marine	Prickly Bay	473-444-4257	spiceislandmarine.com

Trinidad

Coral Cove Marina	Chaguaramas	868-634-2040	www.coralcovemarina.com/

FACILITY	LOCATION	TELEPHONE #	WEB OR EMAIL ADDRESS
CrewsInn Marina	Chaguaramas	868-634-4828	crewsinn@tstt.net.tt
IMS Yacht Services	Chaguaramas	868-625-2104	ims@imsyacht.com
Peake Yacht Services	Chaguaramas	868-634-4423	peakeyachts.com
Power Boats	Chaguaramas	868-634-4303	powerboats.com
TTSA	Carenage	868-634-4519	ttsailing.org

Jamaica

FACILITY	LOCATION	TELEPHONE #	WEB OR EMAIL ADDRESS
Barham's Wharf	Savanna-la-Mar	876-955-3171	
Black's Drydock	Port Royal		
Errol Flynn Boatyard	Port Antonio	876-715-6044	www.errolflynnmarina.com
Royal Jamaica YC	Kingston	876-924-8685	

Cayman Islands

FACILITY	LOCATION	TELEPHONE #	WEB OR EMAIL ADDRESS
Barcadere (Scott's)	GT, Grand Cayman	345-949-3743	Barcadere.com
Harbour House Marina	GT, Grand Cayman	345-947-1307	harbourhousemarina.com
ProYacht	GT, Grand Cayman	345-916-0697	proyacht.ky

Honduras

FACILITY	LOCATION	TELEPHONE #	WEB OR EMAIL ADDRESS
Eagle Marine Dry.	La Ceiba	504-440-7648	Eaglemarine@gmail.com
FH Marine Railway	French Hrbr., Roatán	504-382-2172	
La Ceiba Shipyard	La Ceiba	504-441-9426	laceibashipyard.com/
Navy Base	Puerto Cortés		bahacortes@yahoo.com
Seth Archer's	French Hrbr., Roatán		
Shipyard*	Oak Ridge, Roatán		
Shipyard*	Savannah Bight, Guanaja		

Guatemala

FACILITY	LOCATION	TELEPHONE #	WEB OR EMAIL ADDRESS
Astilleros Magdalena	Río Dulce	502-7930-5059	astillerorio@yahoo.com
Carlos Welding	Río Dulce		
Nana Juana Marina	Río Dulce	502 7930 5230	info@hotelmarinananajuana.com
RAM Marina	Río Dulce	502-7930-5408	rammarina.com

Belize

FACILITY	LOCATION	TELEPHONE #	WEB OR EMAIL ADDRESS
Cucumber Beach	Belize City	501-222-4129	oldbelize.com
Hakim's Boatyard	Belize City	501-615-9341	
Thunderbird's	Placencia	501-670-3737	thunderbirdsmarinebz@gmail.com

Mexico

FACILITY	LOCATION	TELEPHONE #	WEB OR EMAIL ADDRESS
V&V	Cancun	998-234-0100	www.marinavv.com
Playa Mujeres	Isla Mujeres	998-892-4670	info@laamadamarina.com
Puerto Isla Mujeres	Isla Mujeres	998-887-0330	www.puertoislamujeres.com

Cuba

FACILITY	LOCATION	TELEPHONE #	WEB OR EMAIL ADDRESS
Marina Gaviota	Varadero	534-566-4115	gaviota-grupo.com

* These shipyards are used primarily by the shrimping fleet, they can haul out a yacht of large size, but you will have to make arrangements well beforehand.

Chapter 27

List of Marinas

Some folks want to hide in a mangrove lined creek, others want to haul out, still others want to tie up their boat in a slip and perhaps get a hotel room for the duration of the hurricane. With that in mind we offer this (probably incomplete) list of marinas in The Bahamas and Caribbean. You should contact the marina of your choosing extremely early to find out if they have a hurricane policy and will accept transients, not everybody will do that. Some require you to vacate their marina in case of a named storm. Follow the marina's instructions and good luck!

Bahamas

Northern Bahamas

FACILITY	LOCATION	TELEPHONE #	WEB OR EMAIL ADDRESS
Abacos			
Abaco Yacht Services	Green Turtle Cay	242-365-4033	abacoyachtservices@gmail.com
Baker's Bay Marina	Great Guana Cay		bakersbayclub.com
(Baker's Bay Marina is a Private, Members-Only Marina)			
Black Sound Marina	Green Turtle Cay	242-365-4531	blacksoundmarinagreenturtle.com
Bluff House	Green Turtle Cay	242-365-4247	bluffhouse.com
Boat Harbour Mar.	Marsh Harbour	242-367-2158	abacobeachresort.com
Conch Inn Marina	Marsh Harbour	242-367-4000	www.diveabaco.com/conch.htm
Donny's Marina	Green Turtle Cay	242-577-1339	www.donnysboatrentalsgtc.co
Edwin's Boat Yard	Man-O-War Cay	242-365-6007	edwinsboatyard.com
Green Turtle Club	Green Turtle Cay	242-365-4271	greenturtleclub.com
Guana Hideaway	Great Guana Cay	242-365-5070	www.everythingguana.com
Harbour View Marina	Marsh Harbour	242-367-3910	www.harbourviewmarina.com
Hopetown Hideaways	Elbow Cay	242-366-0434	www.hopetown.com/company
Hopetown Marina	Elbow Cay	242-366-0003	www.hopetownmarina.com
Leeward YC & Mar.	Green Turtle Cay	242-365-4191	www.leewardyachtclub.com
Lighthouse Marina	Hope Town	242-366-0154	www.htlighthousemarina.com
Long's Landing	Marsh Harbour		

FACILITY	LOCATION	TELEPHONE #	WEB OR EMAIL ADDRESS
Man-O-War Marina	Man-O-War Cay	242-365-6008	www.manowarmarina.com
Mangoe's Marina	Marsh Harbour	242-367-4255	www.mangoesabaco.com
Marsh Harbour Boat	Marsh Harbour	242-367-5025	mhby.com
Marsh Harbour Mar.	Marsh Harbour	242-367-2700	www.jibroom.com
Orchid Bay Marina	Great Guana Cay	242-365-5175	orchid-bay-marina.com
Other Shore Club	Green Turtle Cay	For Sale	(no website)
Rosie's Place	Grand Cay	242-353-1200	www.rosiesplace.com
Schooner Bay	Great Abaco	888-275-1639	schoonerbaybahamas.com
Sea Spray Resort	Elbow Cay	242-366-0065	www.seasprayresort.com
Spanish Cay Marina	Spanish Cay	242-365-0083	www.spanishcay.com
Treasure Cay Marina	Treasure Cay	242-361-8250	treasurecay.com

Grand Bahama

FACILITY	LOCATION	TELEPHONE #	WEB OR EMAIL ADDRESS
Bahama Bay	Lucaya		info@bahamiaservices.com
Bradford Ship	Freeport	242-352-7711	bradford-marine.com
Flamingo Bay Marina	Lucaya	242-373-5640	www.flamingobaymarina.com
Grand Bahama YC	Lucaya	242-373-7616	www.grandbahamayachtclub.com
Hawksbill YC	Lucaya	242-373-1513	hawksbillyachtclub@yahoo.com
Knowles Marine	Lucaya	242-351-2769	www.knowles-marine.com
Laurie Sykes Dock	Lucaya	242-373-1344	
Ocean Reef YC	Lucaya	242-373-4661	www.oryc.com
Old Bahama Bay	West End	242-602-5171	www.oldbahambay.com/marina
Port Lucaya	Lucaya	242-373-9090	www.portlucayamarina.com
Sea Breeze Marina	Lucaya	242-351-1186	seabreezevacationvillas.com
Sunrise Marina	Lucaya	242-352-6834	www.sunriseresortandmarina.com
Xanadu Beach	Lucaya		xanadubeachhotelandmarina.com

Central Bahamas

Andros

FACILITY	LOCATION	TELEPHONE #	WEB OR EMAIL ADDRESS
Kamalame Cove	Staniard Creek	242-368-6281	www.kamalame.com
Lighthouse Marina	Fresh Creek	242-368-2305	

Berry Islands

FACILITY	LOCATION	TELEPHONE #	WEB OR EMAIL ADDRESS
Berry Islands Club	Chub Cay	504-655-8464	
Chub Cay Marina	Chub Cay	242-325-1490	www.chubcay.com
Great Harbour Cay	Great Harbour Cay	242-367-8005	greatharbourcay.com

Biminis

FACILITY	LOCATION	TELEPHONE #	WEB OR EMAIL ADDRESS
Brown's	North Bimini	242-347-3116	www.brownsmarinabimini.com
Bimini Beach Club	South Bimini	242-359-8228	
Bimini Big Game	North Bimini	242-347-3391	biggameclubbimini.com
Bimini Blue Water	North Bimini	242-347-3166	biminibluewater@hotmail.com
Bimini Sands	South Bimini	242-347-3500	www.thebiminisands.com
Cat Cay Yacht Club	Cat Cay	242-347-3565	catcayyachtclub.com
Fisherman's Village	North Bimini		rwbimini.com
Mega Yacht Marina	North Bimini		rwbimini.com
Seacrest Marina	North Bimini	242-347-3071	www.seacrestbimini.com
Weech's Bimini Dock	North Bimini	242-347-3028	www.weechsbiminidock.com

FACILITY	LOCATION	TELEPHONE #	WEB OR EMAIL ADDRESS

New Providence
Albany Marina	Adekaude	242-676-6010	albanybahamas.com
Atlantis	Nassau	242-363-3000	www.atlantisresort.com
Bayshore Marina	Nassau	242-393-8232	bayshoremarina@hotmail.com
Bay Street Marina	Nassau	242-676-7000	baystreetmarina.com
Brown's Boat Basin	Nassau	242-393-3331	
Harbour Central	Nassau	242-323-2172	harbourcentralmarina.com
Hurricane Hole Mar.	Nassau	242-363-3600	www.hurricaneholemarina.com
Lyford Cay Marina	Lyford Cay	242-362-4271	lyfordcay.com
Nassau Harbour Club	Nassau	242-393-0771	www.nassauharbourclub.com
Nassau Yacht Haven	Nassau	242-393-8173	www.nassauyachthaven.com
Palm Cay Marina	Nassau	242-676-8554	www.palmcay.com
Paradise Harbour	Nassau	242-363-2992	
Texaco Fuel Dock	Nassau		

Eleuthera
Cape Eleuthera	Powell Point	242-334-8500	capeeleuthera.com
Davis Harbour	Weymss Bay	242-334-6303	
French Leave Mar.	Governor's Harbour	242-332-3778	frenchleaveeleuthera.com
Harbour Isl. Club	Harbour Island	242-333-2427	www.harbourislandmarina.com
Romora Bay Marina	Harbour Island	242-333-2325	www.romorabay.com
Runaway Bay Mar.	Runaway Bay	242-332-1744	
Spanish Wells YH	Spanish Wells	242-333-4255	swyachthaven.com
Valentine's YC	Harbour Island	242-333-2142	www.valentinesresort.com

Cat Island
Hawk's Nest Resort		242-342-7050	www.hawks-nest.com
Flamingo Hills	Springfield Bay	242-464-6404	flamingohills.com

Exumas
Compass Cay Marina	Compass Cay	242-422-7300	www.compasscaymarina.com
Emerald Bay Marina	Great Exuma	242-336-6100	www.marinaemeraldbay.com
Exuma YC	George Town	242-336-2578	http://exumayachtclub.wixsite.com
Farmer's Cay YC	Little Farmer's Cay	242-355-4017	
George Town Marina	George Town	242-345-5116	
Highborne Cay Mar.	Highborne Cay	242-355-1008	highbournecaybahamas.com
Kevalli House Marina	George Town	242-357-0118	www.kevallihouse.com/marina
Master Harbour Mar.	George Town	242-422-7310	
Safe Harbour Marina	Cave Cay	242-357-0143	http://www.cavecay.com
Staniel Cay Marina	Staniel Cay	242-355-2024	www.stanielcay.com

Southern Bahamas

Long Island
Flying Fish Marina	Clarence Town	242-337-3430	www.flyingfishmarina.com
Stella Maris	Stella Maris	242-338-2050	www.stellamarisresort.com

San Salvador
Riding Rock Marina	Cockburn Town	242-331-2631	ridingrock.com

| FACILITY | LOCATION | TELEPHONE # | WEB OR EMAIL ADDRESS |

Turks and Caicos Islands

Facility	Location	Telephone	Web/Email
Blue Haven	Providenciales, Caicos	649-946-9900	www.bluehaventci.com
Caicos Marina	Providenciales, Caicos	649-946-5600	caicosmarinashp@tciway.tc
Seaview Marina	South Caicos		
South Side Marina	Providenciales, Caicos	649-946-3417	southsidemarina-tci.com
Turtle Cove	Providenciales, Caicos	649-941-3781	
Walkin & Son	Providenciales, Caicos		
West Caicos Marina	West Caicos	Closed, due to be refinanced soon	
Windward Marina	Ambergris Cay	Still under construction-lack of financing	

Dominican Republic

Facility	Location	Telephone	Web/Email
Cap Cana Marina	Punta Cana	809-227-2262	capcana.com
Casa de Campo Mar.	La Romana	809-523-2111	www.marinacasadecampo.com.do
Club Náutico	Haina	809-537-3961	info@nauticohaina.cjb.net
Club Náutico S.D.	Boca Chica	809-683-2582	www.clubnautico.com.do
Luperón Marina	Luperón		
Marina Punta Cana	Punta Cana	809-959-2714	
Marina Rio Ozama	Santo Domingo		
Marina Tropical	Luperón	809-315-1940	marinatropicalnautica.com
Marina Zar Par	Boca Chica	809-523-5858	www.marinazarpar.com
Ocean World Mar.	Cofresi	809-291-1000	www.oceanworldmarina.com
Puerto Bahía	Samaná	809-503-6363	www.puertoBahíasamana.com
Puerto Blanco	Luperón	809-739-2010	www.puertoblancomarina.com
Salinas Marina	Las Salinas		

Puerto Rico

Facility	Location	Telephone	Web/Email
Cangrejos YC	Carolina	787-791-1015	www.cangrejosyachtclub.com
Cayo Obispo Mar.	Cayo Obispo	787-863-0370	
Club Deportiva	Punta Ostiones	787-851-8880	www.clubdeportivodeloeste.com
Club Nautico	Boqueron	787-851-1336	
Club Nautico	Dorado		
Club Nautico	La Parguera	787-899-5590	
Club Nautico	San Juan	787-722-0177	nauticodesanjuan.com
El Conquistador	Fajardo	787-863-1000	
Island Marine	Cayo Obispo	787-382-3051	
Marina de Salinas	Salinas	787-824-5973	marinadesalinas.com
Marina Pescaderia	Puerto Real	787-717-3638	marinapescaderia.com
Palmas Del Mar	Humacao	787-656-7300	www.palmasdelmar.com
Ponce Yacht Club	Ponce	787-842-9003	ponceyachtandfishingclub.com
Puerto Chico	Fajardo	787-863-0834	www.marinapuertochico.com
Puerto Del Rey Mar.	Ceiba	787-860-1000	www.puertodelrey.com
Puerto Real Marina	Fajardo	787-863-2188	
Punta Pozuelo	Punta Pozuelo		
San Juan Bay Mar.	San Juan	787-721-8062	www.sjbaymarina.com
Sea Lovers Marina	Fajardo	787-863-3762	
Sun Bay Marina	Fajardo	787-863-0313	www.sunbaymarina.com

FACILITY	LOCATION	TELEPHONE #	WEB OR EMAIL ADDRESS
Veradero Fajardo	Las Croabas	787-656-7300	
Veradero Puerto V.	La Parguera	787-899-5588	
Villa Marina	Fajardo	787-863-5131	www.villamarinapr.com
Villa Pesquera	Puerto Real	787-458-4332	

United States Virgin Islands

FACILITY	LOCATION	TELEPHONE #	WEB OR EMAIL ADDRESS
American Yacht Hrb.	Red Hook, St. Thomas	340-775-6454	www.igy-americanyachtharbor.com
Avery's Boathouse	Charlotte Amalie, St. T.	340-776-0113	
Boater's Haven	Charlotte Amalie, St. T.	340-775-6144	
Caneel Bay Ship	Cruz Bay, St. John	340-693-8771	
Compass Point Mar.	Benner Bay, St. Thomas	340-775-4166	https://compasspointmarina.com
Crown Bay Marina	Crown Bay, St. Thomas	340-774-2255	www.crownbay.com
Fish Hawk Marina	Benner Bay, St. Thomas		
Frenchtown Harbour	Charlotte Amalie, St. T.		
Green Cay Marina	Christiansted, St. Croix	340-718-4455	www.tamarindreefresort.com/marina
Haulover Marine	Subbase, S. Thomas	340-776-2078	www.subbasedrydock.com
Independent Boatyard	Benner Bay, St. T.	340-776-0466	ibyvi.com
Jones Maritime	Christiansted, St. Croix		jonesmaritime.com
La Vida Marina	Benner Bay, St. Thomas		
Oasis Cove Marina	Benner Bay, St. Thomas		www.oasiscove.com
Saga Haven	Benner Bay, St. Thomas	340-775-0520	www.sagahaven.com
Salt River Marina	Salt River, St. Croix	340-778-9650	
Sapphire Beach	Sapphire Beach, St. T.		www.sapphirebeachmarina.com
Silver Bay Dock	Gallows Bay, St. Croix	340-773-4709	
St. Croix Marine	Gallows Bay, St. Croix	340-773-0289	www.stcroixmarine.com
St. Croix YC	Teague Bay, St. Croix	340-773-9531	www.stcroixyc.com
Tropical Marine	Benner Bay, St. Thomas	340-775-6595	
Vessup Point Mar.	Red Hook, St. Thomas	340-244-3464	
Yacht Haven Grande	St. Thomas	340-775-6454	www.igy-americanyachtharbor.com

British Virgin Islands

FACILITY	LOCATION	TELEPHONE #	WEB OR EMAIL ADDRESS
Baugher's Bay	Road Town, Tortola	284-494-2393	mcconsultants@surfbvi.com
Bitter End YC	North Sound, V. G.	800-872-2392	www.beyc.com
Bletiner's Marina	Baugher's Bay	284-494-1000	
Fort Burt Marina	Road Harbour, Tortola	284-494-2587	fortburt.com
Foxy's Taboo	Long Bay, Jost Van Dyke	284-495-9258	foxysbar.com
Hannah Bay Marina	Nanny Cay, Tortola		
Harbour View Mar.	Fat Hogs Bay, Tortola	284-495-0165	
Hodges Creek Mar.	Maya Cove, Tortola	284-494-5000	sunyacht@caribsurf.com
Inner Harbour Mar.	Road Town, Tortola	284-494-4502	
Joma Marina	Road Town, Tortola	888-615-4006	bviyachtcharters.com
Leverick Bay	Virgin Gorda	284-495-7421	leverickbay.com
Lighthouse Marina	West End, Tortola	284-495-3445	admiralbvi.com
Little Harbour Mar.	Little Harbour, J. V. D.	284-495-9835	
Manuel Reef Marina	Sea Cows Bay, Tortola	282-495-2066	www.manuel-reef-marina.com
Moorings	Road Harbour, Tortola	284-393-2331	moorings.com
Nanny Cay Marina	Nanny Cay, Tortola	284-345-2657	nannycay.com

FACILITY	LOCATION	TELEPHONE #	WEB OR EMAIL ADDRESS
North Latitude Mar.	Great Harbour, J. V. D.	284-495-9930	northlatitudemarina.com
Penn's Landing Mar.	Fat Hogs Bay, Tortola	284-495-1134	www.pennslandingbvi.com
Peter Island Yacht Hr.	Sprat Bay, Peter Island	284-495-2000	peterisland.com/marina
Prospect Reef	Road Town, Tortola	284-494-3773	resort@prospectreef.com
Pusser's Leverick Bay	North Sound, V. G.		
Pusser's Marina Cay	Road Town Tortola	284-494-2174	marinacay@pussers.com
Road Reef Marina	Road Harbour, Tortola	284-494-2751	
Saba Rock	North Sound, V. Gorda	284-495-7711	www.sabarock.com
Scrub Island Marina	Scrub Island	877-890-7444	www.scrubisland.com
Soper's Hole	Soper's Hole, Tortola	284-495-4589	www.sopersholemarina.com
Sunsail	Frenchman's Cay, Tor.	888-350-3568	Sunsail.com
Virgin Gorda Yacht	Virgin Gorda	284-495-5500	www.virgingordayachtharbour.com
Village Cay Marina	Road Town, Tortola	294-494-2771	villagecaybvi.com
YC Costa Smeralda	Biras Creek	284-393-2000	marina.yccs.com

Leeward Islands

Antigua

FACILITY	LOCATION	TELEPHONE #	WEB OR EMAIL ADDRESS
Antigua Slipway	English Harbour	268-460-1056	www.antiguaslipway.com
Antigua Yacht Club	Falmouth Harbour	268-460-1799	antiguayachtclub.com
Bailey's Boatyard	Falmouth Harbour	268-460-1503	catamaranmarina.com
Catamaran Marina	Falmouth Harbour	268-460-1503	catamaranmarina.com
Falmouth Harbour	Falmouth Harbour	268-460-6054	antigua-marina.com
Great House Marina	Oyster Pond		www.greathousemarina.com
Jolly Harbour	Jolly Harbour	268-462-6042	www.jolly-harbour-marina.com
Nelson's Dockyard	English Harbour	268-481-5033	info@nelsonsdockyardmarina.com
North Sound Marina	Parham	268-562-3499	northsoundmarine.com
Shell Beach Marina	Parham Harbour	268-562-0185	shellbeachmarinaantigua.com
St. James YC	Mamora Bay	866-237-2071	www.stjamesclubantigua.com

Guadeloupe

FACILITY	LOCATION	TELEPHONE #	WEB OR EMAIL ADDRESS
Marina Bas du Fort	Pointe-à-Pitre	0590-93-66-20	www.marinaguadeloupe.com/en
Mar. Rivière Sens	Basse Terre	0590-86-79-43	www.marina-rivieresens.com/en
Mar. St.-François	St.-François		

Isle des Saintes

FACILITY	LOCATION	TELEPHONE #	WEB OR EMAIL ADDRESS
Bourgeois des Saintes Baie de Marigot		0590-81-53-57	lessaintesmultiservices@gmail.com

St. Christopher (St. Kitts)

FACILITY	LOCATION	TELEPHONE #	WEB OR EMAIL ADDRESS
Christophe Harbour	Ballast Bay	869-466-4557	www.christopheharbour.com
Port Zante	Basseterre	869-466-5021	www.portzantemarina.com
Telca Marina	St. Kitts Boat Works	869-662-8930	www.skmw.net

St. Martin/Sint Maarten

FACILITY	LOCATION	TELEPHONE #	WEB OR EMAIL ADDRESS
Bobby's Marina	Philipsburg	721-542-2366	www.bobbysmarina.com
Captain Oliver's	Oyster Pond	0590-87-40-26	www.captainolivershotel.com
Crown Boatyard	Philipsburg	721-580-7114	http://crownboatyard.com
Dock Marten Marina	Philipsburg	721 542-5705	www.dockmaarten.com
Gateway Marina	Simpson Bay	721-554-2324	www.gatewaymarinasxm.com
Great House Marina	Oyster Pond		www.greathousemarina.com

FACILITY	LOCATION	TELEPHONE #	WEB OR EMAIL ADDRESS
Island Water World	Cole Bay	599-544-5310	sales@islandwaterworld.com
Lagoon Marina	Cole Bay	721-544-2611	www.lagoon-marina.com
La Samana	Simpson Bay	Under Construction	
Marina Fort-Louis	Baie du Marigot	0590-511-11-11	www.marinafortlouis.com
Palapa	Simpson Bay	721-545-2735	www.palapamarinasxm.com
Port la Royale	Marigot		www.marinaportlaroyale.com
Porto Cupecoy	Simpson Bay Lagoon	721-546-4900	www.portocupecoy.com
Radisson Marina	Anse Marcel		
Simpson Bay YC	Simpson Bay	721-544-2309	www.igy-simpsonbay.com
St. Maarten Shipyard	Simpson Bay Lagoon	721-545-3740	www.stmaartenshipyard.com
YC Isla del Sol	Simpson Bay	721-544-2408	www.igy-isledesol.com
YC Port de Plais	Simpson Bay	721-544-4565	www.portdeplaisancemarina.com

St. Barths

Port of Gustavia	Gustavia	0590-27-66-97	www.portdegustavia.fr

Windward Islands

Barbados

Port St. Charles	Six Man's Bay	246-419-1000	www.portstcharles.com

Martinique

Carenantilles	Fort-de-France	0596-63 76 74	www.carenantilles.com
Carenantilles	Le Marin	0596-74 77 70	www.carenantilles.com
Club Nautique	Le Marin	0596-74 92 48	clubnautiquedumarin.com
La Marina du Marin	Port de Plaisance	0596-74-83-83	www.marina-martinique.fr
La Marina du Robert	Le Robert	407 9633 1790	
Marina Port Cohé	Cohé Lamentin		
Marina La Neptune	Cohé Lamentin		
Somatras Marina	Pointe du Bout	0596-66-07-74	www.marina3ilets.com

St. Lucia

Marina at Marigot	Marigot Bay	758-458-5300	
Rodney Bay Marina	Rodney Bay	758-458-4892	www.igy-rodneybay.com/index.php
Waterside Landings	Rodney Bay	758-458-7300	www.landingsstlucia.com

St. Vincent

Barefoot Yacht	Calliaqua	784-456-9526	barefootyachts.com/barefoot-base
Lagoon Marina	Blue Lagoon	784-458-4308	www.bluelagoonsvg.com
Ottley Hall Marina	Kingstown	784-457-2178	ottleyhall@spiceisle.com

Bequia

Bequia Marina	Admiralty Bay	784-458-3272	
YC at Bequia	Admiralty Bay	784-457-3407	www.yachtclub.vc

Canouan

Glossy Bay Marina	Charlestown Bay	784-431-2828	www.glossybay.com

Union Island

Anchorage YC	Clifton	784-458-8221	aycunionisland.com
Bougainvilla	Clifton	784-458-8678	bougainvilla@spiceisle.com

FACILITY	LOCATION	TELEPHONE #	WEB OR EMAIL ADDRESS

Carriacou
Carriacou Marine	Tyrrel Bay	473-443-6292	www.carriacoumarine.com

Grenada
Clarke's Court	Clarke's Court	473-439-3939	www.clarkescourtmarina.com
Grenada Marine	St. David's	473-443-1667	grenadamarine.com
Grenada Yacht Club	St. George's	473-440-6826	www.grenadayachtclub.com
Le Phare Bleu	Phare Bleu Bay	473-444-2400	www.lepharebleu.com
Martin's Marina	Mt. Hartman	473-444-4449	martinsmarina@spiceisle.com
Port Louis Marina	St. George's	473-439-0000	www.portlouisgrenada.com
Prickly Bay Marina	Prickly Bay	473-439-5265	pricklybaymarina.com
Secret Harbour	St. George's	473-444-4449	www.secretharbourgrenada.com
Spice Island Marine	Prickly Bay	473-444-4342	spiceislandmarine.com
True Blue Resort	True Blue	473-443-8783	www.truebluebay.com
Whisper Cove Mar.	Clarke's Court	473-444-5296	whispercovemarina.com

Trinidad
Bay View	Gaspar Grande		
Coral Cove Marina	Chaguaramas	868-634-2040	www.coralcovemarina.com
CrewsInn Marina	Chaguaramas	868-634-4000	crewsinn@tstt.net.tt
Hummingbird	Chaguaramas	868-634-2773	
IMS Yacht Services	Chaguaramas	868-625-2104	ims@imsyacht.com
Peake Yacht Services	Chaguaramas	868-634-4420	peakeyachts.com
Power Boats	Chaguaramas	868-634-4303	powerboats.co.tt
PPYC	Point-a-Pierre		
SFYC	San Fernando	For shallow draft powerboats	
Tardieu Marine	Chaguaramas	868-634-4534	
TTYC	Bayshore	868-270-4141	

Jamaica
Errol Flynn Marina	Port Antonio	876-715-6044	www.errolflynnmarina.com
Fisherman's Inn	Falmouth	876-954-3427	www.fishermannsinnjamaica.ca
Glistening Waters	Falmouth	876-954-3229	www.glisteningwaters.com
Montego Bay YC	Montego Bay	876-979-8038	www.mobayyachtclub.com
Ocho Rios Marina	Ocho Rios		
Pier One	Montego Bay	876-952-2452	pieronejamaica.com
Royal Jamaica YC	Kingston	876-924-8685	www.rjyc.org.jm

Cayman Islands
Barcadere (Scott's)	GT, Grand Cayman	345-949-3743	www.barcadere.com
Camana Bay Marina	Grand Cayman	345-640-3500	camanabay.com
Cayman Islands YC	GT, Grand Cayman	345-747-2492	www.ciyachtclub.ky
Harbour House Mar.	GT, Grand Cayman	345-947-1307	ww.harbourhousemarina.com
Kaibo YC	GT, Grand Cayman	345-947-9975	www.kaibo.ky

Honduras
Barefoot Cay Mar.	Brick Bay, Roatán	504-9967-3652	barefootcay.com/marina
Billares Marina	Utila		

FACILITY	LOCATION	TELEPHONE #	WEB OR EMAIL ADDRESS
Brick Bay Marina	Brick Bay, Roatán		
Brooksy Point Mar.	French Harbour, Roatán	504-9455-2330	brooksypointyachtclub.com
CoCo View Resort	French Harbour, Roatán	304-948-7506	www.cocoviewresort.com
Fantasy Island	French Harbour, Roatán	504-9705-6131	www.fantasyislandresort.com
Gibson Bight Mar.	Gibson Bight, Roatán	970-290-0654	www.gibsonbightmarina.com
Jonesville Point Mar.	Jonesville Bight, Roatán	504-9967-3803	jonesvillepointmarina.com
La Ceiba Shipyard	La Ceiba	504-3370-6442	laceibashipyard.com
Oak Ridge Marina	Oak Ridge, Roatán	504-2435-2163	
Parrot Tree Plant	2nd Bight, Roatán	713-234-1477	www.parrottree.com
Roatán YC	French Harbour, Roatán		www.roatanyachtclubhn.com
Turtlegrass Marina	Calabash, Roatán		www.turtlegrass.net

Guatemala

FACILITY	LOCATION	TELEPHONE #	WEB OR EMAIL ADDRESS
Amatique Bay Mar.	Puerto Barrios	502-7931-0000	www.amatiquebay.net
Bruno's Marina	Río Dulce	502-7930-5721	brunoshotel.com
Burnt Key Marina	Bahía de Tejano	502-5747-9717	www.burntkeymarina.com
Calypso Marina	Laguna La Joya	507-6440-3585	www.calypsomarina.com
Capt. John's Mar.	Río Dulce	502-5732-0219	www.riodulcemarina.com
Catamaran	Río Dulce	502-4145-3901	www.catamaranisland.com
Crowbar Marina	Río Dulce		
El Relleno	Río Dulce		
Freddie's Marina	Río Dulce		
Hacienda Tijax	Río Dulce	502-7930-5505	www.tijax.com
La Joya del Rio Marina	Río Dulce	502-7930-5594	
Mango's Marina	Río Dulce	502-4032-4444	mango@riodulcemangomarina.com
Mansion del Rio	Río Dulce	502-7930-5020	www.mansiondelrio.com.gt/en
Mar Marina	Río Dulce	502-7930-5090	info@marmarine.com
Monkey Bay Mar.	Río Dulce	502-5368-9604	www.monkeybaymarina.com
Nana Juana Marina	Río Dulce	5023253-3206	hotelmarinananajuana.com
RAM Marina	Río Dulce	502-7930-5408	www.rammarina.com
Tortugal Marina	Río Dulce	502-5306-6432	tortugal.com
Vista Rio	Río Dulce	502-7930-5665	
Xalaha Marina	Río Dulce		

Belize

FACILITY	LOCATION	TELEPHONE #	WEB OR EMAIL ADDRESS
Belize Yacht Club	San Pedro	501-226-4340	
Cucumber Beach	Belize City	501-222-4129	oldbelize.com
Laru Beya Marina	Placencia	501-523-3476	larubeya.com
Placencia Marina	Placencia	800-810-8567	www.theplacencia.com/marina
Princess Hotel	Belize City	501-223-2670	princessbelize.com
Roberts Grove	Placencia	501-523 3565	robertsgrove.com
Sanctuary Belize	Sapodilla Bay	501-533-7565	www.SanctuaryBelize.com
Sittee River Marina	Sittee River	501-533-7888	www.sitteerivermarina.com
Thunderbird's	Placencia	501-670-3737	thunderbirdsmarinebz@gmail.com

Mexico (dial +52 before each number)

FACILITY	LOCATION	TELEPHONE #	WEB OR EMAIL ADDRESS
Club Nautico	Cozumel	919-872-1113	

FACILITY	LOCATION	TELEPHONE #	WEB OR EMAIL ADDRESS
El Milagro Marina	Isla Mujeres	998-877-1708	www.elmilagrobeachhotelandmarina.com
Marina del Mar	Cancún	800-501-3030	marinahaciendadelmar.com
Marina del Sol	Isla Mujeres	998-888-0929	
Marina El Cid	Cancún	998-871-0185	www.elcid.com
Marina Paraiso	Isla Mujeres	998-877-0252	www.marinaparaiso-islamujeres.com
Marina Silcer	Progreso	969-934-0491	www.marinasilcer.com
Marina Sureste	Progreso	999-911-0417	www.marinasureste.com
Puerto Aventuras	Chetumal	984-873-5108	www.puertoaventuras.com
Puerto Isla Mujeres	Isla Mujeres	998-887-0330	www.puertoislamujeres.com
Terminal Maritima	Cancún	919-883-0775	www.marinavv.com
V&V Marina	Cancún	998-234-0100	www.marinavv.com

Cuba

FACILITY	LOCATION	TELEPHONE #	WEB OR EMAIL ADDRESS
Gaviota	Varadero	534-566-4115	www.gaviota-grupo.com
Marina Cayo Blanco	Casilda	534-199-6205	marinastdad@enet.cu
Marina Cayo Largo	Cayo Largo	534-454-8213	www.cayolargo.net/marina.html
Marina Dársena	Varadero	534-566-8060	
Marina Hemingway	Havanna	537-204-5088	www.hemingwaycuba.com
Marina Internacional	Bahía de Vita	533-243-0445	marvita@enet.cu
Marina Puertosol	Cienfuegos	534-325-1241	rrpp@nautica.cfg.tur.cu
Marina Puertosol	Havanna	537-917-1462	
Marina Puertosol	Isla de la Juventud	533-619-8181	
Marina Puertosol	Trinidad	533-330-1738	psol@cayo.cav.cyt.cu
Marina Punta Gorda	Santiago de Cuba	532-269-1446	admin@marlin.seu.tur.cu
Marina Varedero	Varadero	534-566-8060	Puerto@marlinv.var.cyt.cu
Marlin Marina	Cayo Largo	534-566-8060	comercial@marlin.mtz.tur.cu
Marlin Marina	Cienfuegos	534-566-8060	comercial@marlin.mtz.tur.cu
Marlin Marina	Santiago de Cuba	534-566-8060	comercial@marlin.mtz.tur.cu
Marlin Marina	Varadero	534-566-8060	comercial@marlin.mtz.tur.cu
Puerto de Vita	Bahía de Vita	532-430-4751	

Index

A

Admiralty Bay 222
Albany Marina 55
Alliance 134, 135, 136
Angelfish Point 35
Anse Marcel 222
Arthur's Town 64
Athol Island 56

B

Bacaye Harbour 151
Bahía de Amatique 173
Bahía de Boqueron 98
Bahía de Demajagua 105
Bahía de Guanica 101
Bahía de Guayanilla 101
Bahía la Graciosa 173
Baie de Fort-de-France 138
Baie de Marigot 214, 221
Baie des Flamands 138
Baie des Tourelles 138
Baie du Marigot 222
Bajo Amarillo 107
Bajo Enmedio 98, 100
Bajo Villedo 174
Baker's Bay Marina 39
Banc du Ft. St. Louis 138
Barcadere 212–215, 223
Barefoot Cay Marina 165
Barracouta Rocks 35
Barracouta Rocks Channel 35
Basin Harbour Cay 32
BASRA 200, 202
Basseterre 221
Basse Terre 134, 221
Batelco 50, 51, 60
Bay Islands 159, 166
Beef Island 122
Belize 201, 203
Bellamy Cay 122
Bell Channel 26–29
Bell Channel Bay 29
Benner Bay 113, 213, 220
Bennett's Harbour 63
Bernard Bay 90

Between The Majors 70
Big Cay 163
Big Cay Channel 163
Big French Cay 164
Big Major's Spot 72
Bight of Acklins 83
Bight of Eleuthera 60
Bimini Sands Marina 49
Bishop Shoal 129
Black Sound 37
Blue Lagoon 222
Bluff Cay 37
Bluff Harbour 37
Boat Harbour 82
Boat Harbour Marina 42
Boatswain Point 169
Boca del Infierno 103
Boca de Monos 152
Bocas de Bujajal 179
Bodden Bight 162
Bogue Lagoon 156, 158
Bonacca 159, 160
Bon Accord Lagoon 152, 154
Boqueron 98
Bradford Yacht and Ship 26
Brick Bay 165, 223
Brick Bay Marina 165
Bridgetown 142, 143, 214
Brimstone Hill 132
Brooksy Point Marina 164
Buck Island 121
Bullock's Harbour 49, 50

C

Caicos Marina and Shipyard 88
Calabash Bight 161
Calivigny Harbour 151
Calivigny Island 150
Calliaqua 222
Camel Bay 82
Canal de la Mona 98
Canal Norte 98
Caneel Bay 213, 220
Cape Eleuthera 58, 59
Cape Eleuthera Marina 58
Capesterre 214
Capesterre Belle Eau 214
Carenage 142, 147, 215
Carriacou 146, 203
Carters Cays 35
Castries 144
Cave Cay 72

Cave Cay Cut 72
Cayman Islands 170, 172
Cayman Islands Yacht Club 170
Cayo Mata 102
Cayo Obispo 105, 213
Cayos de Ratones 103
Channel Cay 80, 81
Charlotte Amalie 220
Chetumal 201
Chicken Cay 70
Christiansted 116–118
Christiansted Harbor 116
Chub Cay Marina 49, 51, 52
Chub Point 51, 52
Chub Rock 57, 58
Clarke's Court Bay 150, 223
Clifton 222
Clifton Point 55
Cockburn Harbour 84
CoCo View Marina 163, 164
Cole Bay 214, 222
Columbus Point 119
Company Point 90
Compass Cay 70
Compass Cay Cut 70
Compass Cay Marina 70
Conch Cut 70
Conch Point 169
Conch Shell Point 122
Coral Bay 114, 116
Coral Harbour 55, 202
Cortés 215
Crab Cay 75, 76
Cruz Bay 213, 220
Cul de Sac Bay 144
Cul-de-Sac du Marin 140, 141
Culebra 107

D

Darby Island 74
Davis Harbour 57
Davis Harbour Marina 57
Deep Sea Cay 45
Defence Force 55
Dellis Cay 87
Dickie's Cay 42
Dollar Cay 80
Dollar Harbour 79, 80
Double Breasted Cay 35
Dove Cay 78, 79
Duck Lake 48
Duck Pond Bight 172

Dunbar Rock 159
Duncan Town 82

E

Eagle Rock 43
East Caicos 84
Eastern Harbour 42
East Harbour 156, 159
El Bight 159
Emerald Bay 74
Emerald Bay Marina 74
English Cut 84
English Harbour 129, 130, 202, 213, 221
English Harbour Radio 202
Enseñada Boca Ancha 181
Ensenada Honda 107
Enseñada Laguna 181
Enseñada Los Lagartos 181

F

Fajardo 105, 202, 213, 220
Falmouth Harbour 128, 129, 213, 221
False Channel 154
Fantasy Island 164
Farmer's Hill 74
Felix Cay 33
Fief Hill 126
Flamingo Bay Hotel and Marina 29
Flamingo Hills Resort and Marina 61
Folly Point 156
Fort-de-France 138, 214, 222
Fort George Cay 87
Fort George Cut 86, 87
Fowl Cay 70
Fox Town 35
Freeman Bay 129
Freeport 26, 212
Freeport Peninsula 158
French Cay Harbour 164
French Harbour 164, 165
French Wells 83
Fresh Creek 53, 54
Frozen Cay 51
Ft. Berkeley Point 129, 130

G

Gabarre 134, 135, 136
Galleon Beach 129, 130
Galliot Cut 72
Gallows Bay 213, 220
George Town 74, 75, 169

Gibson Channel 154
Gosier 134
Governor's Creek 170, 172
Governor's Harbour 170
Goyave 213
Grand Bahama Yacht Club 29
Grand Cayman 169, 215, 223
Grand Cul-de-Sac Marin 135
Grand Lucayan Waterway 29, 30
Grande Terre 134
Grassy Cay 113
Great Abaco 39
Great Harbour 49, 50
Great Harbour Cay 49, 50
Great Harbour Cay Marina 49, 50
Great Sale Cay 35
Green Cay 118
Green Island 131
Grier Channel 153, 154
Gringo Bay 177
Guana Cay 39, 51
Guana Cay Harbour 39
Guanaja 159, 215
Guanica 101
Gulf of Paria 152
Gulf Stream 202
Gully Cay 35, 36
Gun Point 82
Gustavia 137, 214

H

Harbor Point 116
Harbour Cut 60
Harbour House Marina 171, 172
Harris 213, 221
Hatchet Bay 59, 60
Hâvre du Robert 141
Hawksbill Creek 26
Hawksbill Yacht Club 27, 28
Hawk's Nest Marina 62
Highborne Cay Marina 65
Hill's Creek 40
Hog Cay 69, 82
Hog Island 149, 150
Hog Pen Bight 161
Hogsty Harbour 35, 36
Hope Town 43
Hurricane Hole 36

I

Îles des Saintes 213

Indian Creek 130, 131
Indian Creek Point 131
Inner Point Cay 41
Izabal 179

J

Jackson Hole 38
Jesse Arch Cay 165
Joe Cay 70
Joe Sound 77, 78
John Devine Cay 75, 76
Jolly Harbour 128, 213
Jonesville Bight 162
Jonesville Cay 162

K

Kaibo Yacht Club 169, 170
Kamalame Cay 52
Kamalame Cove 52
Kingston 203, 215
Knowles Marina and Boatyard 26
Knowles Marine 27

L

La Ceiba 159, 166, 167, 215, 224
La Ceiba Shipyard 159, 166, 167, 215, 224
Lago Izabal 179
Lagoon Bleu 134
Lagoon Point 116
Laguna Calix 175
Laguna el Diamente 159, 167, 168
Laurie Sykes Docks 29
Le Bourg 213
Leduck Island 116
Leeward Going Through 86
Le François 222
Leisure Lee 39, 40
Le Marin 140, 141, 214, 222
Le Robert 141, 222
Les Gros Îlets 137
Lighthouse Marina 44, 53
Little Darby Island 74
Little Harbour 45, 46, 49–51
Little Harbour Bar 45
Little Harbour Cay 50, 51
Little Major's Spot 72. *See* Between The Majors
Little Ragged Island 82
Little Sound 169, 170
Little Stirrup Cay 49, 50
Livingston 177
Long Bay 52, 89

Long Bay Cay 52
Long Bay Hills 89
Long Cay 53
Long Point 116
Los Galafatos 107
Lower Deadman's Cay 81
Lower Hill Cay 40
Lubber's Quarters 45
Lucaya 29
Luperón 77, 93, 95
Lyford Cay 54
Lyford Cay Marina 54

M

Main Channel 169
Mama Rhoda Reef 51
Mama Rhoda Rock 51
Mamora Bay 130, 131
Mangrove Bush 81
Mangrove Cay 31
Man-O-War Cay 42, 75, 212, 216
Man-O-War Marina 43
Marigot 214, 221, 222
Marigot Bay 144, 145, 222
Marina de Port Cohé 139, 140
Mario's Marina 178
Mariposa 38
Marsh Harbour 39, 41, 42, 212, 217
Maya Cove 122
Middle Channel 45
Middle Reef 131
Mitchell's Creek 171
Montego Bay 156
Moon Cay 72
moorings 112, 122
Moraine Cay 35
Mores Island 33
Morris Bay 128
Mosquito Cove 128
Mouth of Harbour Cay 43
Mt. Hartman Bay 149
Muddy Hole 60
Musha Cay 72, 73

N

Nanny Cay 123, 213, 220
Navy Island 156
New Plymouth 37
Nixon's Harbour 49
No Name Harbour 58, 59
Norman's Cay 66

Norman's Spit 66
North Bimini 48
North Caicos 84
North Creek 91
Northeast Point 91
North Harbour 42, 43
North Sound 169–172
North Sound Estates 172
Northwest Point 169
Nuevitas Rocks 79

O

Oak Ridge 215
Ocean Reef Marina 28
Ocean Reef Yacht Club 27
Ocean Reef Yacht Club and Marina 27, 28
Ocean World Marina 94
Old French Harbour 164
Old Port Royal 160
Old Yankee Cay 35, 36
Orange Creek 62
Ordnance Bay 129
Otter Creek 114
Ottley Hall Marina 145, 214, 222
Oyster Pond 126, 221

P

Palm Cay Marina 56
Paraquita Bay 121
Parham 132, 213, 221
Parrot Cays 43
Parrot Tree Plantation 162
Parrot Tree Plantation Marina 162
Passe Champagne 136
Passe du Marin 141
Pearns Point 128
Phare Bleu Bay 223
Phillipsburg 214, 221
Pigeon Cay 82
Pigeon Point 154
Pine Cay 87
Pipe Cay 70
Pipe Creek 70
Point-a-Pierre 223
Point Charlotte 129
Pointe-à-Pitre 213, 214, 221
Pointe des Carrières 138
Pointe des Sables 138
Point Set Rock 43
Polochic 179, 181
Porgee Rock 56

Port Antonio 215, 223
Port du Plaisance 134
Port Egmont 150, 151
Port Lucaya 29
Port Lucaya Marina 29
Port of Spain 152, 153
Port Royal 160, 215
Port St. Charles 142
Port Zante 221
Powell Point 58, 59
Prickly Bay 148, 149, 203, 214, 223
Prickly Point 148
Princess Bay 114
Puerto Barrios 173, 175
Puerto Cortés 203, 215
Puerto de Cabotaje 159
Puerto Real 98
Puerto Refugio 179
Punta Arenas 102, 103
Punta Caballos 168
Punta Carenero 107
Punta Chapin 179
Punta Gotay 101
Punta La Mela 98
Punta Manglar 173
Punta Pepillo 101
Punta Sal 159

R

Ragged Island 82
Ragged Island Harbour 82
RAM Marine 179
Randall's Creek 31
R&B Boatyard 60
Red Hook 112
Red Hook Point 112
Red Shanks 75, 76
Resorts World Bimini 48
Río Ciénaga 178
Río Dulce 173, 177, 178, 215, 224
Río Oscuro 179
Río Pichilingo 175
Rivière Salée 134, 135
Rivière Sens 133, 134, 221
Roatán 159–165, 215, 223, 224
Rodney Bay 143, 203, 214, 222
Roker's Point 74
Rose Island 56
Rotto Cay 113
Round Reef 117
Round Rock 49
Royal Island 60

Rudder Cut 73, 74
Rudder Cut Cay 73, 74
Rum Point 169
Rum Point Channel 169
Runnin Mon Marina. *See* Sunrise Marina
Russell Island 60

S

Safety Harbour 35, 36
Sainte-Anne 136
Salinas 102
Salt Cay 82
Salt Pond 79–81
Salt River Bay 118
Salt River Bay National Park 118
Salt River Point 118
Sanders Bay 116
Sandy Cay 80
San Fernando 223
Sapphire Beach Marina 112
Savannah Bight 159, 215
Schooner Channel 117
Scotch Bank 116
Scotland Bay 152
Sea Lots Channel 153
Sea Spray Resort 44
Second Bight 162, 163
Sellar's Cut 84, 85
Shearpin Cay 45
Shell Bay 179
Silver Cove 27
Silver Point 27
Simms 79
Simpson Bay Lagoon 127, 222
Snake Cay 45
Snapper Creek Cay 81
Sober Island 131
Soper's Hole 213, 221
South Bimini 48, 49
South Caicos 84, 202
South Mooring Field 69
South Side Basin Marina 88
Spanish Wells Marine 60
Spanish Wells Yacht Haven Marina 60
Staniard Creek 52
St. Croix 117, 202, 213
St. David's 150, 151
St. François 136, 221
St. George's 147, 223
St. George's Cay 60
St. George's Harbour 147
St. Thomas 111, 113

Ste. Anne 141
Steventon 74
Stingray Channel 169, 170
Stirrup Cay 50
Sub Base 213, 220
Sunken Rock 130
Sunrise Marina 26, 27

T

Tank Bay 129, 130
The Holes at Stocking Island 74, 75
Tilloo Bank 45
Tilloo Cay 44
Tilloo Cut 44
Titchfield Peninsula 156
Tom Snooch Reef 142
Top Cay 35, 36
Tortola 202
Treasure Cay 38
Treasure Cay Marina 38, 39
Trellis Bay 122
True Blue Bay 148
TTSA 153, 215
TTYC 223
Turks Island Passage 84
Turner Point 116
Turtle Cove 84, 85
Turtle Cove Marina 84, 85
Turtle Sound 83
Tyrrel Bay 146, 214, 223

U

Union Jack Dock 41
Unnamed Harbour. *See* No Name Harbour
Upper Channel Cay 80, 81
Utila 159

V

Virgin Gorda Yacht Harbour 124
Volleyball Beach 74

W

Warderick Wells 69
Water Bay 114
Water Island 111
Wells Point 80
West Bay 131, 169
West Caicos 89, 90
West Caicos Marina 89
West Harbour 156

Whale Cay 39
White Horse Reef 118
White Horse Rock 118
White Sound 37
Willoughby Bay 131
Windward 154
Winn's Bay 153

X

Xanadu Channel 27

Y

Young Island 222

About the Authors

Captain Dave Underill and Stephen J. Pavlidis

On his first sailing adventure to the Abacos in the Bahamas, Captain Dave Underill never realized it would turn into a love affair which is now going on 41 years. In addition to the Bahamas, his boats have taken him across the Atlantic from Portugal to the Virgin Islands, from Puerto Rico to St. Lucia, from the Turks and Caicos Islands to Belize and beyond. He enjoys swimming, learning to play guitar, and writing and has published two works of fiction (yes, about a boat captain) on e books under the writer name of Johnny Mayhem: VOYAGER LAWLESS and CUBA, MERMAIDS, and VOYAGER LAWLESS. His current vessel is a Lagoon 440 which reminds him that there's work to be done every day. He has not seen a mermaid... YET!

Captain Dave Underill

Stephen J. Pavlidis

Stephen J. Pavlidis has been cruising and living aboard since the winter of 1989. Starting in The Exumas over 26 years ago, Steve began his writing career with guides to the many fascinating destinations he visited. Many of his books stand alone to this day as the quintessential guides to the areas he covers. His books are different than most other cruising guides in some very significant ways. All of the charts in Steve's books were created using data personally collected while visiting each area using a computerized system that interfaces GPS coordinates with depth soundings. You can learn more about this exceptional author by visiting his Web site, www.Seaworthy.com, where this is current news and information about Steve's latest projects as well as contact information.

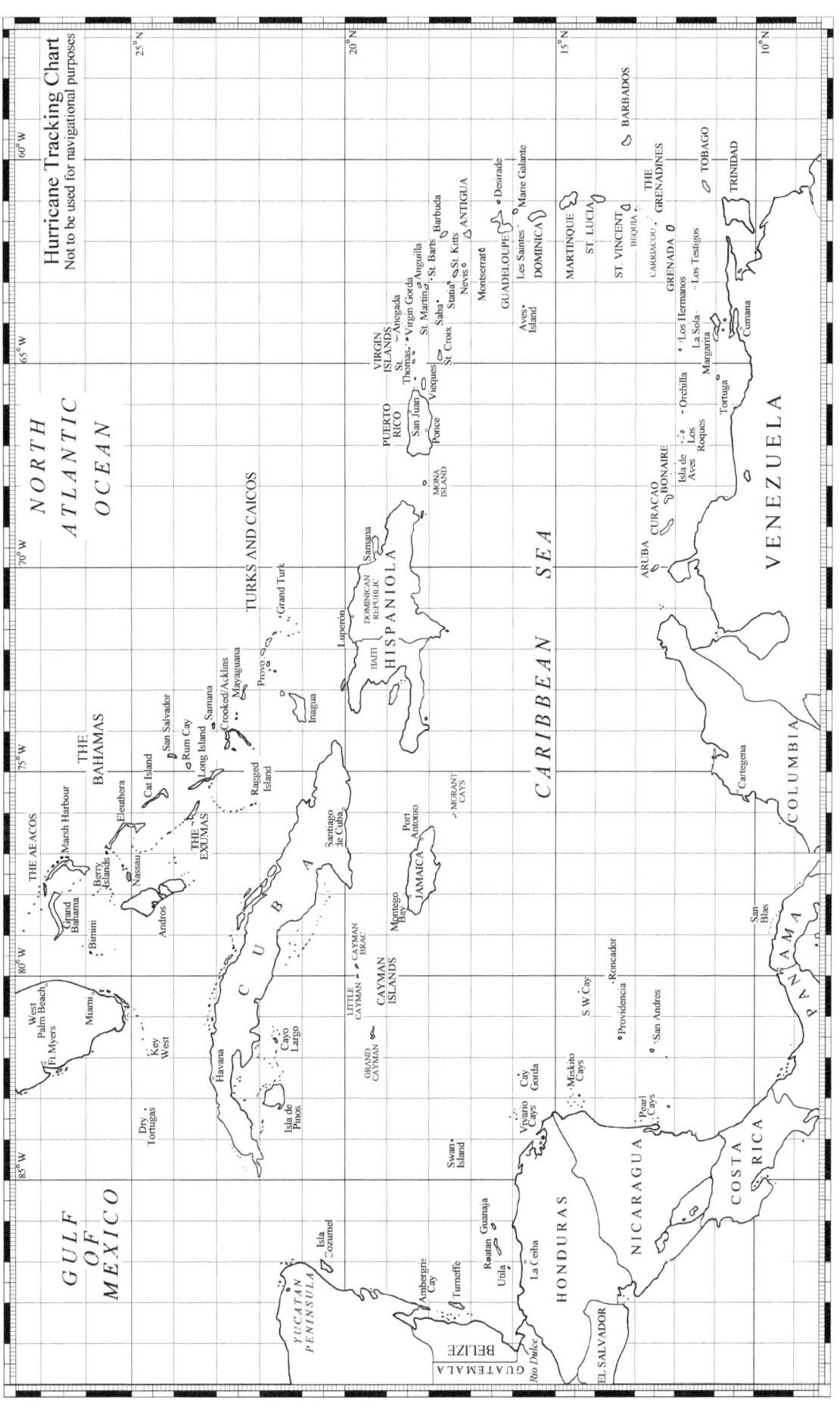

www.ingramcontent.com/pod-product-compliance
Lightning Source LLC
Chambersburg PA
CBHW081809300426
44116CB00014B/2294